Novel Research in Low-Dimensional Systems

Novel Research in Low-Dimensional Systems

Editor

Orion Ciftja

MDPI • Basel • Beijing • Wuhan • Barcelona • Belgrade • Manchester • Tokyo • Cluj • Tianjin

Editor
Orion Ciftja
Physics
Prairie View A&M University
Prairie View
United States

Editorial Office
MDPI
St. Alban-Anlage 66
4052 Basel, Switzerland

This is a reprint of articles from the Special Issue published online in the open access journal *Nanomaterials* (ISSN 2079-4991) (available at: www.mdpi.com/journal/nanomaterials/special_issues/lowdimensional_systems).

For citation purposes, cite each article independently as indicated on the article page online and as indicated below:

LastName, A.A.; LastName, B.B.; LastName, C.C. Article Title. *Journal Name* **Year**, *Volume Number*, Page Range.

© 2023 by the authors. Articles in this book are Open Access and distributed under the Creative Commons Attribution (CC BY) license, which allows users to download, copy and build upon published articles, as long as the author and publisher are properly credited, which ensures maximum dissemination and a wider impact of our publications.
The book as a whole is distributed by MDPI under the terms and conditions of the Creative Commons license CC BY-NC-ND.

Contents

About the Editor

Orion Ciftja

Orion Ciftja is a Professor of Physics at Prairie View A&M University, USA. He received his Ph.D. in the field of Theoretical Condensed Matter Physics from the International School of Advanced Studies (SISSA/ISAS), Trieste, Italy, in 1997. After obtaining his Ph.D., he worked for two years as a Postdoctoral Research Fellow at Ames Laboratory, Iowa State University, USA (1997-1999). Later on, he became a Visiting Assistant Professor at Texas A&M University, College Station, USA (1999-2000), and then a Postdoctoral Research Associate at the University of Missouri, Columbia, USA (2000-2002). He joined Prairie View A&M University, Prairie View, USA, in 2002. Since then, his work has been focused on understanding the properties of strongly correlated electronic systems, fractional/integer quantum Hall effect, semiconductor quantum dots, and low-dimensional devices. His research has been funded by several USA funding agencies, including the National Science Foundation and the Department of Defense. He is the recipient of several teaching and research faculty awards. He is also a reviewer for many journals, where he provides expert advice to various professional organizations, and he serves within several editorial appointments in scientific journals.

 *nanomaterials*

MDPI

Editorial

Novel Research in Low-Dimensional Systems

Orion Ciftja

Department of Physics, Prairie View A&M University, Prairie View, TX 77446, USA; ogciftja@pvamu.edu

Citation: Ciftja, O. Novel Research in Low-Dimensional Systems. *Nanomaterials* **2023**, *13*, 364. https://doi.org/10.3390/nano13020364

Received: 20 December 2022
Accepted: 11 January 2023
Published: 16 January 2023

Copyright: © 2023 by the author. Licensee MDPI, Basel, Switzerland. This article is an open access article distributed under the terms and conditions of the Creative Commons Attribution (CC BY) license (https://creativecommons.org/licenses/by/4.0/).

Low-dimensional systems exhibit unique properties that have attracted considerable attention during the last few decades. Notably, the fabrication of low-dimensional systems and devices with lengths that measure in the nanometer range has opened up investigations in a "new" type of science—nanoscience. A bulk three-dimensional system's properties are typically insensitive to the size (as long as the size is macroscopic). However, all these considerations change when the size of such systems is reduced to the nanometer range. A known fact is that, unlike their bulk counterparts, many low-dimensional systems tend to exhibit novel and unique phenomena of great interest to many scientific disciplines. Furthermore, in the case of nanostructures, many of them manifest size-dependent properties, as well as behaviors that are strongly dictated by the rules of quantum mechanics. Therefore, understanding their properties is both highly interesting and rewarding because of the various possible technological applications. A great deal of progress has been achieved in the field of material science, including the fabrication of novel materials with length scales in the nanometer range [1–6]. Systems such as carbon nanotubes, nanowires, quantum dots, thin films, etc., manifest amazing properties and are already featured in several emerging technologies and advanced applications. The application of new and extraordinary experimental tools in the field has created an urgent need for an improved understanding of the new physical phenomena that occur in such low-dimensional systems. This has drawn the interest of many experimental and theoretical groups around the world [7–12]. The aim of this Special Issue is to provide an overview of the current research in low-dimensional systems by attracting contributions from specialists in the field. This way, we try to provide important insights on the large variety of scientifically fascinating and technologically important phenomena that are being investigated. The covered topics include original research articles on the fundamental and applied aspects of physics in various low-dimensional systems, such as quantum dots, graphene nanosystems, ultrathin films, superconducting nanofilms, novel nanoscale devices, etc. The present Special Issue includes research papers from both theoretical and experimental groups, with many phenomena studied from a multi-disciplinary perspective. There are 10 research papers in this Special Issue, which explore important developments in the field of low-dimensional systems.

The first paper by Metzke et al. [13] illustrates the use of atomic force microscopy (AFM)-based scanning thermal microscopy techniques to characterize the thermal properties of nanoscale systems. Specifically speaking, this work focuses on theoretical studies of ultrathin films with anisotropic thermal properties, such as hexagonal boron nitride (h-BN), and compares the results with a bulk silicon (Si) sample. The second paper by Kapcia [14] investigates the charge order on triangular lattices for fermionic particles that are described by an extended Hubbard model. A triangular lattice is formed by a single layer of graphene or graphite surfaces, as well as the (111) surface of face-centered cubic crystals. The author uses an extension of the lattice gas model for $S = 1/2$ fermionic particles on a two-dimensional triangular (hexagonal) lattice to analyze the system within the mean field approximation. The model's qualitative differences considering hypercubic lattices are also discussed. The third paper by Ciftja [15] represents a theoretical study of a nanocapacitor's electric properties. Such properties can be very different from the expected bulk properties due to the finite-size effects for small length scales. Additionally, a theoretical model for a circular parallel-plate nanocapacitor is considered, and analytic expressions for the nanocapacitor's stored electrostatic energy and capacitance are derived. The obtained results can be easily used to incorporate the effects of a dielectric thin film in case the space between the circular plates of the nanocapacitor is filled with such a film. The fourth paper by Wu et al. [16] considers a graphene nanoribbon gap waveguide as a candidate system for guiding dispersionless gap surface-plasmon polaritons with deep-subwavelength confinement and low loss. An analytical model is developed to analyze the system, in which a reflection phase shift is employed to successfully deal with the influence caused by the boundaries of the graphene nanoribbon. The proposed setup may be of great interest in studying dispersionless and low-loss nanophotonic devices and may have various potential technological applications. The fifth paper by Du et al. [17] focuses on the properties of graphene-based nanocomposite films. Nanocomposite films of this nature are in high demand due to their superior photoelectric and thermal properties; however, their stability and mechanical properties pose challenges. Motivated by these facts, this work illustrates a facile approach that can be used to prepare various nanocomposite films through the non-covalent self-assembly of graphene oxide and biocompatible proteins. Various characterization techniques were employed to characterize the properties of such nanocomposites and to track the interactions between graphene oxide and proteins. It is suggested that this strategy should be facile and effective for fabricating well-designed bio-nanocomposites for universal functional applications. The sixth paper by Wang et al. [18] reports the findings of a study on the influence of ink properties on the morphology of long-wave infrared HgSe quantum-dot films. The main focus of the analysis are the various factors affecting the morphology of the films, including ink surface tension, particle size, and solute volume fraction. This work is important for the morphology control of the filter film arrays, which are core components to many optoelectronic devices and for detecting targets by spectroscopic methods. The various system properties were analyzed in terms of different changing variables. The seventh paper by Alotabi et al. [19] studies the effect of TiO_2 film thickness on the stability of Au_9 clusters with a CrO_x layer. The high-purity TiO_2 films are fabricated via radio-frequency magnetron sputtering techniques, which allow the reliable control of film thickness and uniform morphology. The change in surface roughness upon heating two TiO_2 films with different thicknesses was investigated. Chemically-synthesized phosphine-protected Au_9 clusters covered by a photo-deposited CrO_x layer were used as a probe. It was found that the high mobility of the thick TiO_2 film after heating leads to a significant agglomeration of the Au_9 clusters, even when protected by the CrO_x layer. The eighth paper by Abramkin and Atuchin [20] is a theoretical analysis of hole states energy spectra in novel InGaSb/AlP self-assembled III-V quantum dots. These materials may have possible applications in non-volatile memories. Material intermixing and the formation of strained structures were also taken into account. The authors found that adjusting the values of various parameters allows one to find an optimal device configuration for possible non-volatile memory applications. The search for novel self-assembled quantum dots with hole-localization energy that allow a long charge storage is very important to the field of non-volatile memory applications. The ninth paper by McNaughton et al. [21] studies the causes and consequences of ordering and the dynamic phases of confined vortex rows in superconducting nanostripes. Superconducting nanostripes are a fundamental component in superconducting electronics, and they are crucial components for various applications in the field of quantum technology. Therefore, understanding the behavior of vortices under nanoscale confinement in superconducting circuits is important for the development of superconducting electronics and quantum technologies. Numerical simulations based on the Ginzburg–Landau theory for non-homogeneous superconductivity in the presence of magnetic fields are also carried out. The findings lead to the understanding of how lateral confinement organises vortices in a long superconducting nanostripe. A phase diagram of vortex configurations as a function of the stripe width and magnetic field is also presented. The tenth paper by Sharma et al. [22] sheds light on complex phase-fluctuation effects correlated with granularity in superconducting NbN nanofilms. Superconducting nanofilms are tunable systems that can result in the Berezinskii–Kosterlitz–Thouless superconducting transition when the system approaches the two-dimensional regime. Reducing the dimensionality further to quasi one-dimensional superconducting nanostructures with disorder can generate quantum and thermal phase slips of the order parameter. Experimental studies of these phenomena are difficult. As a result, the characterization of superconducting NbN nanofilms under different conditions carried out in this study can be very useful for future work.

Funding: The author was funded in part by the National Science Foundation (NSF) Grant No. DMR-2001980.

Acknowledgments: I am grateful to all the authors who contributed to this Special Issue. We also express our acknowledgments to the referees for reviewing the manuscripts.

Conflicts of Interest: The author declares no conflict of interest.

References

1. Ashoori, R.C.; Stormer, H.L.; Weiner, J.S.; Pfeiffer, L.N.; Baldwin, K.W.; West, K.W. Single-electron capacitance spectroscopy of discrete quantum levels. *Phys. Rev. Lett.* **1992**, *68*, 3088. [CrossRef] [PubMed]
2. Kastner, M.A. Artificial atoms. *Phys. Today* **1993**, *46*, 24. [CrossRef]
3. Ciftja, O. Classical behavior of few-electron parabolic quantum dots. *Physica B* **2009**, *404*, 1629. [CrossRef]
4. Kim, Y.; Han, H.; Vrejoiu, I.; Lee, W.; Hesse, D.; Alexe, M. Cross talk by extensive domain wall motion in arrays of ferroelectric nanocapacitors. *Appl. Phys. Lett.* **2011**, *99*, 202901. [CrossRef]
5. Hong, N.H.; Raghavender, A.T.; Ciftja, O.; Phan, M.H.; Stojak, K.; Srikanth, H.; Zhang, Y.H. Ferrite nanoparticles for future heart diagnostics. *Appl. Phys. A* **2013**, *112*, 323. [CrossRef]
6. Ciftja, O. Understanding electronic systems in semiconductor quantum dots. *Phys. Scr.* **2013**, *88*, 058302. [CrossRef]
7. Ruiz, F.; Sun, W.D.; Pollak, F.H.; Venkatraman, C. Determination of the thermal conductivity of diamond-like nanocomposite films using a scanning thermal microscope. *Appl. Phys. Lett.* **1998**, *73*, 1802. [CrossRef]
8. Unutmaz, M.A.; Unlu, M. Terahertz spoof surface plasmon polariton waveguides: A comprehensive model with experimental verification. *Sci. Rep.* **2019**, *9*, 8. [CrossRef] [PubMed]
9. Kim, S.J.; Choi, K.; Lee, B.; Kim, Y.; Hong, B.H. Materials for flexible, stretchable electronics: Graphene and 2D materials. *Annu. Rev. Mater. Res.* **2015**, *45*, 63. [CrossRef]
10. Burns, S.E.; Cain, P.; Mills, J.; Wang, J.; Sirringhaus, H. Inkjet printing of polymer thin-film transistor circuits. *MRS Bull.* **2011**, *28*, 829. [CrossRef]
11. Bezryadin, A. Quantum suppression of superconductivity in nanowires. *J. Phys. Cond. Mat.* **2008**, *20*, 043202. [CrossRef]
12. Breznay, N.; Tendulkar, M.; Zhang, L.; Lee, S.C.; Kapitulnik, A. Superconductor to weak-insulator transitions in disordered tantalum nitride films. *Phys. Rev. B* **2017**, *96*, 134522. [CrossRef]
13. Metzke, C.; Kühnel, F.; Weber, J.; Benstetter, G. Scanning thermal microscopy of ultrathin films: Numerical studies regarding cantilever displacement, thermal contact areas, heat fluxes, and heat distribution. *Nanomaterials* **2021**, *11*, 491. [CrossRef]
14. Kapcia, K.J. Charge-order on the triangular lattice: A mean-field study for the lattice S = 1/2 fermionic gas. *Nanomaterials* **2021**, *11*, 1181. [CrossRef]
15. Ciftja, O. Energy stored and capacitance of a circular parallel plate nanocapacitor. *Nanomaterials* **2021**, *11*, 1255. [CrossRef]
16. Wu, Z.; Zhang, L.; Ning, T.; Su, H.; Li, I.L.; Ruan, S.; Zeng, Y.-J.; Liang, H. Graphene nanoribbon gap waveguides for dispersionless and low-loss propagation with deep-subwavelength confinement. *Nanomaterials* **2021**, *11*, 1302. [CrossRef]
17. Du, C.; Du, T.; Zhou, J.T.; Zhu, Y.; Jia, X.; Cheng, Y. Enhanced stability and mechanical properties of a graphene-protein nanocomposite film by a facile non-covalent self-assembly approach. *Nanomaterials* **2022**, *12*, 1181. [CrossRef]
18. Wang, S.; Zhang, X.; Wang, Y.; Guo, T.; Cao, S. Influence of ink properties on the morphology of long-wave infrared HgSe quantum dot films. *Nanomaterials* **2022**, *12*, 2180. [CrossRef] [PubMed]
19. Alotabi, A.S.; Yin, Y.; Redaa, A.; Tesana, S.; Metha, G.F.; Andersson, G.G. Effect of TiO_2 film thickness on the stability of Au_9 clusters with a CrO_x layer. *Nanomaterials* **2022**, *12*, 3218. [CrossRef] [PubMed]
20. Abramkin, D.S.; Atuchin, V.V. Novel InGaSb/AlP quantum dots for non-volatile memories. *Nanomaterials* **2022**, *12*, 3794. [CrossRef] [PubMed]
21. McNaughton, B.; Pinto, N.; Perali, A.; Milošević, M.V. Causes and consequences of ordering and dynamic phases of confined vortex rows in superconducting nanostripes. *Nanomaterials* **2022**, *12*, 4043. [CrossRef] [PubMed]
22. Sharma, M.; Singh, M.; Rakshit, R.K.; Singh, S.P.; Fretto, M.; De Leo, N.; Perali, A.; Pinto, N. Complex phase-fluctuation effects correlated with granularity in superconducting NbN nanofilms. *Nanomaterials* **2022**, *12*, 4109. [CrossRef] [PubMed]

Disclaimer/Publisher's Note: The statements, opinions and data contained in all publications are solely those of the individual author(s) and contributor(s) and not of MDPI and/or the editor(s). MDPI and/or the editor(s) disclaim responsibility for any injury to people or property resulting from any ideas, methods, instructions or products referred to in the content.

nanomaterials

MDPI

Article

Scanning Thermal Microscopy of Ultrathin Films: Numerical Studies Regarding Cantilever Displacement, Thermal Contact Areas, Heat Fluxes, and Heat Distribution

Christoph Metzke [1,2], Fabian Kühnel [1], Jonas Weber [1,3,4] and Günther Benstetter [1,*]

1   Department of Electrical Engineering and Media Technology, Deggendorf Institute of Technology, Dieter-Görlitz-Platz 1, 94469 Deggendorf, Germany; christoph.metzke@th-deg.de (C.M.); fabian.kuehnel@th-deg.de (F.K.); jonas.weber@th-deg.de (J.W.)
2   Department of Electrical Engineering, Helmut Schmidt University/University of the Federal Armed Forces Hamburg, Holstenhofweg 85, 22043 Hamburg, Germany
3   Institute of Functional Nano and Soft Materials, Collaborative Innovation Center of Suzhou Nanoscience & Technology, Soochow University, 199 Ren-Ai Road, Suzhou 215123, China
4   Department of Applied Physics, University of Barcelona, Martí i Franquès 1, 08028 Barcelona, Spain
*   Correspondence: guenther.benstetter@th-deg.de

Abstract: New micro- and nanoscale devices require electrically isolating materials with specific thermal properties. One option to characterize these thermal properties is the atomic force microscopy (AFM)-based scanning thermal microscopy (SThM) technique. It enables qualitative mapping of local thermal conductivities of ultrathin films. To fully understand and correctly interpret the results of practical SThM measurements, it is essential to have detailed knowledge about the heat transfer process between the probe and the sample. However, little can be found in the literature so far. Therefore, this work focuses on theoretical SThM studies of ultrathin films with anisotropic thermal properties such as hexagonal boron nitride (h-BN) and compares the results with a bulk silicon (Si) sample. Energy fluxes from the probe to the sample between 0.6 µW and 126.8 µW are found for different cases with a tip radius of approximately 300 nm. A present thermal interface resistance (TIR) between bulk Si and ultrathin h-BN on top can fully suppress a further heat penetration. The time until heat propagation within the sample is stationary is found to be below 1 µs, which may justify higher tip velocities in practical SThM investigations of up to 20 µms^{-1}. It is also demonstrated that there is almost no influence of convection and radiation, whereas a possible TIR between probe and sample must be considered.

Keywords: scanning thermal microscopy (SThM); numerical study; finite element analysis (FEA); boron nitride; h-BN; ultrathin films; heat transfer; thermal contact; penetration depth; stationary time

Citation: Metzke, C.; Kühnel, F.; Weber, J.; Benstetter, G. Scanning Thermal Microscopy of Ultrathin Films: Numerical Studies Regarding Cantilever Displacement, Thermal Contact Areas, Heat Fluxes, and Heat Distribution. *Nanomaterials* **2021**, *11*, 491. https://doi.org/10.3390/nano11020491

Academic Editor: Orion Ciftja

Received: 20 January 2021
Accepted: 15 February 2021
Published: 16 February 2021

Publisher's Note: MDPI stays neutral with regard to jurisdictional claims in published maps and institutional affiliations.

Copyright: © 2021 by the authors. Licensee MDPI, Basel, Switzerland. This article is an open access article distributed under the terms and conditions of the Creative Commons Attribution (CC BY) license (https://creativecommons.org/licenses/by/4.0/).

1. Introduction

Since the thermal properties of thin films vary significantly from those of the corresponding bulk materials [1–4], new promising materials for micro and nanoscale devices require a detailed investigation with advanced techniques. Moreover, due to several atomic and molecular effects, such as grain boundaries, or the transition to ballistic heat transport, the thermal characterization becomes increasingly challenging. Methods for thin-film thermal characterization are also limited by various factors such as film thickness or anisotropic material properties. One possible method is SThM, which is specifically designed to characterize the local thermal properties of thin films. SThM is applied in an AFM system together with additional measurement equipment. It is a method in which a cantilever is in direct physical contact with a sample. The sample is scanned in a special pattern to obtain local thermal properties. SThM thermal images are likely to be influenced by the sample's topography, which has been explained in literature in recent years [5–12]. To ensure a correct interpretation of the recorded measurement results, a deep understanding of heat transfer during SThM measurements is essential to comprehend the origin and impact of those and further effects. This work aims to provide these insights.

In this work, the goal is to accurately predict the heat transfer process during realistic SThM measurements of ultrathin films such as h-BN and a bulk Si sample. Therefore, theoretical calculations and finite element analysis (FEA) are performed. FEA is a versatile tool to simulate, e.g., heat transfer or mechanical problems that can be described by mathematical equations. Similar SThM measurements with h-BN have been performed in recent works [12]. We tried to create theoretical measurement setups that comply with real scenarios so that the theoretical results may be adopted by other researchers to compare them with practical measurements or to explain certain effects of SThM applied to ultrathin films.

The literature in recent years especially focused on qualitative local thermal properties. Leitgeb, Fladischer et al. investigated 500 nm tungsten films employing SThM based on the 3ω method and compared the results to the time domain thermoreflectance technique. Said tungsten films would have a thermal conductivity between 151.4 $Wm^{-1}K^{-1}$ and 156.0 $Wm^{-1}K^{-1}$ depending on the heat treatment [13,14]. Chen et al. estimated the thermal conductivity of a single SiO_2 nanoparticle at 300 K to 0.95 $\pm$ 0.08 $Wm^{-1}K^{-1}$ using SThM. They also found out that the TIR between the probe and the nanoparticle accounts for 70% of the total thermal resistance. Therefore, TIRs would have to be considered in thermal conductivity measurements [15]. Existing TIRs from metal/non-metal and non-metal/non-metal contacts within a sample were studied by Park et al. using ultrahigh vacuum SThM. They suggest the presented method to be used actively for nanoscale TIR measurements and show the significant contribution of TIRs to the entire heat dissipation at the nanoscale [16]. Chirtoc et al. conclude that the heat management of nanofabricated thermal probes might be optimized by decreasing the TIRs between tip and sample [17]. To obtain local thermal properties using SThM, a tip calibration is necessary. A new calibration method for local temperature measurements is introduced by Nguyen et al., whose active thermal microchips could be used for different SThM probes. In addition, they estimate the TIRs between tip and sample, which would have a great impact on the results [18]. Recent literature demonstrates the necessity of detailed insights into the heat transfer process of SThM measurements, but also the great possibilities of SThM. This work aims to provide values regarding energy fluxes, heat distribution and the influence of possible TIRs between tip and sample. Said values can hardly be found in the literature, especially for h-BN or Si samples, which are interesting in our field of research. We want to support other researchers to fully understand and interpret upcoming effects in their practical SThM investigations.

The measurement and simulation procedure applied in this research work is a sequence of interdependent steps. First, the geometry of a used SThM probe was investigated via SEM (Section 4.1). The information of the SEM images provided the fundamentals to model the probe in SolidWorks (SW). The geometry was then imported into COMSOL Multiphysics (CM). Here, the cantilever displacement under specific forces (Section 4.2) was simulated. After a theoretical calculation of the thermal contact area (TCA) between tip and sample (Section 4.3), we were able to model realistic measurement scenarios using a sample with an ultrathin top surface, which exhibits anisotropic thermal conductivity (Section 4.4). Parametric sweep studies enabled the simulation of different probe temperatures and various anisotropic material properties. Subsequently, these simulations were compared to those with a standard bulk Si sample with isotropic thermal properties (Section 4.5). Both simulation setups deliver particular results concerning the heat distribution and the heat fluxes from tip to sample. Henceforth, the time until the heat transfer process becomes stationary was investigated (Section 4.6). The influence of convection and radiation on the present simulations was also taken into account (Section 4.7), as well as the influence of a possible TIR between tip and sample (Section 4.8). Finally, the applied meshing strategy is verified to demonstrate the reliability of our results (Section 4.9).

2. Materials and Methods

AFM is a method to characterize surfaces according to topographical, mechanical, and electrical properties on a nanometer scale. One of the key parts is a microfabricated probe with an ultrasharp tip to contact the sample surface. Tip radii are in the small nm range (depending on the desired application), e.g., Bruker VITA-DM-NANOTA-200 with a tip radius of up to 30 nm and a tip height of 3–6 μm for a new tip [19,20]. The sample is scanned with a predefined number of lines and readings per line, which results in a line-by-line image. SThM is a method to qualitatively map local thermal conductivities and temperatures as a subcategory of AFM that requires additional measuring equipment. In standard applications, SThM measurements are performed in contact mode. Here, the tip is in direct physical contact with the surface under investigation. The force between tip and sample is held constant in most cases and can be defined by the user during the measurement. Typically, the force of contact mode measurements is within the range of 10 nN to 100 nN [21]. Moreover, a thermal signal is measured and assigned simultaneously to the corresponding topography. Thereunto, a thermal resistive probe is heated with a specific heating power. The temperature of the probe depends on the heat exchange between tip and sample and, therefore, on the sample's local thermal conductivity. If the local thermal conductivity of the sample is high, more heat can spread into the sample, causing a temperature decrease of the probe. The AFM relies on a feedback algorithm and a Wheatstone bridge to evaluate the local thermal conductivities. As a result, SThM measurements create two images simultaneously, one topography and one thermal image of the same position. Such images can be found in [12]. For more details regarding the SThM measurement process, please refer to [22].

In our field of research, special focus is on the thermal characterization of ultrathin films such as h-BN with thicknesses of approximately 10 nm, which are supposed to have anisotropic thermal properties. Recent works demonstrated the possibility of such SThM measurements [12]. This work continues said research and compares it to "standard" SThM measurements of bulk Si with isotropic thermal properties. We deploy the following SThM probe, soft- and hardware:

- Thermal probe: SThM probe Bruker VITA-DM-NANOTA-200 (Bruker Corporation, Billerica, MA, USA) [19,20]; The thermal probe is made from crystalline Si and can be heated repeatedly and reliably up to temperatures of 350 °C, according to manufacturer information. Consecutively, we use the term "probe" for the whole thermal probe (as it can be purchased), the term "cantilever" for the flexible mechanical part of the probe, and the term "tip" only for the sharp area on the front side of the probe, which is in direct contact with the surface of the samples.
- SEM investigation: Zeiss ULTRA 55 (Carl Zeiss AG, Oberkochen, Germany);
- Modeling process: SolidWorks 2020 (SW; Dassault Systemes Deutschland GmbH, Stuttgart, Germany);
- Simulations: COMSOL Multiphysics 5.5 (CM; Comsol Multiphysics GmbH, Göttingen, Germany), which is a versatile FEA tool to simulate, e.g., heat transfer or mechanical problems that can be described by mathematical equations.

3. Theoretical Background

3.1. Cantilever Displacement

The displacement of a beam w(l), which is fixed at one end and stressed by a single force F at the other end, can be calculated using Equation (1), where l is the cantilever length and E Young's modulus. The moment of inertia I of a cantilever with a rectangular cross-section can be calculated with $I = \left(b \cdot h^3\right)/12$, in which b is the width and h is the height of the cantilever [23].

$$w(l) = \frac{1}{3} \cdot \frac{F \cdot l^3}{E \cdot I} \text{ in } [m] \tag{1}$$

The bending stress σ_b can be calculated according to Equation (2) by the division of the bending moment $M_b = F \cdot l$ and the moment of resistance $W = \left(b \cdot h^2\right)/6$ (in case of a rectangular beam) [23].

$$\sigma_b = \frac{M_b}{W} \text{ in } \left[\text{Nm}^{-2}\right] \tag{2}$$

In Section 4.2, Equations (1) and (2) are used to calculate the cantilever displacement and the bending stress in SThM measurements.

3.2. Hertzian Surface Pressure

Hertzian surface pressure occurs when two rigid bodies touch each other. In the special case of a sphere with radius R, Poisson's ratio ν_1 and Young's modulus E_1 touching a flat surface with Poisson's ratio ν_2 and Young's modulus E_2 under a force F, the indentation depth d can be calculated using Equation (3) [24].

$$d = \sqrt[3]{\left(\frac{3F}{4\,E' \cdot \sqrt{R}}\right)^2} \text{ in } [\text{m}] \tag{3}$$

E' is the combined Young's modulus and can be calculated according to Equation (4) [24].

$$E' = \frac{E_1 \cdot E_2}{\left(1 - \nu_1^2\right) \cdot E_2 + \left(1 - \nu_2^2\right) \cdot E_1} \text{ in } \left[\text{Nm}^{-2}\right] \tag{4}$$

The touching radius r can then be calculated with $r = \sqrt{R \cdot d}$ [24]. It must be stated that using the touching radius r does not lead to the mathematical exact contact area as the contact area is not a flat circle but a small section of a sphere. However, the calculated indentation depths are by far smaller than the touching radius ($d(r) = r^2/R$; e.g., d = 0.1 nm and r = 5.7 nm with r_{tip} = 300 nm and tip force = 100 nN) and can thus be neglected. We, therefore, use the touching radius r for the calculation of the TCAs in SThM measurements (Section 4.3) as an adequate approximation for the simulations in Section 4.4 to Section 4.9.

3.3. Heat Transfer

Equation (5) represents the general heat conduction equation in the three-dimensional case without inner heat sources, in which the quotient $k/(\rho \cdot c)$ is expressed by the thermal diffusivity a [25].

$$\frac{dT}{dt} = \frac{k}{\rho \cdot c} \cdot \left(\frac{\partial^2 T}{\partial x^2} + \frac{\partial^2 T}{\partial y^2} + \frac{\partial^2 T}{\partial z^2}\right) = a \cdot \Delta T \tag{5}$$

If there are inner heat sources h(t) (e.g., joule heating or chemical reactions), the inhomogeneous heat conduction Equation (6) follows (5) [25]. In this work, thermal conduction is applied to the entire geometry of the simulations in Section 4.4 to Section 4.9. CM solves said equations using numerical methods.

$$\frac{dT}{dt} = a \cdot \Delta T + h(t) \tag{6}$$

Convection is a mass bound energy transport caused by macroscopic particle movement. Considering free convection, this flow mainly results from the temperature-dependent particle movement and the thermal buoyancy. The reason for this is that fluids normally expand with increasing temperatures and therefore exhibit lower densities. Forced convection occurs if, e.g., a ventilator causes the flow field. Forced convection then overlaps with free convection. The influence of free convection is studied in Section 4.7 using Equation (7), where α is the heat transfer coefficient, which depends on the materials, surfaces and ambient conditions, A_{conv} the convective area and T_2 and T_1 the temperatures of the material and the surrounding fluid (e.g., air), respectively [25]. In this work, the influence of convection can be neglected. This is further discussed in Section 4.7.

$$\dot{Q}_{conv} = \alpha \cdot A_{conv} \cdot (T_2 - T_1) \text{ in } [W] \tag{7}$$

Thermal radiation is not mass bound. Therefore, it occurs even under high vacuum conditions. Electromagnetic waves run from one body surface to another. Each body absorbs and emits radiation. In the special case of a small body with an area A_{rad} and temperature T_2 surrounded by a much greater emissive area with a temperature T_1, the net radiative heat flow of the body can be calculated with Equation (8), using the Stefan-Boltzmann law [25].

$$\dot{Q}_{rad} = -\epsilon_{rad} \cdot \sigma \cdot A_{rad} \cdot \left(T_2^4 - T_1^4\right) \text{ in } [W] \tag{8}$$

The negative radiation constant $-\epsilon_{rad}$ depends on the considered body, and σ is the Stefan-Boltzmann constant. When T_2 is greater than T_1 the sign of $\dot{Q}_{rad}$ is negative which means, that the considered body loses energy. In this work, the influence of radiation of the heated SThM cantilever can be neglected. This is further discussed in Section 4.7.

4. Results and Discussion

4.1. SEM Investigation of a Used SThM Tip

To obtain vital information about the geometry of the cantilever and tip, we performed an SEM investigation of the thermal SThM probe Bruker VITA-DM-NANOTA-200 [20], which was already in use. This facilitated modeling the probe in detail in SW and paved the way for the simulations. SEM investigations require a conductive connection between the sample and the holder to avoid electrical overcharging caused by trapped electrons from the electron beam. Therefore, the probe was fixed on a sample holder using conductive silver and gold in a sputtering process. Figure 1a–c shows the probe before and after the preparation process, as well as inside the sample chamber of the SEM. To collect precise geometry data, the probe was investigated in two positions, one side view and one top view. The results are depicted in Figure 1d–i. Therefore, the tip radius was estimated to be approximately 300 nm. The geometry data for the simulations were created upon these images and are depicted in Figure 1j.

We assume a tip radius of 300 nm to represent an "average" used probe. However, an "average" used probe can hardly be defined as it depends on various factors such as the number of measurements, tip velocities and forces, sample materials, and many more. Mostly, SThM probes are used as long as they deliver reliable results, and from our experience, such tips exhibit tip radii within the region of 300 nm. Furthermore, new tips may degrade much faster than used tips, which is the reason why larger tip radii of around 300 nm may occur more often in practical SThM measurements. As the main residue is not located directly at the TCA (see Figure 1f), we assume it to have no influence. However, residues may increase the TIR between tip and sample. This effect is further discussed in Section 4.8.

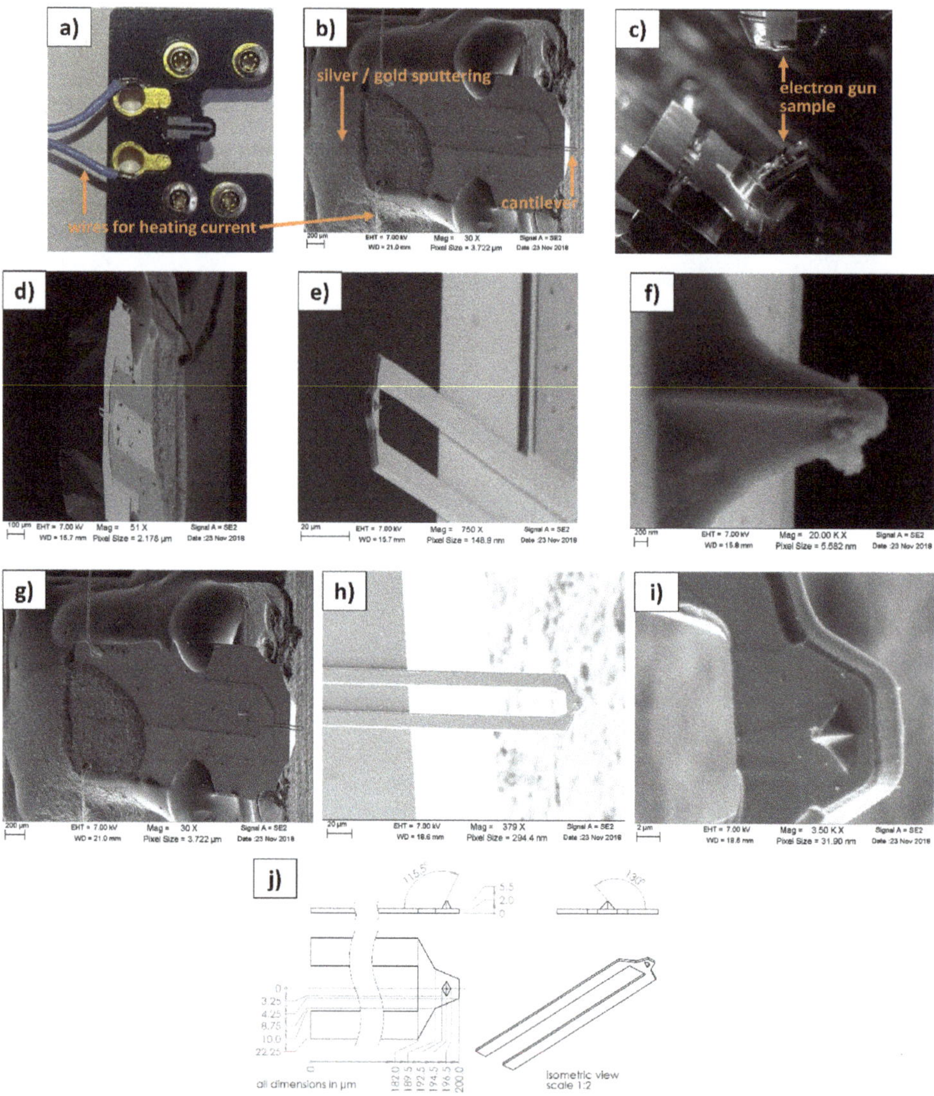

Figure 1. (**a**) Photography image of the scanning thermal microscopy (SThM) probe Bruker VITA-DM-NANOTA-200 before preparation for SEM. (**b**) SEM image of a fully prepared tip. The sputtering area, wires for the heating current, and the cantilever are visible. (**c**) Image of the sample fixation inside the SEM. (**d**) Side-view close-up of the tip in (**e**,**f**). (**g**) Top-view close-up of the tip in (**h**,**i**). (**j**) Dimensions of the probe used for the simulations and isometric view. This SEM investigation made it possible to estimate the geometry of the cantilever and the radius of the used tip, which is approximately 300 nm.

4.2. Cantilever Displacement and Von Mises Stress

To study the displacement and the von Mises stress of the cantilever, a 3D-simulation was set up. The cantilever (geometry see Figure 1j) is fixed by the two connectors on one end. The simulated load is applied directly to the cantilever tip on the other end. All other edges and areas are set to be free. Moreover, gravity is set to a value of 9.81 ms^{-2} in the z-direction. For the cantilever, the material bulk Si was used with Young's modulus of 131 GPa and a Poisson's ratio of 0.221 [26]. The mesh was built with the option *Physics-*

controlled extremely fine, which provides the finest automatically built mesh in CM. The study was performed in the stationary regime.

Figure 2a illustrates the total displacement in µm, (b) represents the von Mises stress in Nm^{-2} under a tip force of 100 nN, which acts directed towards the tip in the negative z-direction. Logically, the total displacement reaches its maximum value at the largest x-coordinate, whereas the von Mises stress achieves its maximum directly at the clamping. Here are some important simulation results for a tip force of 100 nN. The simulation results for 10 nN and 1000 nN can be derived by the multiplication of the following values with 0.1 and 10, respectively:

- Total displacement: 0.114 µm at the center point of the tip and 0.119 µm at the free end of the cantilever;
- Maximum von Mises stress at the fixed end of the cantilever: $1.26 \times 10^6 \ \text{Nm}^{-2}$.

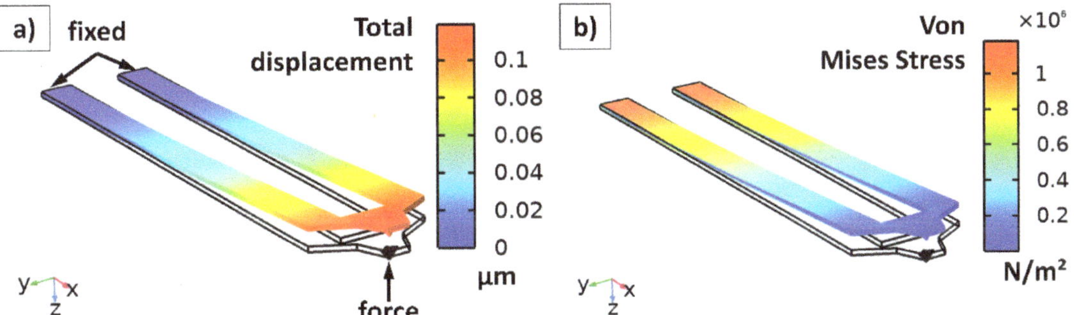

Figure 2. Simulation results under a tip force of 100 nN. (**a**) Total displacement in µm. (**b**) Von Mises stress in Nm^{-2}.

These values were also calculated manually for a simplified cantilever with a rectangular cross-section (b = 24.5 µm, h = 2 µm and l = 194.5 µm) fixed on only one end using the Equations (1) and (2). The results of the calculations are w (l = 194.5 µm) = 0.115 µm and $\sigma_b = 1.19 \times 10^6 \ \text{Nm}^{-2}$. Compared to the simulation results above (0.114 µm and $1.26 \times 10^6 \ \text{Nm}^{-2}$), it is obvious that the exact geometry around the tip has just little influence on the displacement and the maximum von Mises stress. The breaking strength of Si is assumed to be in the range from $5 \times 10^7 \ \text{Nm}^{-2}$ to $20 \times 10^7 \ \text{Nm}^{-2}$ [27]. Compared to the maximum von Mises stress at the fixed end of the cantilever ($1.26 \times 10^6 \ \text{Nm}^{-2}$), there is a safety factor of approximately 40 to 160 before the cantilever breaks. Based on this calculation, the maximum tip forces that can be applied to the tip without breaking the cantilever are in the range from 4 µN to 16 µN. However, such values will not be reached in reasonable SThM measurements.

4.3. Calculation of TCAs in Realistic SThM Investigations

To estimate TCAs between an SThM probe and different samples, we performed a mathematical evaluation regarding Equations (3) and (4). We calculated six different cases: a new Si probe with a tip radius of 30 nm and an "average" used to probe with a tip radius of 300 nm each combined with the Si sample, SiO_2 sample and h-BN sample. The tip radius (300 nm) for an "average" used probe was determined by SEM investigations (Section 4.1). For the calculations, we used the material properties in Table 1. The calculated thermal contact areas are depicted in Figure 3.

Table 1. Material properties are used for the theoretical calculation of the thermal contact areas (TCAs) in Figure 3. Values for SiO_2 are approximated from the predefined material in COMSOL Multiphysics (CM).

Material	Si	SiO_2	h-BN
Young's modulus (GPa)	131 [26]	70	850 [28]
Poisson's ratio (–)	0.221 [26]	0.157	0.2 [29]

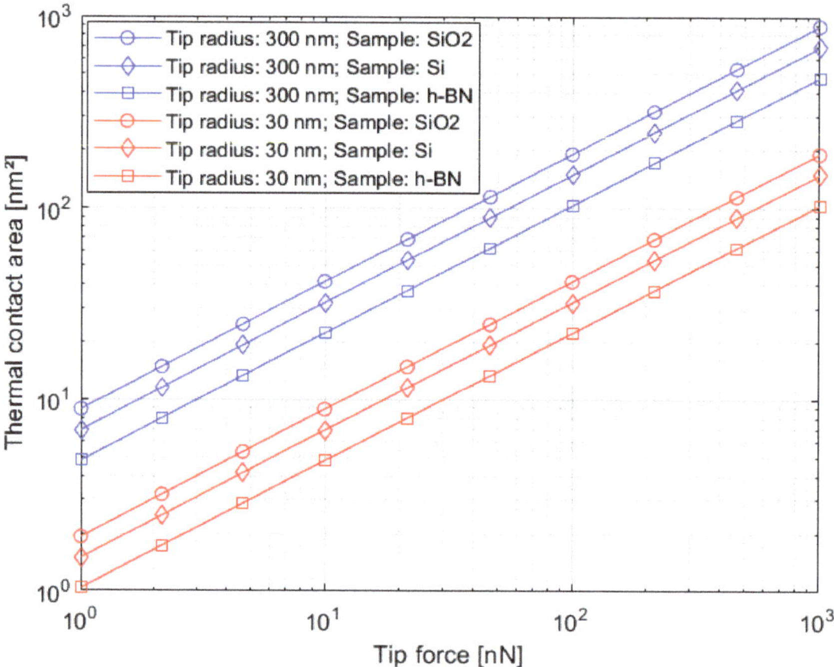

Figure 3. Calculated TCAs in realistic SThM measurements for different probe-sample combinations. The red curves represent a new probe with a tip radius of 30 nm, and the blue curves represent an "average" used probe with a tip radius of 300 nm. A similar investigation can be found in [30].

We chose the above described tip-sample combinations for reasons of reusability as these configurations are comparatively common and might also occur in similar SThM investigations of other researchers. We also used the calculated TCAs of the h-BN curves to calculate the thermal contact radius (TCR) for the simulations in Section 4.4 to Section 4.9.

Appropriate material properties for h-BN are not easy to estimate, as in ultrathin (2D) materials, they are also strongly dependent on factors such as material orientation and exact film thickness. However, in our calculations concerning the h-BN sample, the Poisson's ratio has just little influence on the calculated contact area. The main factor is Young's modulus, as it is assumed to be extremely high in comparison to Si and SiO_2. Consequently, if other researchers might find different values of Young's modulus of h-BN appropriate for their individual case, the contact areas can be approximated to lie in between the h-BN and SiO_2 curves in Figure 3. Prerequisites are a similar measurement setup and Young's modulus between 70 and 850 GPa. Nevertheless, it must be stated that the lower Young's modulus, the higher becomes the influence of different values for the Poisson's ratio.

For our simulations, we assume the calculated ideal contact areas in this section to be realistic, as the impact of surface roughness decreases with lower surface roughnesses. We also assume roughness to be neglectable as we work with samples exhibiting surface roughnesses in the sub-nm area. In the following sections, we use the calculated TCAs to simulate the heat transfer in SThM measurements.

4.4. Heat Spreading in SThM Measurements Regarding Ultrathin h-BN Film

To study the heat spreading in SThM measurements on a sample with an ultrathin h-BN film on top, this simulation was set up.

Simulation setup: For the cantilever (dimensions see Figure 1j), the CM-predefined material bulk Si was set. The rectangular sample (footprint 100 µm × 100 µm) consists of the CM-predefined material h-BN with a thickness of 10 nm on top of 10 µm thick bulk Si. Regarding the anisotropic thermal properties of h-BN, we assume a cross-plane thermal conductivity of h-BN of 1 $Wm^{-1}K^{-1}$. The in-plane thermal conductivity of h-BN was defined using the parameter k, which varies depending on the simulations [31]. The thermal contact resistance between h-BN and bulk Si is assumed to be 4×10^{-8} Km^2W^{-1}. However, this value can only be estimated and will vary in practical measurements depending on material quality, ambient conditions and manufacturing processes. We chose this value inspired by investigations on similar material stacks such as graphene/Si [32] since the atomic structure of graphene is similar to h-BN. Villaroman et al. estimate the thermal contact resistance between graphene and Si to $3.1–4.9 \times 10^{-8}$ Km^2W^{-1} [32]. Values in the same order of magnitude can be found in [33,34].

The initial value for the temperature was set to 293.15 K for all boundaries. Between tip and sample, an ideal thermal contact was defined. We calculated the TCR of the TCA as follows: With Figure 3 (blue squared h-BN curve), we assumed a realistic TCA of ~103.4 nm^2 for r_{tip} = 300 nm and a tip force of 100 nN, which leads to a TCR of ~5.7 nm (TCA = $TCR^2 \cdot \Pi$). This appears to be a realistic value for an ideal thermal contact between a used SThM probe and a sample with an ultrathin h-BN film on top. The temperature of the topside of the cantilever is defined using the parameter *temp*, which varies between 50 °C and 200 °C depending on the simulations. These are appropriate cantilever temperatures for practical SThM investigations [12]. The remaining outer areas of the cantilever were defined as thermal isolating as well as the top surface of the sample, while the remaining outer areas of the sample take over the ambient temperature of exactly 293.15 K. An overview of the simulation setup and boundary conditions is given in Figure 4. The center point of the TCA defines the origin of the coordinate system.

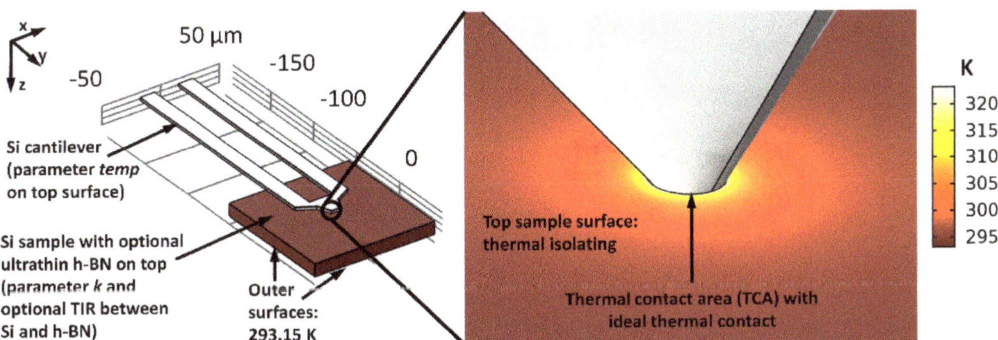

Figure 4. Simulation setup and boundary conditions. Left: The general setup consists of the heated SThM probe and the sample. Right: zoom into the TCA with ideal thermal contact between tip and sample.

Meshing strategy: To simulate the thermal contact with sufficient resolution, a user-controlled mesh was created. We defined the maximum dimension of a triangular mesh element for the smallest area of the entire simulation, the TCA of the cantilever, to 0.2 nm. Moreover, a circle with a radius of 500 nm around the TCA was defined, in which the maximum dimension of a mesh element is 5 nm. For the remaining geometry of the cantilever and sample, we used the predefined meshing strategy with element size normal. In the present simulations, this meshing strategy represents a good tradeoff between accuracy and simulation time and delivers trustworthy results. The meshing strategy is evaluated in Section 4.9, in which the reliability of our results is demonstrated. Figure 5 illustrates the applied meshing strategy in more detail.

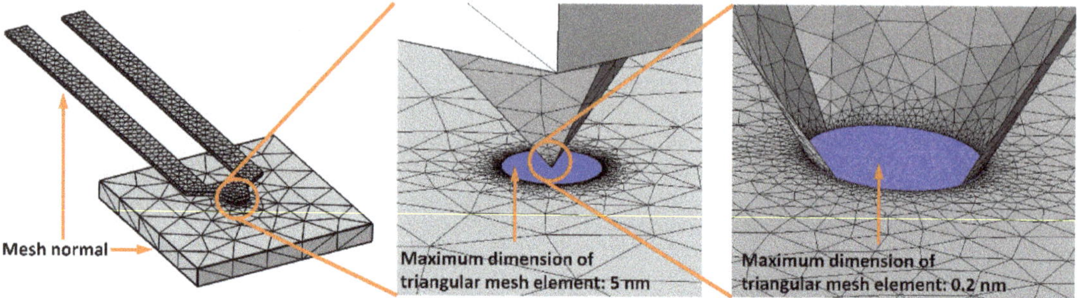

Figure 5. Meshing strategy zooming from the geometry overview to the small TCA.

Parametric studies: To investigate the cross- and in-plane heat distribution for different cases, a parametric sweep study in the stationary regime was performed. Every single combination of the parameters k and *temp* was calculated, which are defined as follows:

- k: Ratio between in-plane and cross-plane thermal conductivity $k = \lambda_\parallel / \lambda_\perp$ with $\lambda_\perp = 1\,\mathrm{Wm^{-1}K^{-1}}$ [31]. In the parametric sweep studies in this section, k resembles the values 1, 2, 5, 10, 20, 50, and 100;

- *temp*: Temperature of the top surface of the SThM cantilever (boundary condition in Figure 4). In the parametric sweeps, *temp* takes on the values from 50 °C to 200 °C in steps of 25 °C. For reasons of better differentiation and easier comparison with practical measurements, *temp* is always specified in °C, whereas all other temperatures are specified in K.

Simplifications: Simulations can never represent the real world as the results are only as good as the chosen simulation model. We tried to create the simulation models as realistic as possible. However, with regards to simulation time and the reusability of our results, we also had to introduce simplifications. The chosen simplifications do not decisively influence our results and are therefore justified. The main simplifications are listed below:

- We assume an ideal thermal contact between tip and sample in Section 4.4 to Section 4.7. TIRs in practical SThM measurements are quite hard to estimate as they depend on numerous factors such as surface roughness, material combination, vertical steps, tip material and geometry, contaminations, tip force, and some more. Due to these numerous influences, TIR will also vary greatly during a single SThM measurement. In literature, values in the range of $10^{-8}\,\mathrm{Km^2W^{-1}}$ to $10^{-10}\,\mathrm{Km^2W^{-1}}$ for different probes, samples, and measurement scenarios can be found [35–37]. Indeed, we assume the TIR to vary in a broader range due to the great number of possible measurement scenarios. To realize a comparison between the simulations in Sections 4.4 and 4.5, we assume an ideal thermal contact in said sections. The influence of a possible TIR is further studied in Section 4.8. It must also be stated that the present investigation in Section 4.4 focuses on ultrathin h-BN samples, which in fact are super flat as they are built of a stack of single h-BN layers. Roughnesses of high-quality h-BN are assumed to be less than 0.4 nm [38], which is significantly lower compared to a tip radius of 300 nm. As surface roughness has a great impact on TIR, values for h-BN samples should be lower compared to samples with higher roughnesses;

- We neglect a possible water meniscus around the tip. Other researchers also propose that a water meniscus can often be neglected in SThM measurements [22]. On one hand, a present water drop could cause heat conduction and may increase the amount of heat flux between tip and sample slightly. On the other hand, SThM measurements can also be performed under high vacuum conditions, where the appearance of a water meniscus can be excluded. We, therefore, focus on simulations without a possible water meniscus;
- In practical SThM measurements, the probe is scanning over the surface with a specific velocity. This is not presentable in simulations in the stationary regime, where a motionless probe is assumed. However, we justify such simulations in Section 4.6 through the investigation of the time until heat distribution is stationary;
- We also neglect radiation and convection as the influence on our simulated cases is neglectable. This is further discussed in Section 4.7.

Results: To compare single measurements, it is necessary to define useful measurable parameters. Hence, we define the thermal active radius (TAR) located on the top surface of the sample starting at the geometric center point of the tip. The temperature until the TAR is greater than 294.15 K, which represents a temperature rise of 1 K in comparison to ambient conditions. In Figure 6b, the heat flow in the z-direction is investigated. It is obvious that with increasing *temp*, the temperature at TCA also increases. However, an increasing k results in a greater heat spreading in x and y-direction, which leads to a stronger temperature decrease within the ultrathin h-BN film and the tip (compare curves of the same color each in Figure 6b. It can also be observed that the heat penetration in z-direction ends at the z-coordinate 10 nm, which is exactly where the h-BN ends. The thermal contact resistance between h-BN and Si causes this rapid temperature decrease. Thus, no heat is transferred into the bulk Si. The green curve in Figure 6a illustrates the isothermal line of the TAR. With increasing k and *temp*, this line moves to greater values of x, which can also be seen in Figure 7b–d.

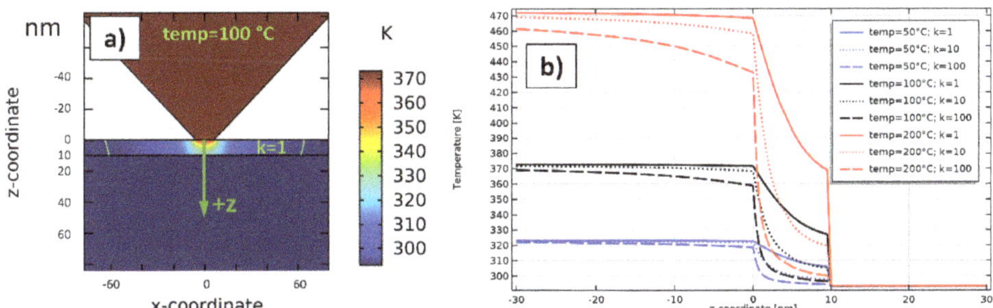

Figure 6. Simulation regarding ultrathin (10 nm) h-BN film on top of Si. (**a**) Stationary heat distribution for $k = 1$ and *temp* = 100 °C on the x/z-plane through the center of TCA (y = 0). (**b**) Temperature curves along the z-coordinate through the center point of TCA for various simulated cases.

Figure 7a illustrates the heat distribution along the x-direction (y = z = 0). It can be seen that higher values of k result in a greater TAR. This can also be seen in Figure 7b–d for the special case *temp* = 100 °C. With an increasing k, the green isothermal circle line of the TAR moves to greater values (64 nm @ $k = 1$ and 232 nm @ $k = 100$), indicating that heat spreading on the x/y-plane of h-BN increases. Furthermore, we can see that the heat spreading effect between $k = 1$ and $k = 10$ is greater than between $k = 10$ and $k = 100$ (compare curves of the same color in Figure 7a. There seems to be a kind of saturation effect for an increasing k. Surely, the temperature of TCA increases with increasing *temp* (Figure 7a). A higher k leads to an effective greater thermal conductivity of h-BN. The small TCA then has an increased proportion of the entire thermal resistance, which is the reason for lower temperatures at the TCA with *temp* = const. and k increasing.

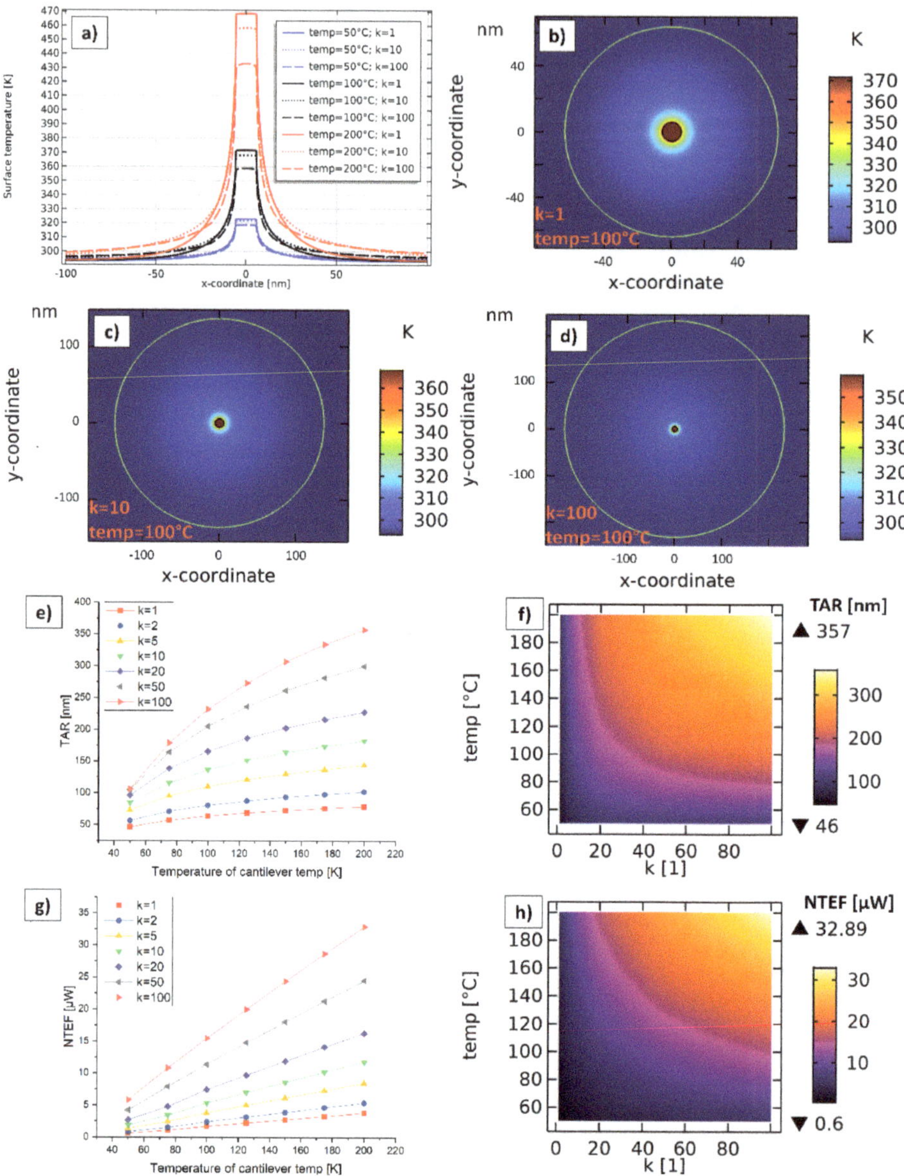

Figure 7. Simulation regarding ultrathin (10 nm) hexagonal boron nitride (h-BN) film on top of Si. (**a**) Temperature curves along the x-coordinate on the top surface (y = z = 0) for different simulated cases. (**b**) to (**d**) Heat distribution and thermal active radius (TAR) on x/y-plane for *temp* = 100 °C and *k* = 1 in (**b**), *k* = 10 in (**c**) and *k* = 100 in (**d**). (**e**) TAR in dependency of *temp* and *k* for all simulated cases. (**f**) Interpolated color plot of the simulated TARs in (**e**). (**g**) normal total energy flux (NTEF) through the TCA in dependency of *temp* and *k* for all simulated cases. (**h**) Interpolated color plot of the simulated NTEFs in (**g**).

Figure 7e shows the simulated values for the TAR as a graphical representation. The TAR increases with an increasing *temp*, but there seems to be a kind of saturation effect. The TAR logically also increases with higher *k* as heat spreading is greater there. Values for the TAR lie in between 46 nm (@ *temp* = 50 °C and k = 1) and 357 nm (@ *temp* = 100 °C and k = 100). Figure 7f delivers a color plot of the TARs in dependency of k (x-axes) and *temp* (y-axes). This color plot was created using the simulated values in Figure 7e). Intermediate points are interpolated. Depending on the case, other researchers can estimate the TAR for their research using Figure 7a–f. The thermal penetration depths for all cases is exactly 10 nm.

We also simulated the normal total energy flux (NTEF) in the stationary state, which flows through the small TCA for every case. Figure 7g illustrates the simulated cases. The NTEF seems to increase nearly linearly for an increasing *temp*. The NTEF also increases with greater k. Values for the NTEF vary from 0.6 μW (@ *temp* = 50 °C and k = 1) to 32.9 μW (@ *temp* = 100 °C and k = 100). Figure 7h shows a color plot of the NTEF in dependency of k (x-axes) and *temp* (y-axes). This color plot was created using the simulated values in Figure 7g. Intermediate points are interpolated. It, therefore, allows a rough estimation for other research for different cases.

A comparison of the NTEFs to measured values of other researchers is quite hard. There is an almost infinite number of possible measurement setups, and our specific setup could not be found in literature so far, to the best of our knowledge. However, the following references used different setups but may verify our simulated range of the NTEFs. Hwang et al. measured heat fluxes during null-point SThM in the range of approximately 1 μW (Teflon-coated surface) and 6 μW (SiO_2 surface) using a thermocouple probe [39]. Assy and Gomès used a Wollaston wire probe at 140 °C and a Kelvin nanotechnology probe at 65 °C on germanium and Si samples. They calculated the probe Joule power relative difference $\Delta P/P = (P_c - P_a)/P_c$, where P_c and P_a represent the probe Joule power in contact and out of contact, respectively. Unfortunately, only relative values of $\Delta P/P$ ranging from 0.003 to 0.058 are presented [40].

4.5. Heat Spreading in SThM Measurements Regarding a Bulk Si Sample

To compare the results of Section 4.4 to a realistic SThM measurement with a bulk Si sample, this simulation was set up. The only difference to the simulation in Section 4.4 is that the 10 nm thick h-BN layer is removed. The main results are presented in Figure 8.

Figure 8a shows the heat distribution for *temp* = 100 °C on the x/z-plane, whereas (c) shows the same heat distribution on the x/y-plane. Since the Si sample features isotropic thermal conductivity, the heat spreading is more or less circular around the tip. The green line is the TAR-line, where the temperature rise regarding the ambient temperature of 293.15 K is 1 K. As a reason of the mesh density, this line is not exactly a semicircle. The ideal semicircle is represented by the red dotted line, which overlaps the green semicircle. It can be seen that there is only a little deviation from the ideal semicircle caused by the mesh density. Compared to the h-BN sample (Section 4.4), we can see that the penetration depth is not limited by a TCR at 10 nm, and therefore, the penetration depth is much larger than for the h-BN sample. Figure 8b illustrates the temperature drop in z-direction exactly through the center of the TCA for all simulated cases. It is obvious that with increasing *temp*, the temperature at the TCA and in general also increases. Temperatures at z = 0 nm, which is directly at the physical center point of the TCA, are in between 301.8 K (@ *temp* = 50 °C) and 336.1 K (@ *temp* − 200 °C). Compared to the h-BN sample, we can see that the temperature drop within the cantilever is much higher. This is because Si has a higher overall thermal conductivity than h-BN, whereby the thermal resistance of the small TCA creates a greater proportion of the entire thermal resistance. Figure 8d shows the temperature curves along the x-direction (y = z = 0). As expected, with an increasing *temp* the curves also rise.

Figure 8e illustrates the TAR and the NTEF in dependency of *temp* for all simulated cases of the Si sample and compares them to the simulations of the isotropic case (k = 1) of the h-BN sample in Section 4.4. It is obvious that the NTEFs are significantly greater for the Si sample and increase with increasing *temp*. In general, the NTEFs for the Si sample are in between 28.3 μW (@ *temp* = 50 °C) and 126.8 μW (@ *temp* = 200 °C), the TARs are ranging from 29 nm (@ *temp* = 50 °C) to 132 nm (@ *temp* = 200 °C). The TARs of the Si sample are greater compared to the h-BN sample from *temp* ≈ 90 °C upwards. Values for the NTEF are also significantly greater compared to the maximum curves of the h-BN sample with k = 100 in Figure 7g. The root cause for this is the greater overall thermal conductivity of Si.

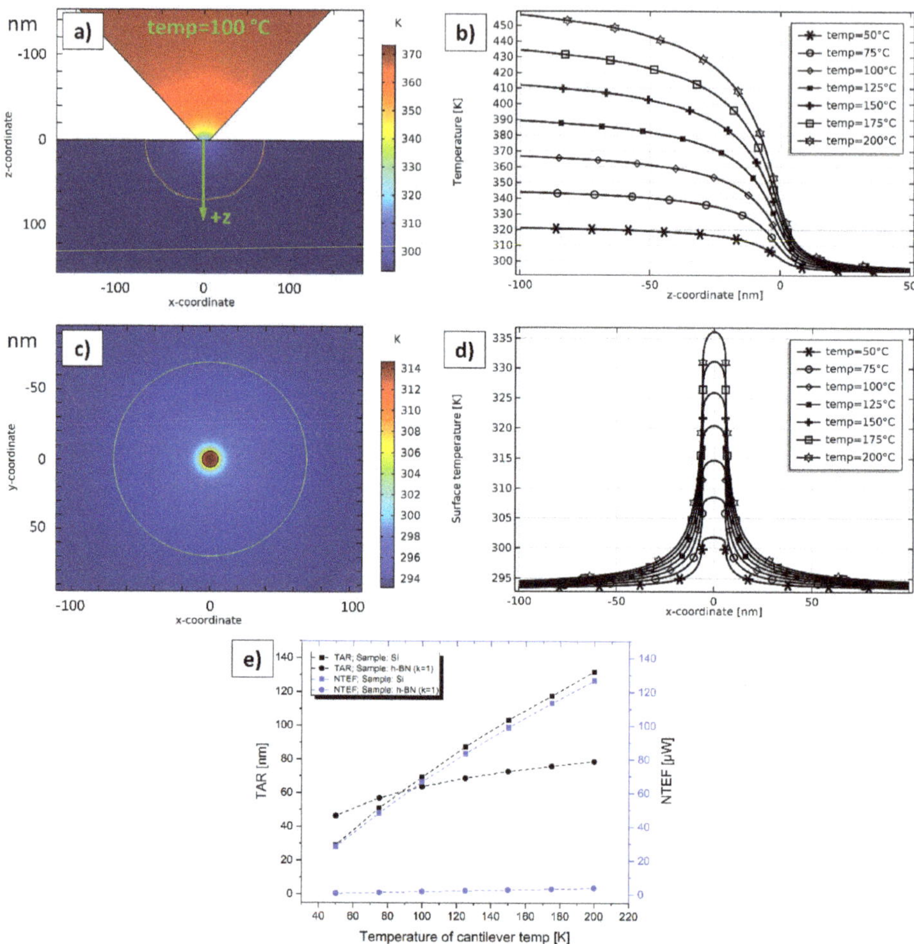

Figure 8. Simulation regarding bulk Si sample. (**a**) Stationary heat distribution for *temp* = 100 °C on the x/z-plane through the center of the TCA (y = 0). (**b**) Temperature curves along the z-coordinate through the center point of the TCA for different simulated cases. (**c**) Heat distribution and TAR on x/y-plane for *temp* = 100 °C. (**d**) Temperature curves along the x-coordinate on the top surface (y = z = 0) for different simulated cases. (**e**) TAR and NTEF in dependency of *temp* for all simulated cases of the Si sample and comparison to simulations with the h-BN sample (isotropic case with k = 1).

4.6. Stationary Time of Heat Distribution

To study the time until heat dissipation is stationary (t_{stat}), we performed the subsequent simulations. The simulations are based on Sections 4.4 and 4.5, with the difference being time-dependent instead of stationary. We consider the minimum and maximum case regarding the NTEF through the TCA for both the h-BN and the Si sample. We consider the time-dependent temperature curve for the geometric center point of the TCA. Those four curves are depicted in Figure 9.

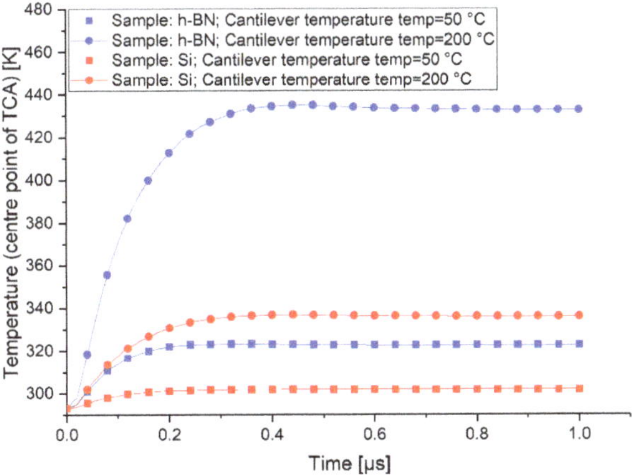

Figure 9. Time-dependent simulation of the temperature rise at the center point of the TCA for the minimum and maximum cases of the h-BN and Si sample, respectively.

As a result, we can say that the t_{stat} for every simulated case is below 1 µs. In general, it seems that higher cantilever temperatures lead to a slightly greater t_{stat}. In practical SThM measurements, the cantilever is usually moving over the sample with a specific velocity. This is not representable in simulations. The tip velocity in practical SThM measurements is normally below 20 µms^{-1} [12], which means that in 1 µs, the tip moves less than 20 pm. This is by far smaller than the atomic radius of Si or the lattice constant of h-BN. Therefore, simulations in the stationary regime in Sections 4.4 and 4.5, which assume a motionless cantilever, are justified.

4.7. Influence of Convection and Radiation

To study the influence of radiation, we set up the same simulation as in Section 4.5, with the only difference of surface radiation being enabled for all areas which were isolated in Section 4.5. Surface emissivity ϵ_{rad} of Si was set to 0.67 [41]. We also consider the minimum and maximum cases regarding the NTEF. By comparing the NTEF and the TAR for the minimum and maximum cases, we obtain the same results whether radiation being enabled or disabled (28.3 µW and 29 nm @ *temp* = 50 °C; 126.8 µW and 132 nm @ *temp* = 200 °C; compare also Figure 8e). Therefore, we can say that the influence of radiation can be neglected in our simulations. To roughly estimate the amount of radiative heat losses, a radiative area A_{rad} with a radius of 100 nm and a constant surface temperature T_2 of 393.15 K with an ambient temperature T_1 of 293.15 K and an ϵ_{rad} of 0.67 is considered. Using Equation (8), we calculate radiative heat losses of ~19.7 pW.

To roughly estimate the influence of convection a manual calculation with a convective heat transfer coefficient α of −5 Wm^{-2} K^{-1} [42], a convective area A_{conv} with a radius of 100 nm, a constant surface temperature T_2 of 393.15 K and an ambient temperature T_1 of 293.15 K is considered. Using Equation (7), we reach a loss of power caused by free convection of ~15.7 pW.

Finally, we may state that estimated convective and radiative heat losses in this work are extremely low compared to the simulated NTEF in Figures 7g and 8e. A comparative simulation also shows no differences regarding the NTEF and the TAR with radiation being enabled or disabled. Therefore, we neglect the influence of convection and radiation in our simulations. Furthermore, in the literature, the influence of radiation and convection is neglected in special cases of SThM measurements [22].

4.8. Influence of the TIR at the TCA regarding the TAR and the NTEF

To study the influence of a possible TIR between probe and sample directly at the TCA, these simulations were set up. TIRs in practical SThM measurements are quite hard to estimate as they depend on numerous factors such as surface roughness, material combination, vertical steps, tip material and geometry, contaminations, residue, tip force, and some more. Due to these numerous influences, the TIR will also vary significantly during a single SThM measurement. In literature, values in the range of 10^{-8} Km^{2}W^{-1} to 10^{-10} Km^{2}W^{-1} for different probes, samples, and measurement cases can be found [35–37]. Indeed, we assume the TIR to vary in a broader range due to the great number of possible measurement scenarios. To enable a comparison of the simulations in Sections 4.4 and 4.5, we assume an ideal thermal contact in said sections. In contrast, here, the influence of a varying TIR is investigated. Figure 10 illustrates the simulation results. We created the same simulations as in Section 4.4 (with $k = 1$) and Section 4.5 for *temp* = 50 °C and *temp* = 200 °C with the difference of varying the TIR from 5×10^{-7} Km^{2}W^{-1} to 1×10^{-13} Km^{2}W^{-1}.

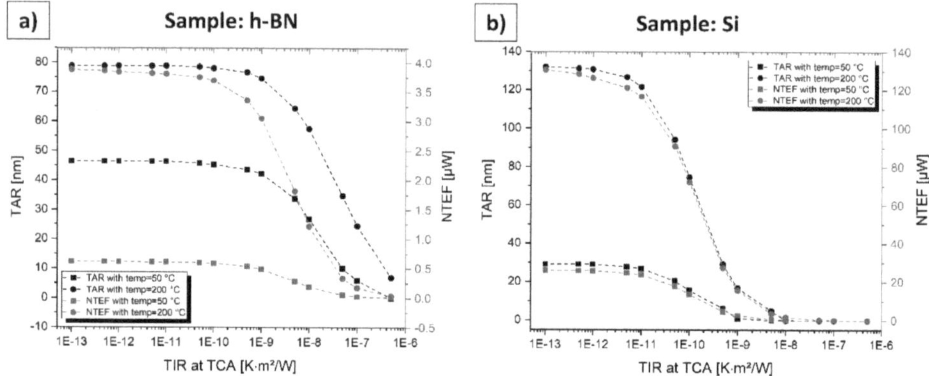

Figure 10. TAR and NTEF in dependency of the TIR at the TCA. (**a**) Sample: h-BN (compare Section 4.4). (**b**) Sample: Si (compare Section 4.5).

The case with an ultrathin h-BN film on the top surface is demonstrated in Figure 10a, representing ultrathin films with low thermal conductivities. It can be deduced that both, the TAR and the NTEF increase with decreasing TIR and converge exactly at the same values for the corresponding case in Section 4.4 (46 nm and 0.6 µW @ *temp* = 50 °C; 79 nm and 3.8 µW @ *temp* = 200 °C). This convergence starts approximately from 1×10^{-10} Km^{2}W^{-1} downwards. Figure 10b represents the case of the bulk Si sample as a sample with high thermal conductivity. Values for the TAR and the NTEF also converge at similar values compared to the corresponding cases in Section 4.5 (29 nm and 28.3 µW @ *temp* = 50 °C; 132 nm and 126.8 µW @ *temp* = 200 °C). Compared to the h-BN sample, this convergence effect starts with significantly lower TIR values at approximately 1×10^{-12} Km^{2}W^{-1}. However, it seems that in samples with a higher thermal conductivity, the influence of the TIR also increases. Nevertheless, it must be stated that SThM measurements with different material combinations cannot be compared directly, as the TIR depends on numerous influence factors as listed above. As a result, we may conclude that a TIR at the TCA has some influence and will reduce the ideal simulated values of the TAR and the NTEF in Sections 4.4 and 4.5. This effect should be considered by other researchers assuming specific values for the TIR.

4.9. Mesh Verification

To demonstrate the reliability of the results, we set up two parametric simulations. They are based on the simulations in Section 4.4 (h-BN sample) and Section 4.5 (Si sample). For the h-BN sample, the minimum case regarding the NTEF with *temp* = 50 °C and $k = 1$

was simulated, whereas for the Si sample, the maximum case regarding the NTEF with *temp* = 200 °C was performed. The sweep parameter in these simulations is the maximum size of a triangular mesh element within the TCA (see Figure 5 right). This parameter was swept between 0.09 nm and 10 nm to obtain information about the dependency of the NTEFs and the TARs on the mesh density and to confront them with the simulation time. Additionally, the maximum size in the circle around the TCA (see Figure 5 center) is defined to be 25 times larger than the maximum size within the TCA. Figure 11 illustrates the findings.

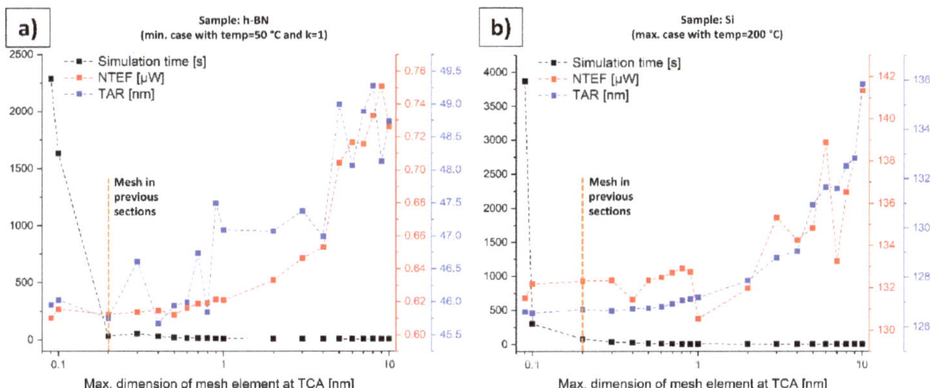

Figure 11. Dependency of the NTEFs and the TARs on the mesh density and confrontation with the simulation time. (**a**) Sample: h-BN (compare Section 4.4). (**b**) Sample: Si (compare Section 4.5).

It is obvious that with a finer mesh (left part of the *x*-axis), the values for the NTEFs and the TARs show a kind of saturation effect. For the h-BN sample, there is only little change for x values smaller than 0.2 nm, which was the chosen mesh density in all previous simulations. For the Si sample, this saturation effect starts below 0.4 nm on the *x*-axis. On the other hand, the simulation times increase enormously with x values below 0.2 nm (h-BN sample) and 0.1 nm (Si sample). The final results of the present work were obtained by performing 60 single simulations, excluding the significantly larger number of "pre-simulations". Therefore, the chosen meshing strategy with a maximum size of a triangular mesh element at the TCA of 0.2 nm provides a good compromise between accuracy and simulation time.

5. Conclusions

This work provides detailed insights into the heat transfer process during realistic SThM measurements. An SEM investigation of a used SThM probe made it possible to model an "average" used probe and calculate realistic values for the TCAs. Realistic values for thermal penetration depths, TARs, and NTEFs through the tip in contact with an h-BN or a Si sample, respectively, are provided. This allows other researchers to estimate said values for their special measurement setup and may help to interpret practical SThM measurements correctly or to explain occurring effects such as topography influences. In addition, the presented values for the TARs may help to evaluate the lateral resolution of SThM measurements as the TARs help to interpret the effect of adjacent layers. Similar to the proposal of other researchers [22], it could be shown that the influence of convection and radiation may be neglected in such studies.

From the authors' point of view, one of the most interesting findings of this study is the great impact of possible TIRs, which may not be neglected. This work may also justify higher tip velocities in practical SThM measurements as t_{stat} is estimated below 1 µs. As a single SThM measurement can take more than 1 h, a possible increase of the tip velocities may accelerate practical measurements without a loss of thermal accuracy.

However, t_{stat} only represents the stationary time of the heat propagation of the tip-sample contact. The time constant of the entire probe may slow down the sensing mechanism. In recent practical measurements, we tried to use higher tip velocities and compared the thermal images to a tip velocity of 1 μms^{-1} without significant differences. Said practical observation, therefore, agrees with the theoretical findings in this work and may justify tip velocities of up to 20 μms^{-1}. Nevertheless, the values obtained in this work are only theoretical results, which could hardly be verified as almost no comparative results can be found in the literature so far. In the future, similar practical measurements need to be performed to verify the presented values.

Author Contributions: Conceptualization, C.M.; methodology, C.M.; software, C.M.; validation, C.M., F.K., J.W. and G.B.; formal analysis, C.M.; investigation, C.M.; resources, G.B.; data curation, C.M.; writing—original draft preparation, C.M.; writing—review and editing, C.M., F.K., J.W. and G.B.; visualization, C.M.; supervision, G.B.; project administration, G.B.; funding acquisition, G.B. All authors have read and agreed to the published version of the manuscript.

Funding: This research was funded by "Bayerisches Staatsministerium für Wirtschaft, Landesentwicklung und Energie StMWi", grant number DIE-2003-0004//DIE0111/02. The APC was funded by Technische Hochschule Deggendorf.

Institutional Review Board Statement: Not applicable.

Informed Consent Statement: Not applicable.

Data Availability Statement: The data presented in this study are available on request from the corresponding author. The data are not publicly available due to the large amount of simulation data and the difficulty to present native modeling and simulation files in a reasonable way.

Acknowledgments: Many thanks to our laboratory engineer Edgar Lodermeier for the support during the SEM investigations. This work contains little parts of the Master Thesis of Christoph Metzke: Scanning Thermal Microscopy—Measurements and Finite Element Method Simulations, Deggendorf, 2019.

Conflicts of Interest: The authors declare no conflict of interest. The funders had no role in the design of the study; in the collection, analyses, or interpretation of data; in the writing of the manuscript, or in the decision to publish the results.

References

1. Cahill, D.G.; Goodson, K.; Majumdar, A. Thermometry and Thermal Transport in Micro/Nanoscale Solid-State Devices and Structures. *J. Heat Transf.* **2002**, *124*, 223–241. [CrossRef]
2. Cahill, D.G.; Ford, W.K.; Goodson, K.E.; Mahan, G.D.; Majumdar, A.; Maris, H.J.; Merlin, R.; Phillpot, S.R. Nanoscale thermal transport. *J. Appl. Phys.* **2003**, *93*, 793–818. [CrossRef]
3. Choi, S.R.; Kim, D.; Choa, S.-H.; Lee, S.-H.; Kim, J.-K. Thermal Conductivity of AlN and SiC Thin Films. *Int. J. Thermophys* **2006**, *27*, 896–905. [CrossRef]
4. Cahill, D.G.; Braun, P.V.; Chen, G.; Clarke, D.R.; Fan, S.; Goodson, K.E.; Keblinski, P.; King, W.P.; Mahan, G.D.; Majumdar, A.; et al. Nanoscale thermal transport. II. 2003–2012. *Appl. Phys. Rev.* **2014**, *1*, 11305. [CrossRef]
5. Price, D.M.; Reading, M.; Hammiche, A.; Pollock, H.M. Micro-thermal analysis: Scanning thermal microscopy and localised thermal analysis. *Int. J. Pharm.* **1999**, *192*, 85–96. [CrossRef]
6. Ruiz, F.; Sun, W.D.; Pollak, F.H.; Venkatraman, C. Determination of the thermal conductivity of diamond-like nanocomposite films using a scanning thermal microscope. *Appl. Phys. Lett.* **1998**, *73*, 1802–1804. [CrossRef]
7. Sadeghi, M.M.; Park, S.; Huang, Y.; Akinwande, D.; Yao, Z.; Murthy, J.; Shi, L. Quantitative scanning thermal microscopy of graphene devices on flexible polyimide substrates. *J. Appl. Phys.* **2016**, *119*, 235101. [CrossRef]
8. Martinek, J.; Klapetek, P.; Campbell, A.C. Methods for topography artifacts compensation in scanning thermal microscopy. *Ultramicroscopy* **2015**, *155*, 55–61. [CrossRef]
9. Shi, L.; Plyasunov, S.; Bachtold, A.; McEuen, P.L.; Majumdar, A. Scanning thermal microscopy of carbon nanotubes using batch-fabricated probes. *Appl. Phys. Lett.* **2000**, *77*, 4295–4297. [CrossRef]
10. Hammiche, A.; Pollock, H.M.; Song, M.; Hourston, D.J. Sub-surface imaging by scanning thermal microscopy. *Meas. Sci. Technol.* **1996**, *7*, 142–150. [CrossRef]
11. Majumdar, A. SCANNING THERMAL MICROSCOPY. *Annu. Rev. Mater. Sci.* **1999**, *29*, 505–585. [CrossRef]
12. Metzke, C.; Frammelsberger, W.; Weber, J.; Kühnel, F.; Zhu, K.; Lanza, M.; Benstetter, A.G. On the Limits of Scanning Thermal Microscopy of Ultrathin Films. *Materials* **2020**, *13*, 518. [CrossRef] [PubMed]

13. Leitgeb, V.; Fladischer, K.; Mitterhuber, L.; Defregger, S. Quantitative SThM Characterization for Heat Dissipation Through Thin Layers. In Proceedings of the 2019 25th International Workshop on Thermal Investigations of ICs and Systems (THERMINIC), Lecco, Italy, 25–27 September 2019; pp. 1–4, ISBN 978-1-7281-2078-2.

14. Fladischer, K.; Leitgeb, V.; Mitterhuber, L.; Maier, G.A.; Keckes, J.; Sagmeister, M.; Carniello, S.; Defregger, S. Combined thermophysical investigations of thin layers with Time Domain Thermoreflectance and Scanning Thermal Microscopy on the example of 500 nm thin, CVD grown tungsten. *Thermochim. Acta* **2019**, *681*, 178373. [CrossRef]

15. Chen, W.; Feng, Y.; Qiu, L.; Zhang, X. Scanning thermal microscopy method for thermal conductivity measurement of a single SiO2 nanoparticle. *Int. J. Heat Mass Transf.* **2020**, *154*, 119750. [CrossRef]

16. Park, J.; Koo, S.; Kim, K. Measurement of thermal boundary resistance in ~10 nm contact using UHV-SThM. *IJNT* **2019**, *16*, 263. [CrossRef]

17. Chirtoc, M.; Bodzenta, J.; Kaźmierczak-Bałata, A. Calibration of conductance channels and heat flux sharing in scanning thermal microscopy combining resistive thermal probes and pyroelectric sensors. *Int. J. Heat Mass Transf.* **2020**, *156*, 119860. [CrossRef]

18. Nguyen, T.P.; Thiery, L.; Euphrasie, S.; Lemaire, E.; Khan, S.; Briand, D.; Aigouy, L.; Gomes, S.; Vairac, P. Calibration Tools for Scanning Thermal Microscopy Probes Used in Temperature Measurement Mode. *J. Heat Transf.* **2019**, *141*. [CrossRef]

19. Bruker. VITA—NanoTA. Available online: https://www.brukerafmprobes.com/c-206-vita-nanota.aspx (accessed on 9 December 2020).

20. Bruker. VITA-DM-NANO-TA-200. Available online: https://www.brukerafmprobes.com/p-3701-vita-dm-nanota-200.aspx (accessed on 9 December 2020).

21. Born, A. Nanotechnologische Anwendungen der Rasterkapazitätsmikroskopie und Verwandter Rastersondenmethoden. Ph.D. Thesis, Universität Hamburg, Hamburg, Germany, 2000.

22. Zhang, Y.; Zhu, W.; Hui, F.; Lanza, M.; Borca-Tasciuc, T.; Muñoz Rojo, M. A Review on Principles and Applications of Scanning Thermal Microscopy (SThM). *Adv. Funct. Mater.* **2019**, *107*, 1900892. [CrossRef]

23. Hering, E.; Modler, K.-H. (Eds.) *Grundwissen des Ingenieurs*; 14.,aktualisierte Aufl; Fachbuchverl. Leipzig im Carl-Hanser-Verl: München, Germany, 2007; ISBN 978-3-446-22814-6.

24. Popov, V.L. *Rigorose Behandlung des Kontaktproblems–Hertzscher Kontakt. Kontaktmechanik und Reibung*; Springer: Berlin/Heidelberg, Germany, 2009; pp. 57–72. ISBN 978-3-540-88836-9.

25. Nitsche, K.; Marek, R. *Praxis der Wärmeübertragung. Grundlagen, Anwendungen, Übungsaufgaben; mit 48 Tabellen, 50 vollständig durchgerechneten Beispielen sowie 93 Übungsaufgaben mit Lösungen*; Fachbuchverl. Leipzig im Carl Hanser Verl.: München, Germany, 2007; ISBN 9783446409996.

26. Korth Kristalle GmbH. Silizium (Si). Available online: https://www.korth.de/index.php/material-detailansicht/items/32.html (accessed on 9 December 2020).

27. Frühauf, J. *Shape and Functional Elements of the Bulk Silicon Microtechnique. A Manual of Wet-Etched Silicon Structures*; Springer: Berlin/Heidelberg, Germany, 2005; ISBN 978-3-540-26876-5.

28. Falin, A.; Cai, Q.; Santos, E.J.G.; Scullion, D.; Qian, D.; Zhang, R.; Yang, Z.; Huang, S.; Watanabe, K.; Taniguchi, T.; et al. Mechanical properties of atomically thin boron nitride and the role of interlayer interactions. *Nat. Commun.* **2017**, *8*, 15815. [CrossRef]

29. Boldrin, L.; Scarpa, F.; Chowdhury, R.; Adhikari, S. Effective mechanical properties of hexagonal boron nitride nanosheets. *Nanotechnology* **2011**, *22*, 505702. [CrossRef]

30. Frammelsberger, W.; Benstetter, G.; Kiely, J.; Stamp, R. C-AFM-based thickness determination of thin and ultra-thin SiO 2 films by use of different conductive-coated probe tips. *Appl. Surf. Sci.* **2007**, *253*, 3615–3626. [CrossRef]

31. Jiang, L.; Shi, Y.; Hui, F.; Tang, K.; Wu, Q.; Pan, C.; Jing, X.; Uppal, H.; Palumbo, F.; Lu, G.; et al. Dielectric Breakdown in Chemical Vapor Deposited Hexagonal Boron Nitride. *ACS Appl. Mater. Interfaces* **2017**, *9*, 39758–39770. [CrossRef]

32. Villaroman, D.; Wang, X.; Dai, W.; Gan, L.; Wu, R.; Luo, Z.; Huang, B. Interfacial thermal resistance across graphene/Al2O3 and graphene/metal interfaces and post-annealing effects. *Carbon* **2017**, *123*, 18–25. [CrossRef]

33. Hong, Y.; Zhang, J.; Zeng, X.C. Thermal contact resistance across a linear heterojunction within a hybrid graphene/hexagonal boron nitride sheet. *Phys. Chem. Chem. Phys.* **2016**, *18*, 24164–24170. [CrossRef]

34. Zhang, J.; Wang, Y.; Wang, X. Rough contact is not always bad for interfacial energy coupling. *Nanoscale* **2013**, *5*, 11598–11603. [CrossRef] [PubMed]

35. Timofeeva, M.; Bolshakov, A.; Tovee, P.D.; Zeze, D.A.; Dubrovskii, V.G.; Kolosov, O.V. Nanoscale Resolution Scanning Thermal Microscopy with Thermally Conductive Nanowire Probes. *arXiv* **2013**, arXiv:1309.2010. Available online: http://arxiv.org/pdf/1309.2010v1 (accessed on 9 December 2020).

36. Kim, K.; Jeong, W.; Lee, W.; Sadat, S.; Thompson, D.; Meyhofer, E.; Reddy, P. Quantification of thermal and contact resistances of scanning thermal probes. *Appl. Phys. Lett.* **2014**, *105*, 203107. [CrossRef]

37. Timofeeva, M.; Bolshakov, A.; Tovee, P.D.; Zeze, D.A.; Dubrovskii, V.G.; Kolosov, O.V. Scanning thermal microscopy with heat conductive nanowire probes. *Ultramicroscopy* **2016**, *162*, 42–51. [CrossRef] [PubMed]

38. Song, Y.; Mandelli, D.; Hod, O.; Urbakh, M.; Ma, M.; Zheng, Q. Robust microscale superlubricity in graphite/hexagonal boron nitride layered heterojunctions. *Nat. Mater.* **2018**, *17*, 894–899. [CrossRef]

39. Hwang, G.; Chung, J.; Kwon, O. Enabling low-noise null-point scanning thermal microscopy by the optimization of scanning thermal microscope probe through a rigorous theory of quantitative measurement. *Rev. Sci. Instrum.* **2014**, *85*, 114901. [CrossRef] [PubMed]

40. Assy, A.; Gomès, S. Heat transfer at nanoscale contacts investigated with scanning thermal microscopy. *Appl. Phys. Lett.* **2015**, *107*, 43105. [CrossRef]
41. Ravindra, N.M.; Sopori, B.; Gokce, O.H.; Cheng, S.X.; Shenoy, A.; Jin, L.; Abedrabbo, S.; Chen, W.; Zhang, Y. Emissivity Measurements and Modeling of Silicon-Related Materials: An Overview. *Int. J. Thermophys* **2001**, *22*, 1593–1611. [CrossRef]
42. Cerbe, G.; Hoffmann, H.-J. *Einführung in die Wärmelehre. Von der Thermodynamik zur technischen Anwendung; mit 30 Tafeln, 111 Beispielen, 119 Aufgaben und 161 Kontrollfragen*; 9., verb. Aufl; Hanser: München, Germany, 1990; ISBN 9783446159525.

 *nanomaterials*

Article

Charge-Order on the Triangular Lattice: A Mean-Field Study for the Lattice $S = 1/2$ Fermionic Gas

Konrad Jerzy Kapcia

Faculty of Physics, Adam Mickiewicz University in Poznań, ulica Uniwersytetu Poznańskiego 2, PL-61614 Poznań, Poland; konrad.kapcia@amu.edu.pl; Tel.: +48-61-829-5051

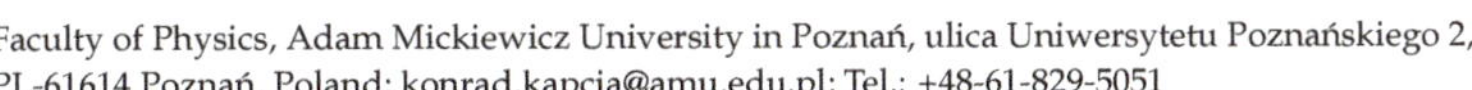

Abstract: The adsorbed atoms exhibit tendency to occupy a triangular lattice formed by periodic potential of the underlying crystal surface. Such a lattice is formed by, e.g., a single layer of graphane or the graphite surfaces as well as (111) surface of face-cubic center crystals. In the present work, an extension of the lattice gas model to $S = 1/2$ fermionic particles on the two-dimensional triangular (hexagonal) lattice is analyzed. In such a model, each lattice site can be occupied not by only one particle, but by two particles, which interact with each other by onsite U and intersite W_1 and W_2 (nearest and next-nearest-neighbor, respectively) density-density interaction. The investigated hamiltonian has a form of the extended Hubbard model in the atomic limit (i.e., the zero-bandwidth limit). In the analysis of the phase diagrams and thermodynamic properties of this model with repulsive $W_1 > 0$, the variational approach is used, which treats the onsite interaction term exactly and the intersite interactions within the mean-field approximation. The ground state ($T = 0$) diagram for $W_2 \leq 0$ as well as finite temperature ($T > 0$) phase diagrams for $W_2 = 0$ are presented. Two different types of charge order within $\sqrt{3} \times \sqrt{3}$ unit cell can occur. At $T = 0$, for $W_2 = 0$ phase separated states are degenerated with homogeneous phases (but $T > 0$ removes this degeneration), whereas attractive $W_2 < 0$ stabilizes phase separation at incommensurate fillings. For $U/W_1 < 0$ and $U/W_1 > 1/2$ only the phase with two different concentrations occurs (together with two different phase separated states occurring), whereas for small repulsive $0 < U/W_1 < 1/2$ the other ordered phase also appears (with tree different concentrations in sublattices). The qualitative differences with the model considered on hypercubic lattices are also discussed.

Keywords: charge order; triangular lattice; extended Hubbard model; atomic limit; mean-field theory; phase diagram; longer-range interactions; thermodynamic properties; fermionic lattice gas; adsorption on the surface

Citation: Kapcia, K.J. Charge-Order on the Triangular Lattice: A Mean-Field Study for the Lattice $S = 1/2$ Fermionic Gas. *Nanomaterials* **2021**, *11*, 1181. https://doi.org/10.3390/nano11051181

Academic Editor: Orion Ciftja

Received: 25 March 2021
Accepted: 29 April 2021
Published: 30 April 2021

Publisher's Note: MDPI stays neutral with regard to jurisdictional claims in published maps and institutional affiliations.

Copyright: © 2021 by the author. Licensee MDPI, Basel, Switzerland. This article is an open access article distributed under the terms and conditions of the Creative Commons Attribution (CC BY) license (https://creativecommons.org/licenses/by/4.0/).

1. Introduction

It is a well known fact that the classical lattice gas model is useful phenomenological model for various phenomena. It has been studied in the context of experimental studies of adsobed gas layers on crystaline substrates (cf., for example pioneering works [1–4]). For instance, the adsorbed atoms exhibit tendency to occupy a triangular lattice formed by periodic potential of the underlying crystal surface. This lattice is shown in Figure 1a. Such a lattice is formed by, e.g., a single layer of graphane or the graphite surface [i.e., the honeycomb lattice; (0001) hexagonal closed-packed (hcp) surface], and (111) face-centered cubic (fcc) surface. Atoms from (111) fcc surface are organized in the triangular lattice, whereas the triangular lattice is a dual lattice for the honeycomb lattice [5]. Note also that arrangements of atoms on (110) base-centered cubic (bcc) surface as well as on (111) bcc surface (if one neglects the interactions associated with other layers under surface) are quite close to the triangular lattice. One should mention that the triangular lattice and the honeycomb lattice are two examples of two-dimensional hexagonal Bravais lattices. Formally, the triangular lattice is a hexagonal lattice with a one-site basis, whereas the honeycomb lattice is a hexagonal lattice with a two-site basis. The classical lattice gas model is equivalent with the $S = 1/2$ Ising model in the external field [1,6–9] (the results for this model on the triangular lattice will be discussed in more details in Section 2).

In the present work, an extension of the lattice gas model to $S = 1/2$ fermionic particles is analyzed. Such a model has a form of the atomic limit of the extended Hubbard model [10], cf. (1). In this model, each lattice site can be occupied not by only one particle as in the model discussed in previous paragraph, but also by two particles. In addition to long-range (i.e., intersite) interactions between fermions, the particles located at the same site can also interact with each other via onsite Hubbard U interaction. For a description of the interacting fermionic particles on the lattice, the single-orbital extended Hubbard model with intersite density-density interactions has been used widely [11–19]. It is one of the simplest model capturing the interplay between the Mott localization (onsite interactions) and the charge-order phenomenon [15–25]. However, in some systems the inclusion of other interactions and orbitals is necessary [11–14,26,27].

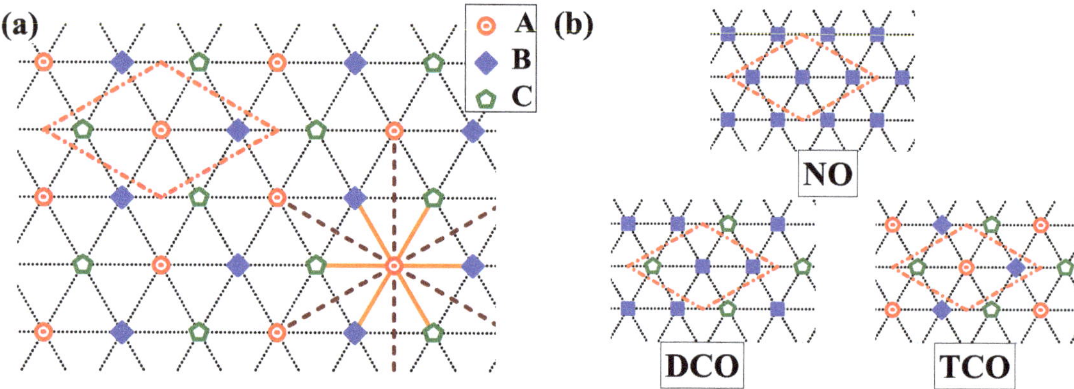

Figure 1. (a) The schema of the triangular lattice on which the extended Hubbard model in the atomic limit is studied in the present work. The lattice is divided into three equivalent sublattices ($\alpha = A, B, C$) denoted by different symbols. The dash-dotted line denotes the boundaries of $\sqrt{3} \times \sqrt{3}$ unit cell. By solid and dashed lines all nearest neighbors and all next-nearest neighbors of a chosen site from sublattice A are indicated, respectively. (b) There different types of particle arrangements in $\sqrt{3} \times \sqrt{3}$ unit cells (i.e., the tri-sublattice assumption) corresponding to NO, DCO, TCO phases (as labeled). Symbol shapes on each panel correspond to respective concentrations at the lattice sites.

This work can be palced among recent theoretical and experimental studies of adsorption of various atoms on (i) the (0001) hcp surface of the graphite [28–32] or of other materials [33,34] and (ii) the (111) fcc surface of metals and semimetals [34–39]. Although in the mentioned works the adsorbed particles on surface are rather classical and the analysis of classical lattice gas can give some predictions, taking into account of the quantum properties of adsorbed particles is necessary for, e.g., a description of experiments with He^4 and He^3 [28,40,41]. Moreover, there is plethora of recent experimental and theoretical studies of quasi-two-dimensional systems, e.g., Na_xCoO_2 [42], $NbSe_2$ [43–47], $TiSe_2$ [48], $TaSe_2$ [49], VSe_2 [50], TaS_2 [51], and other transition metals dichalcogenides [52] as well as organic conductors [53,54], where various charge-ordered patterns have been observed on the triangular lattice. However, for such phenomena the atomic limit of the model studies is less reliable and more realistic description includes also electron hoping term as in the extended Hubbard model [15–19] or coupling with phonons as in the Holstein-Hubbard model [55]. In such cases, results obtained for atomic limit can be treated as a benchmark for models including the itinerant properties of fermionic particles.

The present work is organized as follows. In Section 2 the model and the methods (together with the most important equations) are presented. Section 3 is devoted to the discussion of ground state phase diagrams of the model with non-zero next-nearest neighbor interactions. Next, the finite temperature properties of the model with only the nearest-neighbor interactions are presented in Section 4. Finally, the most important conclusions and supplementary discussion are included in Section 5.

2. The Model and the Method

The extended Hubbard model in the zero-bandwidth limit (i.e., in the atomic limit) with interactions restricted to the second neighbors (or, equivalently, the next-nearest neighbors) can be expressed as:

$$\hat{H} \;=\; U\sum_i \hat{n}_{i\uparrow}\hat{n}_{i\downarrow} + \frac{1}{2}\frac{W_1}{z_1}\sum_{\langle i,j\rangle_1} \hat{n}_i\hat{n}_j + \frac{1}{2}\frac{W_2}{z_2}\sum_{\langle i,j\rangle_2} \hat{n}_i\hat{n}_j - \mu\sum_i \hat{n}_i, \tag{1}$$

where $\hat{n}_i = \sum_\sigma \hat{n}_{i\sigma}$, $\hat{n}_{i\sigma} = \hat{c}_{i\sigma}^\dagger \hat{c}_{i\sigma}$, and $\hat{c}_{i\sigma}^\dagger$ ($\hat{c}_{i\sigma}$) denotes the creation (annihilation) operator of an electron with spin σ at the site i. U is the onsite density interaction, W_1 and W_2 are the intersite density-density interactions between the nearest neighbors (NN) and the next-nearest neighbors (NNNs), respectively. z_1 and z_2 are numbers of NN and NNNs, respectively. μ is the chemical potential determining the total concentration n of electrons in the system by the relation $n = (1/L)\sum_i \langle \hat{n}_i \rangle$, where $0 \leq n \leq 2$ and L is the total number of lattice sites. In this work phase diagrams emerging from this model are inspected. The analyses are performed in the grand canonical ensemble.

In this work the mean-field decoupling of the intersite term is used in the following form

$$\hat{n}_i\hat{n}_j = \langle \hat{n}_i \rangle \hat{n}_j + \hat{n}_i \langle \hat{n}_j \rangle - \langle \hat{n}_i \rangle \langle \hat{n}_j \rangle, \tag{2}$$

which is an exact treatment only in the limit of large coordination number ($z_n \to \infty$; or limit of infinite dimensions) [10–13,56–58]. Thus, for the two-dimensional triangular lattice (with $z_1 = z_2 = 6$) it is an approximation in the general case. It should be underlined that the treatment of the onsite term is rigorous in the present work. Please note that that the interactions U, W_1 and W_2 should be treated as effective parameters for fermionic particles including all possible contributions and renormalizations originating from other (sub-)systems.

Model (1) for $W_2 \neq 0$ has been intensively studied on the hypercubic lattices (see, e.g., [59–68] and references therein). Also the case of two-dimensional square lattice was investigated in detail for $W_2 = 0$ [61–64] as well as for $W_2 \neq 0$ [65–69]. There are also rigorous results for one-dimensional chain for $W_2 = 0$ [70,71] and $W_2 \neq 0$ [72].

In [73] the model with $W_2 = 0$ was investigated on the triangular lattice at half-filling by using a classical Monte Carlo method, and a critical phase, characterized by algebraic decay of the charge correlation function, belonging to the universality class of the two-dimensional XY model with a $\mathbb{Z}_6$ anisotropy was found in the intermediate-temperature regime. Some preliminary results for model (1) on the triangular lattice and for large attractive $U < 0$ and $W_2 = 0$ within the mean-field approximation were presented in [74].

The model in the limit $U \to -\infty$ is equivalent with the $S = 1/2$ Ising model with antiferromagnetic (ferromagnetic) J_n interactions if W_n interaction in model (1) are repulsive, i.e., $W_n > 0$ (attractive, i.e., $W_n < 0$, respectively). The relation between interaction parameters in both models is very simple, namely $J_n = -W_n$. There is plethora of the results obtained for the Ising model on the triangular lattice. One should mention the following works (not assuming a comprehensive review): (a) exact solution in the absence of the external field H, i.e., for $H = 0$ (only with NN interactions, at arbitrary temperature) [5,75–79]; (b) for the model with NNN interactions included: ground state exact results [2], Bethe-Peierls approximation [1], Monte Carlo simulation both for $H = 0$ [80] and $H \neq 0$ [3] (and other methods, e.g., [81,82]); (c) exact ground state results for the model with up to 3rd nearest-neighbor interactions for both $H = 0$ case [83] and $H \neq 0$ case [4,84]. The most important information arising from these analyses is that only for $W_2 \leq 0$ (and arbitrary W_1) one can expect that consideration of $\sqrt{3} \times \sqrt{3}$ unit cells (i.e., tri-sublattice orderings) is enough to find all ordered states (particle arrangements) in the model. The reason is that the range of W_2 interaction is larger than the size of the unit cell. Thus, this is the point for that the present analysis of the model including only $\sqrt{3} \times \sqrt{3}$ unit cell orderings with restriction to $W_2 \leq 0$ is justified. One should not expect occurrence of any other phases beyond the tri-sublattice assumption in the studied range of the model parameters.

Please note that for $W_2 > 0$ it is necessary to consider a larger unit cell to find the true phase diagram of the model even in the $U \to -\infty$ limit (cf., e.g., [2,4,81]). This is a similar situation as for model (1) on the square lattice, where for $W_2 > 0$ and any U not only checker-board order occurs (the two-sublattice assumption), but also other different arrangements of particles are present (the four-sublattice assumption, e.g., various stripes orders) [67,68].

2.1. General Definitions of Phases Existing in the Investigated System

In the systems analyzed only three nonequivalent homogeneous phases can exist (within the tree-sublattice assumption used). They are determined by the relations between concentrations n_α's in each sublattice α ($n_\alpha = (3/L) \sum_{i \in \alpha} \langle \hat{n}_i \rangle$), but a few equivalent solutions exist due to change of sublattice indexes. For intuitive understanding of rather complicated phase diagrams each pattern is marked with adequate abbreviation. The nonordered (NO) phase is defined by $n_A = n_B = n_C$ (all three n_α's are equal), the charge-ordered phase with two different concentrations in sublattices (DCO phase) is defined by $n_A = n_B \neq n_C$, $n_B = n_C \neq n_A$, or $n_C = n_A \neq n_B$ (two and only two out of three n_α's are equal, 3 equivalent solution), whereas in the charge-ordered phase with three different concentrations in sublattices (TCO phase) $n_A \neq n_B$, $n_B \neq n_C$, and $n_A \neq n_C$ (all three n_α's are different, 6 equivalent solutions). All these phases are schematically illustrated in Figure 1b. These phases exist in several equivalent solutions due to the equivalence of three sublattices forming the triangular lattice. Each of these patterns can be realized in a few distinct forms depending on specific electron concentrations on each sublattice (cf. Tables 1 and 2 for $T = 0$). In addition, the degeneracy of the ground state solutions is contained in Table 1 (including charge and spin degrees of freedom).

2.2. Expressions for the Ground State

In the ground state (i.e., for $T = 0$), the grand canonical potential ω_0 per site of model (1) can be found as

$$\omega_0 = \langle \hat{H} \rangle / L = E_D + E_W + E_\mu, \tag{3}$$

where contributions associated with the onsite interaction, the intersite interactions, and the chemical potential, respectively, has the following forms

$$E_D = \frac{U}{6} [n_A(n_A - 1) + n_B(n_B - 1) + n_C(n_C - 1)], \tag{4}$$

$$E_W = \frac{1}{6} W_1(n_A n_B + n_B n_C + n_C n_A) + \frac{1}{6} W_2(n_A^2 + n_B^2 + n_C^2), \tag{5}$$

$$E_\mu = -\tfrac{1}{3}\mu(n_A + n_B + n_C). \tag{6}$$

In the above expressions, concentrations n_α at $T = 0$ take the values from $\{0, 1, 2\}$ set (cf. also Table 1). Please note that the above equations are the exact expressions for ω_0 of model (1) on the triangular lattice.

The free energy per site of homogeneous phases at $T = 0$ within the mean-field approximation is obtained as

$$f_0 = \langle \hat{H} + \mu \sum_i \hat{n}_i \rangle / L = U D_{occ} + E_W, \tag{7}$$

where E_W is expressed by (5). $D_{occ} = (1/L) \sum_i \langle \hat{n}_{i\uparrow} \hat{n}_{i\downarrow} \rangle$ denotes the double occupancy and this quantity is found to be exact, cf. Table 2. One should underline that above expression for f_0 is an approximate result for model (1) on the triangular lattice. Here, it is assumed that concentration n_α are as defined in Table 2 and they are the same in each $\sqrt{3} \times \sqrt{3}$ unit cell in the system. Formally, it could be treated as exact one only if the numbers z_n ($n = 1, 2$) goes to infinity.

The expressions presented in this subsection (for $W_2 = 0$) can be obtained as the $T \to 0$ limit of the equations for $T > 0$ included in Section 2.3.

2.3. Expressions for Finite Temperatures

For finite temperatures ($T > 0$), the expressions given in [59] for the three-sublattice assumption takes the following forms (cf. also these in [68] given for the four-sublattice assumption). In approach used, the onsite U term is treated exactly and for the intersite W_1 term the mean-field approximation (2) is used. For a grand canonical potential ω (per lattice site) in the case of the lattice presented in Figure 1 one obtains

$$\omega = -\frac{1}{6}\sum_\alpha \Phi_\alpha n_\alpha - \frac{1}{3\beta}\sum_\alpha (\ln Z_\alpha). \tag{8}$$

where $\beta = 1/(k_B T)$ is inverted temperature, coefficients Φ_α are defined as $\Phi_\alpha = \mu - \mu_\alpha$,

$$Z_\alpha = 1 + 2\exp(\beta\mu_\alpha) + \exp\left[\beta(2\mu_\alpha - U)\right], \tag{9}$$

and μ_α is a local chemical potential in α sublattice ($\alpha \in \{A, B, C\}$)

$$\mu_A = \mu - \tfrac{1}{2}W_1(n_B + n_C), \ \mu_B = \mu - \tfrac{1}{2}W_1(n_A + n_C), \ \mu_C = \mu - \tfrac{1}{2}W_1(n_A + n_B). \tag{10}$$

For electron concentration n_α in each sublattice in arbitrary temperature $T > 0$ one gets

$$n_\alpha = \frac{2}{Z_\alpha}\{\exp\left(\beta\mu_\alpha\right) + \exp\left[\beta(2\mu_\alpha - U)\right]\} \quad (\text{for } \alpha \in \{A, B, C\}). \tag{11}$$

The set of three Equations (11) for n_A, n_B, and n_C determines the (homogeneous) phase occurring in the system for fixed model parameters U, W_1, and μ. If $n = (1/3)(n_A + n_B + n_C)$ is fixed, one has also set of three equations, but it is solved with respect to μ, n_A, and n_B (the third n_α is obviously found as $n_C = 3n - n_A - n_B$).

The free energy f per site is derived as

$$f = \omega + \frac{1}{3}\mu(n_A + n_B + n_C), \tag{12}$$

where ω and n_α's are expressed by (8)–(11).

2.4. Macroscopic Phase Separation

The free energy f_{PS} of the (macroscopic) phase separated state (as a function of total electron concentration n; and at any temperature $T \geq 0$) is calculated from

$$f_{PS}(n) = \frac{n - n_-}{n_+ - n_-}f_+(n_+) + \frac{n_+ - n}{n_+ - n_-}f_-(n_-), \tag{13}$$

where $f_\pm(n_\pm)$ are free energies of separating homogeneous phases with concentrations $n_\pm$. The factor before $f_\pm(n_\pm)$ is associated with a fraction of the system, which is occupied by the phase with concentration $n_\pm$. Such defined phase separated states can exist only for n fulfilling the condition $n_- < n < n_+$. For $n_\pm$ only the homogeneous phase exists in the system (one homogeneous phase occupies the whole system). Concentrations $n_\pm$ are simply determined at the ground state, whereas for $T > 0$ they can be found as concentrations at the first-order (discontinuous) boundary for fixed μ or by minimizing the free energy f_{PS} [i.e., (13)] with respect to n_+ and n_- (for n fixed). For more details of the so-called Maxwell's construction and macroscopic phase separations see, e.g., [59,68,85,86]. The interface energy between two separating phases is neglected here.

Table 1. Homogeneous phases ($z_n \to \infty$, $n = 1, 2$) or $\sqrt{3} \times \sqrt{3}$ unit cells (the triangular lattice) at $T = 0$ (for fixed μ). Star "$*$" in superscript indicates that the phase is obtained by the particle-hole transformation (i.e., $n_\alpha \to 2 - n_\alpha$; the NO_1 and TCO phases are invariant under this transformation). In the brackets also an alternative name is given. The degeneration $d_c \times d_s$ of the unit cells (equal to the degeneration of the ground state for $z_n \to \infty$ limit) and degeneration $D_c \times D_s$ of the ground state phases constructed from the corresponding unit cells for the triangular lattice is given (with respect to charge and spin degrees of freedom).

Phase	n_A	n_B	n_C	$d_c \times d_s$	$D_c \times D_s$	ω_0
NO_0 (NO_2^*)	0	0	0	1×1	1×1	0
NO_1 (NO_1^*)	1	1	1	1×8	1×2^L	$(-2\mu + W_1 + W_2)/2$
NO_2 (NO_0^*)	2	2	2	1×1	1×1	$-2\mu + U + 2W_1 + 2W_2$
DCO_1	0	0	1	3×2	$3 \times 2^{L/3}$	$(-2\mu + W_2)/6$
DCO_1^*	1	2	2	3×2	$3 \times 2^{L/3}$	$(-10\mu + 4U + 8W_1 + 9W_2)/6$
DCO_2	0	0	2	3×1	3×1	$(-2\mu + U + 2W_2)/3$
DCO_2^*	0	2	2	3×1	3×1	$(-4\mu + 2U + 2W_1 + 4W_2)/3$
DCO_3	0	1	1	3×4	$3 \times 4^{L/3}$	$(-4\mu + W_1 + 2W_2)/6$
DCO_3^*	1	1	2	3×4	$3 \times 4^{L/3}$	$(-8\mu + 2U + 5W_1 + 6W_2)/6$
TCO (TCO*)	0	1	2	6×2	$6 \times 2^{L/3}$	$(-6\mu + 2U + 2W_1 + 5W_2)/6$

Table 2. Homogeneous phases at $T = 0$ (for fixed n) defined by n_α's and D_{occ}. n_s and n_f define the range $[n_s, n_f]$ of n, where the phase is correctly defined. In the last column, the phase separated state degenerated with the homogeneous phase in range (n_s, n_f) for $W_2 = 0$ is mentioned. Star "$*$" in superscript indicates that the phase is obtained by the particle-hole transformation (i.e., $n_\alpha \to 2 - n_\alpha$; TCO_A, TCO_A^*, TCO_B, and TCO_B^* phases are invariant under this transformation).

Phase	n_A	n_B	n_C	D_{occ}	n_s	n_f	PS
DCO_A	0	0	$3n$	$n/2$	0	2/3	NO_0/DCO_2
DCO_B	0	0	$3n$	0	0	1/3	NO_0/DCO_1
DCO_C	0	0	$3n$	$n - 1/3$	1/3	2/3	DCO_1/DCO_2
DCO_D	$3n - 2$	1	1	0	2/3	1	DCO_3/NO_1
TCO_A	0	$3n - 2$	2	$n/2$	2/3	4/3	DCO_2/DCO_2^*
TCO_B	0	$3n - 2$	2	1/3	2/3	1	DCO_2/TCO
TCO_C	0	$3n - 1$	1	0	1/3	2/3	DCO_1/DCO_3
DCO_A^*	$3n - 4$	2	2	$n/2$	4/3	2	DCO_2^*/NO_2
DCO_B^*	$3n - 4$	2	2	$n - 1$	5/3	2	DCO_1^*/NO_2
DCO_C^*	$3n - 4$	2	2	2/3	4/3	5/3	DCO_2^*/DCO_1^*
DCO_D^*	1	1	$3n - 2$	$n - 1$	1	4/3	NO_1/DCO_3^*
TCO_A^*	0	$3n - 2$	2	$n/2$	2/3	4/3	DCO_2/DCO_2^*
TCO_B^*	0	$3n - 2$	2	$n - 2/3$	1	4/3	TCO^*/DCO_2^*
TCO_C^*	1	$3n - 3$	2	$n - 1$	4/3	5/3	DCO_3^*/DCO_1^*

3. Results for the Ground State ($W_1 > 0$ and $W_2 \le 0$)

3.1. Analysis for Fixed Chemical Potential μ

The ground state diagram for model (1) as a function of (shifted) chemical potential $\bar{\mu} = \mu - W_1 - W_2$ is shown in Figure 2. The diagram is determined by comparison of the grand canonical potentials ω_0's of all phases collected in Table 1 [cf. (3)]. It consists of several regions, where the NO phase occurs (3 regions: NO_0, NO_1 and NO_2), the DCO phase occurs (6 regions: DCO_1, DCO_2, DCO_3, DCO_1^*, DCO_2^*, and DCO_3^*) and the TCO phase occurs (1 region).

All boundaries between the phases in Figure 2 are associated with a discontinuous change of at least one of the n_α. The only boundaries associated with a discontinuous jump of two n_α's are: DCO_2–DCO_3 (DCO_2^*–DCO_3^*) and TCO–NO_1. At the boundaries ω_0's of the phases are the same. It means that both phases can coexist in the system provided that a formation of the interface between two phases does not require additional energy. For $W_2 = 0$, only the boundaries DCO$_2$–DCO$_3$ (DCO$_2^*$–DCO$_3^*$) and TCO–NO$_1$ have finite degeneracy (6 and 7, respectively, modulo spin degrees of freedom) and the interface between different types of $\sqrt{3} \times \sqrt{3}$ unit cells increases the energy of the system. Thus, the mentioned phases from neighboring regions cannot coexist at the boundaries. The other boundaries exhibit infinite degeneracy (it is larger than $3 \cdot 2^{L/3}$ modulo spin) and entropy per site in the thermodynamic limit is non-zero. It means that at these boundaries both types of unit cells from neighboring regions can mix with any ratio and the formation of the interface between two phases does not change energy of the system. However, some conditions for arrangement of the cells can exist. For example, the DCO$_2$ phase with $(0,0,2)$ can mix with the DCO$_2^*$ phase with $(0,2,2)$ or $(2,0,2)$, but not with the DCO$_2^*$ phase with $(2,2,0)$. Please note that it is also possible to mix all three unit cells: $(0,0,2)$, $(0,2,2)$, and $(2,0,2)$. In such a case, $(0,2,2)$ and $(2,0,2)$ cells of the DCO$_2^*$ phase cannot be located next to each other, i.e., they need to be separated by $(0,0,2)$ unit cells of the DCO$_2$ phase. Thus, the degeneracy of the DCO$_2$–DCO$_2^*$ boundary is indeed larger than $3 \cdot 2^{L/3}$ modulo spin. This is so-called *macroscopic degeneracy*, cf. [68]). In such a case, we say that the *microscopic phase separation* occurs. For $W_2 < 0$ these degeneracies are removed and all boundaries exhibit finite degeneracy (neglecting spin degrees of freedom). In this case the phases cannot be mixed on a microscopic level.

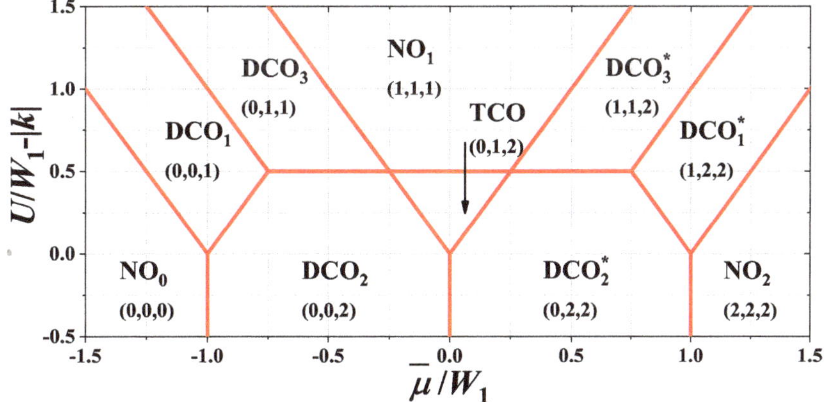

Figure 2. Ground state phase diagram of the model on the triangular lattice as a function of shifted chemical potential $\bar{\mu} = \mu - W_1 - W_2$ for $W_1 > 0$ and $W_2 \leq 0$ ($|k| = |W_2|/W_1$). The regions are labeled by the names of the phases defined in Table 1 and numbers corresponding to concentrations in each sublattice n_A, n_B and n_C.

Please also note that for $W_2 = 0$ as well as for $W_2 < 0$ inside the regions shown in Figure 2, the $\sqrt{3} \times \sqrt{3}$ unit cells of the same type with different orientation cannot mix. It denotes that orientation of one type of the unit cell determines the orientation of other unit cells (of the same type). Thus, the degeneracy of the state of the system is finite (modulo spin) and the system exhibits the long-range order at the ground state inside each region of Figure 2. This is different from the case of two dimensional square lattice, where inside some regions different unit cells (elementary blocks) of the same phase can mix with each other [67,68].

One should underline that the discussed above ground state results for fixed chemical potential are the exact results for model (1) on the triangular lattice. This is due to the fact that the model is equivalent with a classical spin model, namely the $S = 1$ Blume-Cappel model with two-fold degenerated value of $S = 0$ (or the $S = 1$ classical Blume-Cappel with temperature-dependent anizotropy without degeneration), cf. [10,60,63]. For such a model, the mean-field approximation is an exact theory at the ground state and fixed external magnetic field (which corresponds to the fixed chemical potential in the model investigated).

3.2. Analysis for Fixed Particle Concentration n

The ground state diagram as a function of particle concentration n is shown in Figure 3. The rectangular regions are labeled by the abbreviations of homogeneous phases (cf. Table 2). At commensurate filling, i.e., $i/3$ ($i = 0, 1, 2, 3, 4, 5, 6$; but only on the vertical boundaries indicated in Figure 3) the homogeneous phase occurs, which can be found in Table 1 and Figure 2. On the horizontal boundaries the phases from both neighboring regions have the same energies.

For $W_2 = 0$ phase separated states (mentioned in the last column of Table 2) are degenerated with the corresponding homogeneous phases inside all regions of the phase diagram. This degeneracy can be removed in finite temperatures and in some regions the phase separated states can be stable at $T > 0$ (such regions are indicated by slantwise patter in Figure 3, cf. also Section 4). E.g., for $W_2 = 0$, the TCO phase can exist only in the range of $0 < U/W_1 < 1/2$ at $T \neq 0$. For $W_2 < 0$ the phase separated states have lower energies and they occur on the phase diagram (inside the rectangular regions of Figure 3). Obviously, at commensurate filling and for any $W_2 \leq 0$, the homogeneous states can only occur (i.e., solid vertical lines in Figure 3). Please note that the following boundaries between homogeneous states (obtained by comparing only energies of homogeneous phases): (i) the DCO_A and DCO_B phases, (ii) the DCO_A and DCO_C phases, and (iii) the TCO_A and TCO_B phases are located at $U/W_1 = 0$ (and these corresponding for $n > 1$; the dashed line in Figure 3). For $W_2 < 0$ these lines do not overlap with the boundaries between corresponding phase separated states at $U/W_1 - |k| = 0$ (or $U/|W_2| = 1$), but in such a case the homogeneous states have higher energies than the phase separated states. In fact, the homogeneous states for $W_2 < 0$ are unstable (i.e., $\partial \mu / \partial n < 0$) inside the regions of Figure 1. For $W_2 < 0$ they are stable only for commensurate fillings (solid lines in Figure 3).

For the system on the square lattice the similar observation can be made (Figure 1 from [59])—compare HCO_A–LCO_A and HCO_A–HCO_B boundaries at $U/W_1 = 0$ with $PS1_A$–$PS1_B$ and $PS1_A$–$PS1_B$ boundaries at $U/|W_2| = 1$, respectively. In [68] the boundaries between homogeneous phases for $W_2 < 0$ are not shown in Figure 3. Only boundaries between corresponding phase separated states are correctly presented in that figure for $W_2 < 0$. For $U/W_1 > 0$, the CBO_A phase (corresponding to the HCO_A phase from [59]) is not the phase with the lowest energy (among homogeneous phases) in any range of n (but for $U/W_1 < 0$ it has the lowest energy among all homogeneous states). However, the corresponding phase separated state NO_0/CBO_2 (i.e., $PS1_A$ from [59]) can occur for $U/W_1 > 0$ (and for $U/|W_2| < 1$) as shown in Figure 3 of [68].

The vertical boundaries for homogeneous phases (i.e., the transitions with changing n) are associated with continuous changes of all n_α's and D_{occ}, but the chemical potential μ (calculated as $\mu = \partial f / \partial n$) changes discontinuously. Boundaries DCO_A–DCO_B, DCO_A–DCO_C, and TCO_A–TCO_B (and other transitions for fixed n at $U/W_1 = 0$) between homogeneous phases are associated with discontinuous change of only D_{occ}. One should note that it is similar to transition between two checker-board ordered phases on the square lattice, namely CBO_A–CBO_B and CBO_A–CBO_C boundaries, cf. [68] (or the HCO_A–LCO_A and HCO_A–HCO_B boundaries, respectively, from [59]). At the other horizontal boundaries (i.e., transitions for fixed n at $U/W_1 - |k| = 1/2$ in Figure 3) two of n_α's and D_{occ} change discontinuously. At commensurate fillings transitions with changing U/W_1 occur only at points indicated by squares in Figure 3.

All horizontal boundaries between phase separated states (which are stable for $W_2 < 0$) are connected with discontinuous changes of D_{occ}. These boundaries located at $U/W_1 - |k| = 0$ are also associated to a discontinuous change of particle concentration in one of the domains.

The diagram presented in Figure 3 is constructed by the comparison of (free) energies of various homogeneous phases and phase separated states collected in Table 1. The energies of homogeneous phases are calculated from (7), whereas energies of phase separated states are calculated from (13). Please note that it is easy to calculate energies of $f_\pm(n_\pm)$ of separating homogeneous phase (with commensurate fillings) at the ground state by just taking $\mu = 0$ in ω_0's collected in Table 1. Obviously, one can also calculate energies of the phases collected in Table 2 at these fillings (from both neighboring regions). For example, the DCO_B phase and the DCO_C phase at $n = 1/3$ reduce to DCO_1 phase.

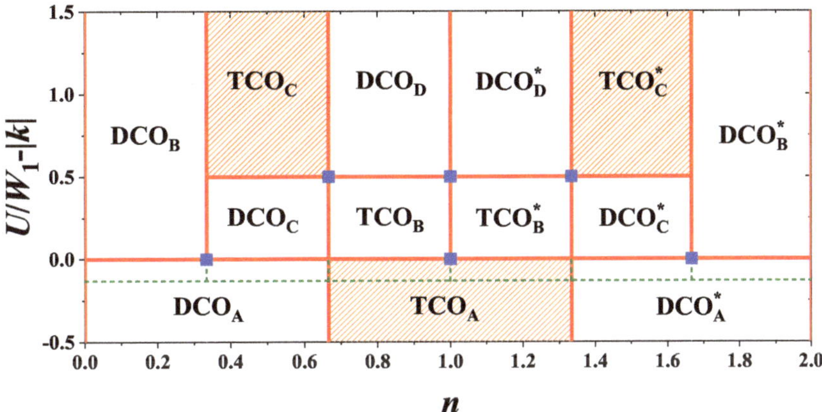

Figure 3. Ground state phase diagram of the model as a function of particle concentration n for $W_1 > 0$ and $W_2 \le 0$ ($|k| = |W_2|/W_1$). The regions are labeled by the names of the homogeneous phases (cf. Table 2). For $W_2 = 0$ all homogeneous phases are degenerated with macroscopic phase separated states indicated in the last column of Table 2. In regions filled by slantwise pattern the phase separated states occurs at infinitesimally $T > 0$ for $W_2 = 0$. For $W_2 < 0$ the phase separated states occur inside the regions, whereas at the vertical boundaries for commensurate filling the homogeneous states (defined in Table 1) still exist. The boundary at $U/W_1 = 0$ (schematically indicated by dashed green line) denotes the boundaries between homogeneous phases, which do not overlap with the boundaries between phase separated states for $W_2 < 0$. Squares denote transitions for fixed n between homogeneous phase at commensurate fillings.

4. Results for Finite Temperatures ($W_1 > 0$ and $W_2 = 0$)

One can distinguish four ranges of U interaction, where the system exhibits qualitatively different behavior, namely: (i) $U/W_1 < 0$, (ii) $0 < U/W_1 < (1/3)\ln(2)$, (iii) $(1/3)\ln(2) < U/W_1 < 1/2$, and (iv) $U/W_1 > 1/2$. In Figures 4–7, the exemplary finite temperature phase diagrams occurring in each of these ranges of onsite interaction are presented. All diagrams are found by investigation of the behavior of n_α's determined by (11) in the solution corresponding to the lowest grand canonical potential [Equation (8), when μ is fixed] or to the lowest free energy [Equations (12) and (13) if n is fixed]. The set of three nonlinear Equations (11) has usually several nonequivalent solutions and thus it is extremely important to find a solution, which has the minimal adequate thermodynamic potential. In Figure 8 the behavior of n_α's as a function of temperature or chemical potential is shown for some representative model parameters. Figure 9 presents the phase diagram of the system for half-filling.

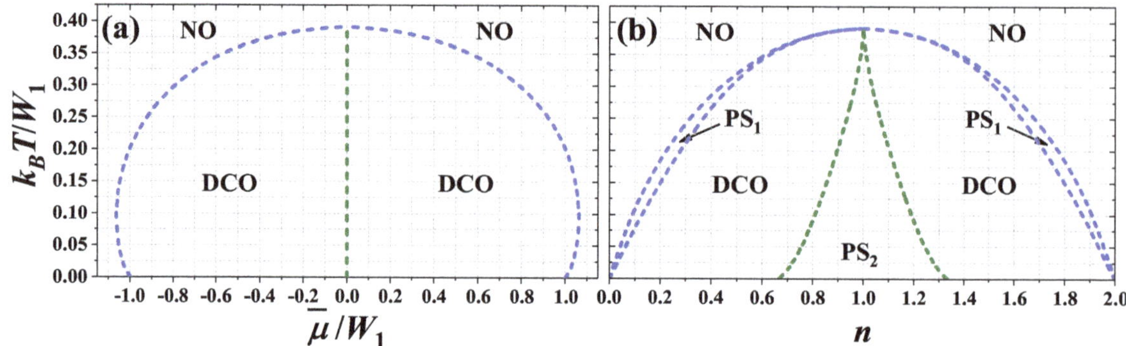

Figure 4. Phase diagrams of the model for $U/W_1 = -1.00$ as a function of (**a**) chemical potential $\bar{\mu}/W_1$ and (**b**) particle concentration n ($W_1 > 0$, $W_2 = 0$). All transitions are first order and regions of phase separated state (PS$_1$:NO/DCO and PS$_2$:DCO/DCO) occurrence are present on panel (**b**). NO and DCO denote homogeneous phases defined in Figure 1b.

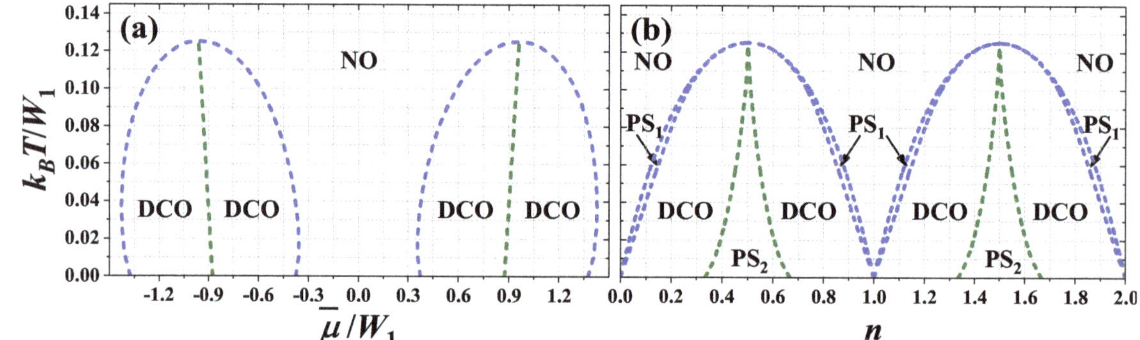

Figure 5. Phase diagrams of the model for $U/W_1 = 0.75$ as a function of (**a**) chemical potential $\bar{\mu}/W_1$ and (**b**) particle concentration n ($W_1 > 0$, $W_2 = 0$). All transitions are first order. Other denotations as in Figure 4.

For $U/W_1 < 0$ and $U/W_1 > 1/2$ the phase diagrams of the model are similar and the DCO phase is only ordered homogeneous one occurring on the diagrams. In the first range, there are two regions of ordered phase occurrence (cf. Figure 4 and [74]), whereas in the second case one can distinguish four regions of the DCO phase stability (cf. Figure 5). The NO–DCO transitions for fixed μ are discontinuous for any values of onsite interaction and chemical potential in discussed range of model parameters and thus phase separated state PS$_1$:NO/DCO occurs in define ranges of n. In this state domains of the NO and the DCO phases coexist.

For $U/W_1 < 0$ the temperature of NO–DCO transition is maximal for $\bar{\mu} = 0$ (i.e., at half-filling)–Figure 4a. Its maximal value T_M monotonously decreases with increasing of U from $k_B T_M / W_1 = 1/2$ for $U \to -\infty$ and at $U = 0$ it is equal to $1/4$. This transition exhibits re-entrant behavior (for fixed $|\bar{\mu}| > 1$). At $T = T_M$ and $\bar{\mu} = 0$ and at only this point, this transition exhibits properties of a second order transition (cf. Figure 8a). In particular, with increasing T for fixed $\bar{\mu} = 0$ n_α's changes continuously at T_M, but two equivalent solutions still exist for any $T < T_M$ (similarly as in the ferromagnetic Ising model at zero field [9]). At $\bar{\mu} = 0$ and $T < T_M$ the discontinuous transition between two DCO phases occurs. In the DCO phase for $\bar{\mu} < 0$ ($n < 1$) [connecting with the DCO$_1$ (DCO$_A$) region at $T = 0$] the relation $n_A = n_B < n_C$ is fulfilled, whereas in the DCO phase for $\bar{\mu} > 0$ ($n > 1$) [connecting with the DCO$_1^*$ (DCO$_A^*$) region at $T = 0$] the relation $n_A < n_B = n_C$ occurs (n_C can be larger than 1 for some temperatures), cf. also Figure 8g,h as well as [74]. Both discontinuous transitions for fixed chemical potential are associated with occurrence of phase separated states. On the diagrams obtained for fixed n three region of phase separated states occurs (Figure 4b). For $W_2 = 0$ the PS_1:NO/DCO phase separated state occurs only for $T > 0$. For $T \rightarrow 0$ the concentrations in both domains of the PS_1 state approach 0 (or 2), whereas for $T \rightarrow T_M$ they approach to 1. Near $n = 1$ the PS_2:DCO/DCO state is stable for $0 \leq T < T_M$. In this state domains of two DCO phases (with different particle concentrations) coexist in the system.

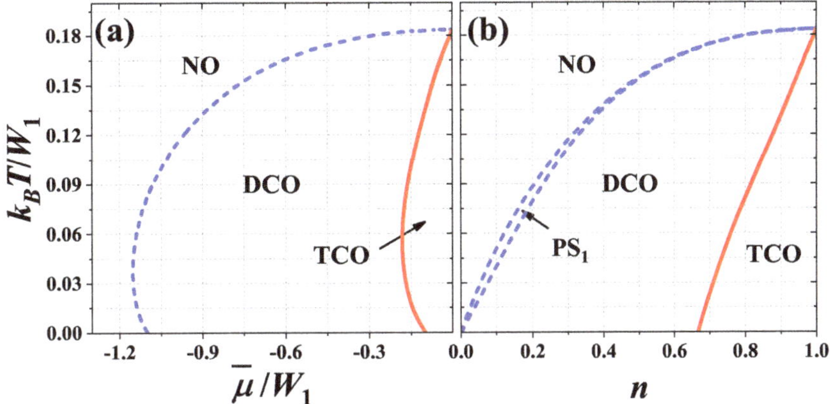

Figure 6. Phase diagrams of the model for $U/W_1 = 0.20$ as a function of (**a**) chemical potential $\bar{\mu}/W_1$ and (**b**) particle concentration n ($W_1 > 0$, $W_2 = 0$). The boundary TCO–DCO is second order, the remaining are first order. Other denotations as in Figure 4. The diagrams are shown only for $\bar{\mu} \leq 0$ and $n \leq 1$, but they are symmetric with respect to $\bar{\mu} = 0$ and $n = 1$, respectively.

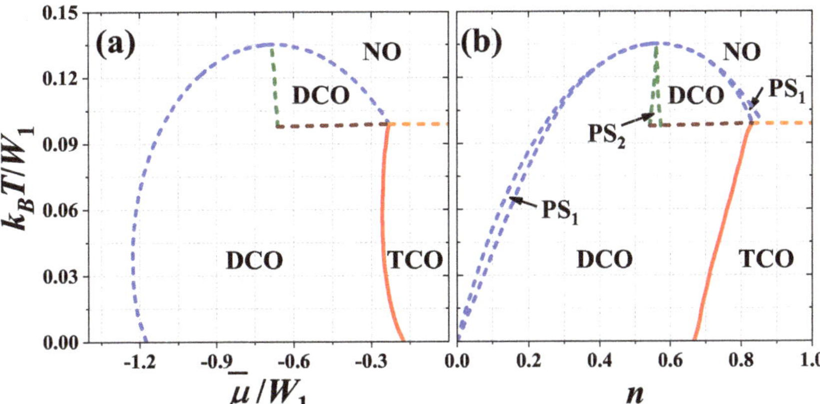

Figure 7. Phase diagrams of the model for $U/W_1 = 0.35$ as a function of (**a**) chemical potential $\bar{\mu}/W_1$ and (**b**) particle concentration n ($W_1 > 0$, $W_2 = 0$). The boundary TCO–DCO is second order, the remaining are first order. Other denotations as in Figure 4. The diagrams are shown only for $\bar{\mu} \leq 0$ and $n \leq 1$, but they are symmetric with respect to $\bar{\mu} = 0$ and $n = 1$, respectively.

For $U/W_1 > 1/2$ the diagrams are similar, but the double occupancy of sites is strongly reduced due to repulsive U (Figure 5). Thus, their structure exhibits two lobs of the DCO phase occurrence in cotrary to the case of $U/W_1 < 0$, where a single lob of the DCO phase is present (as expected from previous studies of the model, cf. [10,59,60]). The maximal value $k_B T_M / W_1$ of NO–DCO transition occurs for $\bar{\mu}/W_1$ corresponding approximately quarter fillings (i.e., near $n = 1/2$ and $n = 3/2$). With increasing U it decreases and finally in the limit $U \rightarrow +\infty$ it reaches $1/8$. At this point DCO–NO boundary exhibits features of continuous transition as discussed previously. In this range, the phase diagrams are (almost) symmetric with respect to these fillings (when one considers only one part of the diagram for $0 < n < 1$ or for $1 < n < 2$).

The most complex diagrams are obtained for $0 < U/W_1 < 1/2$, where the TCO phase appears at $T = 0$ and for finite temperatures near half-filling. For $0 < U/W_1 < (1/3)\ln(2)$ the region of the TCO phase is separated from the NO phase by the region of DCO phase, Figure 6a. The TCO–DCO transition is continuous (cf. Figure 8g,h for $U/W_1 = 0.35$) and its maximal temperature is located for half-filling (at $\bar{\mu} = 0$ or $n = 1$). At this point two first-order NO–DCO and two second-order TCO–DCO boundaries merge (for fixed chemical potential). It is the only point for fixed U/W_1 in this range of model parameters, where a direct continuous transition from the TCO phase to the NO phase is possible (Figure 8b). The continuous TCO–DCO transition temperature can be also found as a solution of (11) and (A4) as discussed in Appendix A. Similarly as for $U/W_1 < 0$, the temperature of NO–DCO transition is maximal at half-filling. For fixed n, the narrow regions of PS_1:NO/DCO states are present between the NO region and DCO regions. Please note that for $T > 0$ there is no signatures of the discontinuous DCO_1–DCO_2 (DCO_1^*–DCO_2^*) boundary occurring at $T = 0$. It is due to the fact that the discontinuous jumps of n_α's occurring for $T = 0$ at these boundaries are changed into continuous evolutions of sublattice concentrations at $T > 0$ and there is no criteria for distinction of these two DCO phases at finite temperatures (cf. also [59–61]). From the same reason, there is no boundary at $T > 0$ for fixed n associated to the DCO_B–DCO_C (DCO_B^*–DCO_C^*) line occurring at $T = 0$ (Figure 6b). However, strong reduction of one n_α from the case where $n_\alpha \approx 2$ to the case of $n_\alpha \approx 1$ is visible (some kind of a smooth crossover inside the DCO region), cf. Figure 8f–h for $U/W_1 = 0.35$.

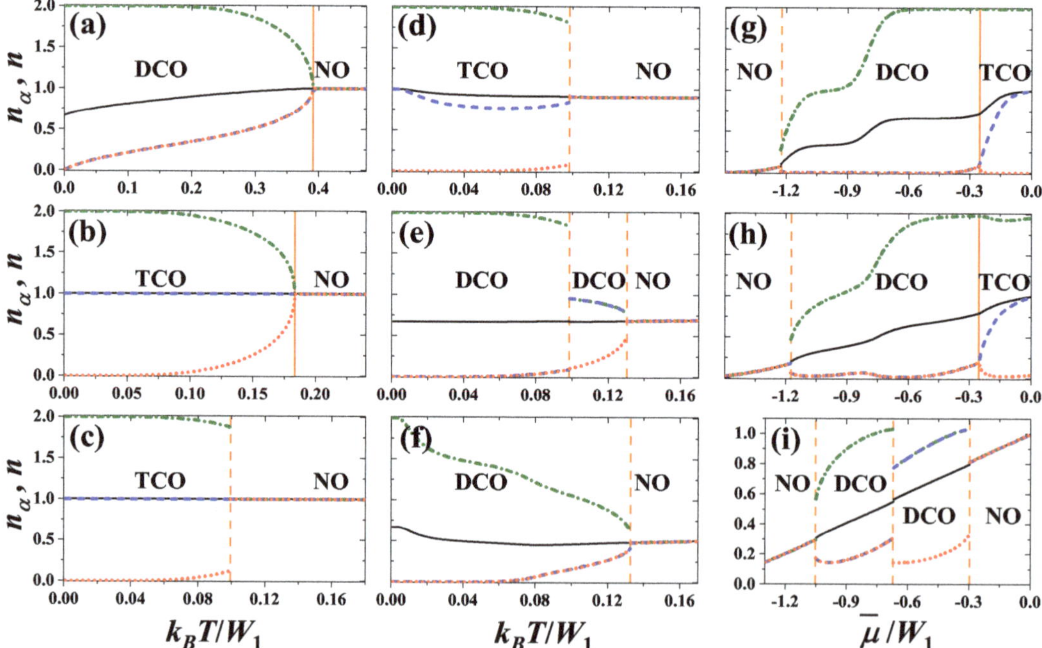

Figure 8. Dependencies of particle concentrations n_α's in the sublattices (red dotted, blue dashed, and green dot-dashed lines) as a function of $k_B T/W_1$ [(**a**–**f**)] and $\bar{\mu}/W_1$ [(**g**–**h**)] for $W_2 = 0$. Black solid lines denote total particle concentration $n = (n_A + n_B + n_C)/3$. They are obtained for: (**a**) $U/W_1 = -1.00$, $\bar{\mu}/W_1 = 0$; (**b**) $U/W_1 = 0.20$, $\bar{\mu}/W_1 = 0$; (**c**) $U/W_1 = 0.35$, $\bar{\mu}/W_1 = 0$; (**d**) $U/W_1 = 0.35$, $\bar{\mu}/W_1 = -0.15$; (**e**) $U/W_1 = 0.35$, $\bar{\mu}/W_1 = -0.5$; (**f**) $U/W_1 = 0.35$, $\bar{\mu}/W_1 = -0.8$; (**g**) $U/W_1 = 0.35$, $k_B T/W_1 = 0.40$; (**h**) $U/W_1 = 0.35$, $k_B T/W_1 = 0.80$; (**i**) $U/W_1 = 0.35$, $k_B T/W_1 = 0.11$. Vertical solid and dashed lines indicate points of continuous and discontinuous transitions, respectively.

For $(1/3)\ln(2) < U/W_1 < 1/2$, the maximum of the NO–DCO transition temperature is shifted towards larger $|\bar{\mu}|/W_1$ (or smaller $|1 - n|$). This is associated with forming of the two-lob structure of the diagram found for $U/W_1 > 1/2$. Inside the regions of the DCO phase occurrence discontinuous transitions between two DCO phases appear—See Figure 7a as well as Figure 8e,i. These new regions of the DCO phase at $T > 0$ [with $n_A < n_B = n_C$ (for $\bar{\mu} < 0$ or $n < 1$); cf. Figure 8e,i] are connected with the DCO_3 and DCO_3^* regions occurring at the ground state. The boundaries DCO–DCO weakly dependent on $\bar{\mu}$ are associated with occurrence of phase separated PS_2:DCO/DCO states (at high temperatures) in some ranges of n, cf. Figure 7b. The other DCO–DCO transitions (which are almost temperature-independent) are not connected with phase separated states. Also the first-order TCO–NO line is present near half-filling, cf. Figure 8d. One should underline that all four lines (three first-order boundaries: DCO–NO, DCO–DCO, TCO–NO and the second-order TCO–DCO boundary) merge at single point with numeric accuracy. However, it cannot be excluded that the DCO–NO and TCO–DCO boundaries connect with the temperature-independent line in slightly different points, what could result in, e.g., the TCO–DCO–NO sequence of transition with increasing temperature for small range of chemical potential $\bar{\mu}$. All of these almost temperature-independent boundaries (i.e., the DCO–DCO and the TCO–NO lines) are located at temperature, which decreases with increasing U/W_1 and approaches 0 at $U/W_1 = 1/2$ [i.e., they connect with the DCO_2–DCO_3 (DCO_2^*–DCO_3^*) and TCO–NO_1 boundaries at $T = 0$ for fixed μ or with the TCO–DCO_D (TCO*–DCO_D^*) lines at $T = 0$ for fixed n]. From the analysis of (11) similarly as it was done in the case of the square lattice [10] (see also Appendix A) one obtains that the point, where the TCO–NO transition changes its order at half-filling, is $k_B T/W_1 = 1/6$ and $U/W_1 = (1/3)\ln(2)$.

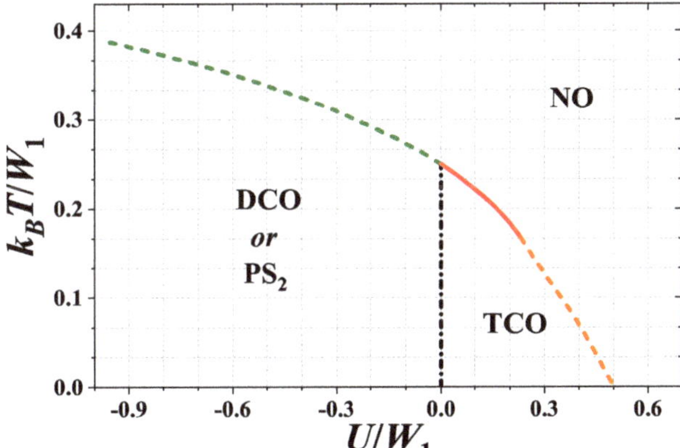

Figure 9. Phase diagram of the model for half-filling ($\bar{\mu} = 0$ or $n = 1$) as a function of onsite interaction U/W_1 ($W_1 > 0$ and $W_2 = 0$). The order-disorder boundary for $U/W_1 < (1/3)\ln(2)$ is a line consisting of some higher-order critical points as discussed in the text.

For better overview of the system behavior, the phase diagram of the model for half-filling ($\bar{\mu} = 0$ or $n = 1$) is presented in Figure 9. The temperature of the order-disorder transition decreases with increasing U/W_1. In low temperatures and for $U/W_1 < 0$, the DCO phases exist in the system (precisely, if μ is fixed—at $\bar{\mu} = 0$ the DCO–DCO discontinuous boundary occurs; whereas if n is fixed—the PS_2:DCO/DCO state is stable at $n = 1$), cf. also Figure 4. For $0 < U/W_1 < 1/2$ the TCO phase is stable below the order-disorder line, but for $(1/3)\ln(2) < U/W_1 < 1/2$ and $k_B T/W_1 < 1/6$ the TCO–NO phase transition is discontinuous (cf. also Figure 7). For $U/W_1 < (1/3)\ln(2)$ the order-disorder boundary presented in Figure 9 is a merging point of several boundaries as presented in Figures 4 and 6, and discussed previously. Thus, formally this order-disorder boundary for $U/W_1 < (1/3)\ln(2)$ occurring at half-filling is a line of some critical points of a higher order.

Please note that the order-disorder transition is discontinuous for any value of onsite interaction and chemical potential [excluding only the TCO–NO boundary for half-filling and $0 < U/W_1 < (1/3)\ln(2)$] in contrast to the case of two- [10,59,60] or four-sublattice [67,68] assumptions, where it can be continuous one for some range of model parameters). In [74] also metastable phases have been discussed in detail for the large onsite attraction limit and the triangular lattice.

5. Final Remarks

In this work, the mean-field approximation was used to investigate the atomic limit of extended Hubbard model [hamiltonian (1)] on the triangular lattice. The phase diagram was determined for the model with intersite repulsion between the nearest neighbors $(W_1 > 0)$. The effects of attractive next-nearest-neighbor interaction $(W_2 < 0)$ were discussed in the ground state. The most important findings of this work are that (i) two different arrangements of particles (i.e., two different charge-ordered phases: the DCO and TCO states) can occur in the system and (ii) attractive $W_2 < 0$ or finite $T > 0$ removes the degeneration between homogeneous phases and phase separated states occurring at $T = 0$ for $W_2 = 0$. It was shown that TCO phase is stable in intermediate range of onsite repulsion $0 < U/W_1 < 1/2$ (for $W_2 = 0$). All transition from the ordered phases to the NO are discontinuous for fixed chemical potential (apart from TCO–NO boundary at half-filling for $0 < U/W_1 < (1/3)\ln(2)$) and the DCO–NO boundaries at single points corresponding to $n = 1/2, 1, 3/2$ as discussed in Section 4), thus the phase separated states occur on the phase diagram for fixed particle concentration.

One should stress that hamiltonian (1) is interesting not only from statistical point of view as a relatively simple toy model for phase transition investigations. Although it is oversimplified for quantitative description of bulk condensed matter systems, it can be useful in qualitative analysis of, e.g., experimental studies of adsorbed gas layers on crystalline substrates.

Additionally, one notes that the mean-field results for model (1) with attractive $W_1 < 0$ and $W_2 \leq 0$ are the same for both two-sublattice and tri-sublattice assumptions. In such a case, three different nonordered phases exist with the discontinuous first-order transition between them (at $\bar{\mu} = 0$ for $U < 0$ or for $|\bar{\mu}| \neq 0$ for $U/(|W_1| + |W_2|) > 1$), and thus for fixed n, several so-called electron-droplet states (phase separations NO/NO) exist (cf. [60,68,87,88], particularly Figure 2 of [60]).

Notice that the mean-field decoupling of the intersite term is an approximation for purely two-dimensional model investigated, which overestimates the stability of ordered phases. For example, the order-disorder transition for the ferromagnetic Ising model is overestimated by the factor two (for the honeycomb, square and triangular lattices rigorous solution gives $k_B T_c/|J|$ as 0.506, 0.568, 0.607, respectively, whereas the mean-field approximation gives $k_B T_c/|J| = 1$) [76]. Moreover, the results for the antiferromagnetic Ising model on the triangular lattice [the limit $U \to \pm\infty$ of model (1)] do not predict long-range order at zero field [1,3,76] and $T > 0$ [corresponding to $n = 1$ or $n = 1/2, 3/2$, respectively, in the case of model (1)]. However, longer-range interactions [3] or weak interactions between adsorbed particles and the adsorbent material occurring in realistic systems could stabilize such an order (such systems are rather quasi-two-dimensional). It should be also mentioned that the charge Berezinskii-Kosterlitz-Thouless-like phase was found in the intermediate-temperature regime between the charge-ordered phase (with long-range order, coresponding to the TCO phase here) and disordered phases in the investigated model [73].

The recent progress in the field of optical lattices and a creation of the triangular lattice by laser trapping [89,90] could enable testing predictions of the present work. The fermionic gases in harmonic traps are fully controllable systems. Note also that the superconductivity in the twisted-bilayer graphene [91–96] is driven by the angle between the graphene layers.

It is associated with an occurrence of the Moiré pattern (the triangular lattice with very large supercell). Hetero-bilayer transition metals dichalcogenides system is the other field where this pattern appears [97,98]. This makes further studies of properties of different models on the triangular lattice desirable.

Author Contributions: Conceptualization, K.J.K.; methodology, K.J.K.; software, K.J.K.; validation, K.J.K.; formal analysis, K.J.K.; investigation, K.J.K.; resources, K.J.K.; data curation, K.J.K.; writing—original draft preparation, K.J.K.; writing—review and editing, K.J.K.; visualization, K.J.K.; supervision, K.J.K.; project administration, K.J.K.; funding acquisition, K.J.K. The author has read and agreed to the published version of the manuscript.

Funding: The support from the National Science Centre (NCN, Poland) under Grant SONATINA 1 no. UMO-2017/24/C/ST3/00276 is acknowledged. Founding in the frame of a scholarship of the Minister of Science and Higher Education (Poland) for outstanding young scientists (2019 edition, no. 821/STYP/14/2019) is also appreciated.

Institutional Review Board Statement: Not applicable. The study does not involve humans or animals.

Informed Consent Statement: Not applicable. The study does not involve humans.

Data Availability Statement: The data presented in this study are available on request from the author. All data presented in Section 4 have been obtained by numerical solving of self-consistent equations given in Section 2.3.

Acknowledgments: The author expresses his sincere thanks to J. Barański, R. Lemański, R. Micnas, P. Piekarz, and A. Ptok for very useful discussions on some issues raised in this work. The author also thanks R. Micnas and I. Ostrowska for careful reading of the manuscript.

Conflicts of Interest: The author declares no conflict of interest. The funders had no role in the design of the study; in the collection, analyses, or interpretation of data; in the writing of the manuscript, or in the decision to publish the results.

Abbreviations

The following abbreviations are used in this manuscript:

NO   Non-ordered phase with the same concentrations in all three sublattices
DCO   Charge-ordered phase with two different concentrations in sublattices
TCO   Charge-ordered phase with three different concentrations in sublattices

Appendix A. Analytic Expressions for Continuous Transition Temperatures

Equations (11) can be written in a different form, namely $n_\alpha = f_\alpha$, where $f_\alpha \equiv 2(t_\alpha + t_\alpha^2 a)/(1 + 2t_\alpha + t_\alpha^2 a)$, $t_\alpha \equiv \exp(\beta\mu_\alpha)$ and $a \equiv \exp(-\beta U)$. One can define $\Delta \equiv (n_A - n_B)/2$ and $\chi \equiv (n_B - n_C)/2$. From (10) one gets:

$$\mu_A = \mu - W_1\left[n - \tfrac{1}{3}(2\Delta + \chi)\right], \tag{A1}$$

$$\mu_B = \mu - W_1\left[n + \tfrac{1}{3}(\Delta - \chi)\right], \tag{A2}$$

$$\mu_C = \mu - W_1\left[n + \tfrac{1}{3}(\Delta + 2\chi)\right]. \tag{A3}$$

Taking the limit $\chi \to 0$ of both sides of the equation $(f_B - f_C)/(2\chi) = 1$ (using de l'Hospital theorem) one gets $(g_B - g_C)/2 = 1$, where $g_\alpha \equiv \frac{\partial f_\alpha}{\partial \chi} = \frac{\partial f_\alpha}{\partial t_\alpha}\frac{\partial t_\alpha}{\partial \mu_\alpha}\frac{\partial \mu_\alpha}{\partial \chi}$. One easily finds that $\partial f_\alpha/\partial t_\alpha = 2(1 + 2t_\alpha a + t_\alpha^2 a)/(1 + 2t_\alpha + t_\alpha^2 a)^2$, $\partial t_\alpha/\partial \mu_\alpha = \beta t_\alpha$ as well as $\partial\mu_A/\partial\chi = -W_1/3$, $\partial\mu_B/\partial\chi = W_1/3$, $\partial\mu_C/\partial\chi = -2W_1/3$. Finally, the equation determining temperature T_c of a continuous transition (at which $n_B \to n_C$) has the form

$$\frac{1}{\beta_c W_1} = \frac{\left(1 + 2t_{BC}\bar{a} + t_{BC}^2\bar{a}\right)t_{BC}}{\left(1 + 2t_{BC} + t_{BC}^2\bar{a}\right)^2}, \tag{A4}$$

where $t_{BC} \equiv \exp(\beta_c \mu_{BC})$, $\mu_{BC} = \mu - W_1(n + \Delta/3)$ (in the considered limit $\mu_B = \mu_C$ and $n_B = n_C$), $\bar{a} \equiv \exp(-\beta_c U)$, $\beta_c \equiv 1/(k_B T_c)$. Concentrations n_A and $n_{BC} \equiv n_B = n_C$ are calculated from (11) for β_c self-consistently. Thus, for fixed μ (or n) one has a set of three equation which is solved with respect to β_c, n (or μ) and Δ.

The solutions of (A4) and (11) with $\Delta \neq 0$ (i.e., $n_A \neq n_B$) correspond to the TCO–DCO boundaries. Such determined temperatures coincide with those found from the analysis of (11) and (8) or (12) and presented in Figures 6 and 7, what supports the findings that the TCO–DCO boundaries are indeed continuous.

The solutions of (A4) and (11) with $\Delta = 0$ (i.e., $n_A = n_B$) correspond to the continuous DCO–NO boundaries. On the diagrams presented in Section 4 such solutions for T_c are located inside the regions of the DCO phase occurrence (and they correspond to the transitions between metastable phases [74] or to vanishing of the NO metastable solution, cf. [88,99]). In the present case of model (1) studied, they coincide with the DCO–NO transitions presented in Figures 4–7 only at $T = 0$ (i.e., for $n = 0,2$ as well as for $n = 1$ and $U/W_1 > 1/2$; or corresponding $\bar{\mu}$) and at $T = T_M$ (i.e., maximal temperature of the DCO–NO transition, occurring for $U/W_1 < (1/3)\ln(2)$ and $n = 1$ or $\bar{\mu} = 0$, as well as for $U/W_1 > 1/2$ and $n \approx 1/2$, $3/2$ or corresponding $\bar{\mu}$; for $(1/3)\ln(2) < U/W_1 < 1/2$ it is located for some intermediate concentrations $1/2 < n < 1$ and $1 < n < 3/2$). For $\Delta = 0$, (A4) and (11) give the following results: (i) for $U \rightarrow -\infty$: $k_B T_c/W_1 = n(2-n)/2$; (ii) for $U = 0$: $k_B T_c/W_1 = n(2-n)/4$; and (iii) for $U \rightarrow +\infty$: $k_B T_c/W_1 = n(1-n)/2$ (if $n < 1$) and $k_B T_c/W_1 = (2-n)(n-1)/2$ (if $n > 1$). Please note that such determined T_c for $\Delta = 0$ is two times smaller than corresponding continuous transitions for the model considered on the hypercubic lattice within the mean-field aprroximation for the intersite term (for the same U/W_1 and n) [10,59,60].

References

1. Campbell, C.E.; Schick, M. Triangular Lattice Gas. *Phys. Rev. A* **1972**, *5*, 1919–1925. [CrossRef]
2. Kaburagi, M.; Kanamori, J. Ordered Structure of Adatoms in the Extended Range Lattice Gas Model. *Japan. J. Appl. Phys.* **1974**, *13* (Suppl. S2), 145–148. [CrossRef]
3. Mihura, B.; Landau, D.P. New Type of Multicritical Behavior in a Triangular Lattice Gas Model. *Phys. Rev. Lett.* **1977**, *38*, 977–980. [CrossRef]
4. Kaburagi, M.; Kanamori, J. Ground State Structure of Triangular Lattice Gas Model with up to 3rd Neighbor Interactions. *J. Phys. Soc. Jpn.* **1978**, *44*, 718–727. [CrossRef]
5. Wannier, G.H. The Statistical Problem in Cooperative Phenomena. *Rev. Mod. Phys.* **1945**, *17*, 50–60. [CrossRef]
6. Ising, E. Beitrag zur Theorie des Ferromagnetismus. *Z. Phys.* **1925**, *31*, 253–258. [CrossRef]
7. Onsager, L. Crystal Statistics. I. A Two-Dimensional Model with an Order-Disorder Transition. *Phys. Rev.* **1944**, *65*, 117–149. [CrossRef]
8. Binney, J.J.; Dowrick, N.J.; Fisher, A.J.; Newman, M.E.J. *The Theory of Critical Phenomena: An Introduction to the Renormalization Group*; Oxford University Press: Oxford, UK, 1992.
9. Vives, E.; Castán, T.; Planes, A. Unified Mean-Field Study of Ferro-and Antiferromagnetic Behavior of the Ising Model with External Field. *Amer. J. Phys.* **1997**, *65*, 907–913. [CrossRef]
10. Micnas, R.; Robaszkiewicz, S.; Chao, K.A. Multicritical Behavior of the Extended Hubbard Model in the Zero-Bandwidth Limit. *Phys. Rev. B* **1984**, *29*, 2784–2789. [CrossRef]
11. Micnas, R.; Ranninger, J.; Robaszkiewicz, S. Superconductivity in Narrow-Band Systems with Local Nonretarded Attractive Interactions. *Rev. Mod. Phys.* **1990**, *62*, 113–171. [CrossRef]
12. Georges, A.; Kotliar, G.; Krauth, W.; Rozenberg, M.J. Dynamical Mean-Field Theory of Strongly Correlated Fermion Systems and the Limit of Infinite Dimensions. *Rev. Mod. Phys.* **1996**, *68*, 13–125. [CrossRef]
13. Imada, M.; Fujimori, A.; Tokura, Y. Metal-Insulator Transitions. *Rev. Mod. Phys.* **1998**, *70*, 1039–1263. [CrossRef]
14. Kotliar, G.; Savrasov, S.Y.; Haule, K.; Oudovenko, V.S.; Parcollet, O.; Marianetti, C.A. Electronic Structure Calculations with Dynamical Mean-Field Theory. *Rev. Mod. Phys.* **2006**, *78*, 865–951. [CrossRef]
15. Davoudi, B.; Hassan, S.R.; Tremblay, A.M.S. Competition Between Charge and Spin Order in the $t-U-V$ Extended Hubbard Model on the Triangular Lattice. *Phys. Rev. B* **2008**, *77*, 214408. [CrossRef]
16. Cano-Cortés, L.; Ralko, A.; Février, C.; Merino, J.; Fratini, S. Geometrical Frustration Effects on Charge-Driven Quantum Phase Transitions. *Phys. Rev. B* **2011**, *84*, 155115. [CrossRef]
17. Merino, J.; Ralko, A.; Fratini, S. Emergent Heavy Fermion Behavior at the Wigner-Mott Transition. *Phys. Rev. Lett.* **2013**, *111*, 126403. [CrossRef] [PubMed]

18. Tocchio, L.F.; Gros, C.; Zhang, X.F.; Eggert, S. Phase Diagram of the Triangular Extended Hubbard Model. *Phys. Rev. Lett.* **2014**, *113*, 246405. [CrossRef] [PubMed]
19. Litak, G.; Wysokiński, K.I. Evolution of the Charge Density Wave Order on the Two-Dimensional Hexagonal Lattice. *J. Magn. Magn. Mater.* **2017**, *440*, 104–107. [CrossRef]
20. Aichhorn, M.; Evertz, H.G.; von der Linden, W.; Potthoff, M. Charge Ordering in Extended Hubbard Models: Variational Cluster Approach. *Phys. Rev. B* **2004**, *70*, 235107. [CrossRef]
21. Tong, N.H.; Shen, S.Q.; Bulla, R. Charge Ordering and Phase Separation in the Infinite Dimensional Extended Hubbard Model. *Phys. Rev. B* **2004**, *70*, 085118. [CrossRef]
22. Amaricci, A.; Camjayi, A.; Haule, K.; Kotliar, G.; Tanasković, D.; Dobrosavljević, V. Extended Hubbard Model: Charge Ordering and Wigner-Mott Transition. *Phys. Rev. B* **2010**, *82*, 155102. [CrossRef]
23. Ayral, T.; Biermann, S.; Werner, P.; Boehnke, L. Influence of Fock Exchange in Combined Many-Body Perturbation and Dynamical Mean Field Theory. *Phys. Rev. B* **2017**, *95*, 245130. [CrossRef]
24. Kapcia, K.J.; Robaszkiewicz, S.; Capone, M.; Amaricci, A. Doping-Driven Metal-Insulator Transitions and Charge Orderings in the Extended Hubbard Model. *Phys. Rev. B* **2017**, *95*, 125112. [CrossRef]
25. Terletska, H.; Chen, T.; Paki, J.; Gull, E. Charge Ordering and Nonlocal Correlations in the Doped Extended Hubbard Model. *Phys. Rev. B* **2018**, *97*, 115117. [CrossRef]
26. Freericks, J.K.; Zlatić, V. Exact Dynamical Mean-Field Theory of the Falicov-Kimball Model. *Rev. Mod. Phys.* **2003**, *75*, 1333–1382. [CrossRef]
27. Kapcia, K.J.; Lemański, R.; Zygmunt, M.J. Extended Falicov–Kimball Model: Hartree–Fock vs DMFT approach. *J. Phys. Condens. Matter* **2020**, *33*, 065602. [CrossRef]
28. Aziz, R.A.; Buck, U.; Jónsson, H.; Ruiz-Suárez, J.; Schmidt, B.; Scoles, G.; Slaman, M.J.; Xu, J. Two- and Three-Body Forces in the Interaction of He Atoms with Xe Overlayers Adsorbed on (0001) Graphite. *J. Chem. Phys.* **1989**, *91*, 6477–6493. [CrossRef]
29. Caragiu, M.; Finberg, S. Alkali Metal Adsorption on Graphite: A Review. *J. Phys. Condens. Matter* **2005**, *17*, R995–R1024. [CrossRef]
30. Petrović, M.; Lazić, P.; Runte, S.; Michely, T.; Busse, C.; Kralj, M. Moiré-Regulated Self-Assembly of Cesium Adatoms on Epitaxial Graphene. *Phys. Rev. B* **2017**, *96*, 085428. [CrossRef]
31. Dimakis, N.; Valdez, D.; Flor, F.A.; Salgado, A.; Adjibi, K.; Vargas, S.; Saenz, J. Density Functional Theory Calculations on Alkali and the Alkaline Ca Atoms Adsorbed on Graphene Monolayers. *Appl. Sur. Sci.* **2017**, *413*, 197–208. [CrossRef]
32. Zhour, K.; El Haj Hassan, F.; Fahs, H.; Vaezzadeh, M. Ab Initio Study of the Adsorption of Potassium on B, N, and BN-Doped Graphene Heterostructure. *Mater. Today Commun.* **2019**, *21*, 100676. [CrossRef]
33. Huang, Y.C.; Zhao, K.Y.; Liu, Y.; Zhang, X.Y.; Du, H.Y.; Ren, X.W. Investigation on Adsorption of Ar and N_2 on α-Al_2O_3(0001) Surface from First-Principles Calculations. *Vacuum* **2020**, *176*, 109344. [CrossRef]
34. Xing, H.; Hu, P.; Li, S.; Zuo, Y.; Han, J.; Hua, X.; Wang, K.; Yang, F.; Feng, P.; Chang, T. Adsorption and Diffusion of Oxygen on Metal Surfaces Studied by First-Principle Study: A Review. *J. Mater. Sci. Technol.* **2021**, *62*, 180–194. [CrossRef]
35. Profeta, G.; Ottaviano, L.; Continenza, A. $\sqrt{3}\times\sqrt{3}$ R30° → 3 × 3 Distortion on the C/Si(111) Surface. *Phys. Rev. B* **2004**, *69*, 241307. [CrossRef]
36. Tresca, C.; Brun, C.; Bilgeri, T.; Menard, G.; Cherkez, V.; Federicci, R.; Longo, D.; Debontridder, F.; D'angelo, M.; Roditchev, D.; et al. Chiral Spin Texture in the Charge-Density-Wave Phase of the Correlated Metallic Pb/Si(111) Monolayer. *Phys. Rev. Lett.* **2018**, *120*, 196402. [CrossRef]
37. Rodríguez, B.C.R.; Santana, J.A. Adsorption and Diffusion of Sulfur on the (111), (100), (110), and (211) Surfaces of FCC Metals: Density Functional Theory Calculations. *J. Chem. Phys.* **2018**, *149*, 204701. [CrossRef]
38. Patra, A.; Peng, H.; Sun, J.; Perdew, J.P. Rethinking CO Adsorption on Transition-Metal Surfaces: Effect of Density-Driven Self-Interaction Errors. *Phys. Rev. B* **2019**, *100*, 035442. [CrossRef]
39. Menkah, E.S.; Dzade, N.Y.; Tia, R.; Adei, E.; de Leeuw, N.H. Hydrazine Adsorption on Perfect and Defective FCC Nickel (100), (110) and (111) Surfaces: A Dispersion corrected DFT-D2 study. *Appl. Sur. Sci.* **2019**, *480*, 1014–1024. [CrossRef]
40. Bretz, M.; Dash, J.G. Ordering Transitions in Helium Monolayers. *Phys. Rev. Lett.* **1971**, *27*, 647–650. [CrossRef]
41. Bretz, M.; Dash, J.G.; Hickernell, D.C.; McLean, E.O.; Vilches, O.E. Phases of He^3 and He^4 Monolayer Films Adsorbed on Basal-Plane Oriented Graphite. *Phys. Rev. A* **1973**, *8*, 1589–1615. [CrossRef]
42. Zhou, S.; Wang, Z. Charge and Spin Order on the Triangular Lattice: Na_xCoO_2 at $x = 0.5$. *Phys. Rev. Lett.* **2007**, *98*, 226402. [CrossRef]
43. Soumyanarayanan, A.; Yee, M.M.; He, Y.; van Wezel, J.; Rahn, D.J.; Rossnagel, K.; Hudson, E.W.; Norman, M.R.; Hoffman, J.E. Quantum Phase Transition from Triangular to Stripe Charge Order in $NbSe_2$. *Proc. Natl. Acad. Sci. USA* **2013**, *110*, 1623–1627. [CrossRef] [PubMed]
44. Ugeda, M.M.; Bradley, A.J.; Zhang, Y.; Onishi, S.; Chen, Y.; Ruan, W.; Ojeda-Aristizabal, C.; Ryu, H.; Edmonds, M.T.; Tsai, H.Z.; et al. Characterization of Collective Ground States in Single-Layer $NbSe_2$. *Nat. Phys.* **2016**, *12*, 92–97. [CrossRef]
45. Xi, X.; Wang, Z.; Zhao, W.; Park, J.H.; Law, K.T.; Berger, H.; Forro, L.; Shan, J.; Mak, K.F. Ising Pairing in Superconducting $NbSe_2$ Atomic Layers. *Nat. Phys.* **2016**, *12*, 139–143. [CrossRef]
46. Ptok, A.; Głodzik, S.; Domański, T. Yu-Shiba-Rusinov States of Impurities in a Triangular Lattice of $NbSe_2$ with Spin-Orbit Coupling. *Phys. Rev. B* **2017**, *96*, 184425. [CrossRef]

47. Lian, C.S.; Si, C.; Duan, W. Unveiling Charge-Density Wave, Superconductivity, and Their Competitive Nature in Two-Dimensional NbSe$_2$. *Nano Lett.* **2018**, *18*, 2924–2929. [CrossRef]
48. Kolekar, S.; Bonilla, M.; Ma, Y.; Diaz, H.C.; Batzill, M. Layer- and Substrate-Dependent Charge Density Wave Criticality in 1T-TiSe$_2$. *2D Mater.* **2018**, *5*, 015006. [CrossRef]
49. Ryu, H.; Chen, Y.; Kim, H.; Tsai, H.Z.; Tang, S.; Jiang, J.; Liou, F.; Kahn, S.; Jia, C.; Omrani, A.A.; et al. Persistent Charge-Density-Wave Order in Single-Layer TaSe$_2$. *Nano Lett.* **2018**, *18*, 689–694. [CrossRef]
50. Pásztor, A.; Scarfato, A.; Barreteau, C.; Giannini, E.; Renner, C. Dimensional Crossover of the Charge Density Wave Transition in Thin Exfoliated VSe$_2$. *2D Mater.* **2017**, *4*, 041005. [CrossRef]
51. Zhao, J.; Wijayaratne, K.; Butler, A.; Yang, J.; Malliakas, C.D.; Chung, D.Y.; Louca, D.; Kanatzidis, M.G.; van Wezel, J.; Chatterjee, U. Orbital Selectivity Causing Anisotropy and Particle-Hole Asymmetry in the Charge Density Wave Gap of $2H-$TaS$_2$. *Phys. Rev. B* **2017**, *96*, 125103. [CrossRef]
52. Chhowalla, M.; Shin, H.S.; Eda, G.; Li, L.J.; Loh, K.P.; Zhang, H. The Chemistry of Two-Dimensional Layered Transition Metal Dichalcogenide Nanosheets. *Nat. Chem.* **2013**, *5*, 263–275. [CrossRef]
53. Kaneko, R.; Tocchio, L.F.; Valentí, R.; Gros, C. Emergent Lattices with Geometrical Frustration in Doped Extended Hubbard Models. *Phys. Rev. B* **2016**, *94*, 195111. [CrossRef]
54. Kaneko, R.; Tocchio, L.F.; Valenti, R.; Becca, F. Charge Orders in Organic Charge-Transfer Salts. *New J. Phys.* **2017**, *19*, 103033. [CrossRef]
55. Han, Z.; Kivelson, S.A.; Yao, H. Strong Coupling Limit of the Holstein-Hubbard Model. *Phys. Rev. Lett.* **2020**, *125*, 167001. [CrossRef] [PubMed]
56. Müller-Hartmann, E. Correlated Fermions on a Lattice in High Dimensions. *Z. Phys. B Condens. Matter* **1989**, *74*, 507–512. [CrossRef]
57. Pearce, P.A.; Thompson, C.J. The Anisotropic Heisenberg Model in the Long-Range Interaction Limit. *Commun. Math. Phys.* **1975**, *41*, 191–201. [CrossRef]
58. Pearce, P.A.; Thompson, C.J. The High Density Limit for Lattice Spin Models. *Commun. Math. Phys.* **1978**, *58*, 131–138. [CrossRef]
59. Kapcia, K.; Robaszkiewicz, S. The Effects of the Next-Nearest-Neighbour Density-Density Interaction in the Atomic Limit of the Extended Hubbard Model. *J. Phys. Condens. Matter* **2011**, *23*, 105601. [CrossRef]
60. Kapcia, K.J.; Robaszkiewicz, S. On the Phase Diagram of the Extended Hubbard Model with Intersite Density-Density Interactions in the Atomic Limit. *Phys. A* **2016**, *461*, 487–497. [CrossRef]
61. Borgs, C.; Jedrzejewski, J.; Kotecký, R. The Staggered Charge-Order Phase of the Extended Hubbard Model in the Atomic Limit. *J. Phys. A Math. Gen.* **1996**, *29*, 733–747. [CrossRef]
62. Fröhlich, J.; Rey-Bellet, L.; Ueltschi, D. Quantum Lattice Models at Intermediate Temperature. *Commun. Math. Phys.* **2001**, *224*, 33–63. [CrossRef]
63. Pawłowski, G. Charge Orderings in the Atomic Limit of the Extended Hubbard Model. *Eur. Phys. J. B* **2006**, *53*, 471–479. [CrossRef]
64. Ganzenmüller, G.; Pawłowski, G. Flat Histogram Monte Carlo Sampling for Mechanical Variables and Conjugate Thermodynamic Fields with Example Applications to Strongly Correlated Electronic Systems. *Phys. Rev. E* **2008**, *78*, 036703. [CrossRef] [PubMed]
65. Jędrzejewski, J. Phase Diagrams of Extended Hubbard Models in the Atomic Limit. *Phys. A* **1994**, *205*, 702–717. [CrossRef]
66. Rademaker, L.; Pramudya, Y.; Zaanen, J.; Dobrosavljević, V. Influence of Long-Range Interactions on Charge Ordering Phenomena on a Square Lattice. *Phys. Rev. E* **2013**, *88*, 032121. [CrossRef] [PubMed]
67. Kapcia, K.J.; Barański, J.; Robaszkiewicz, S.; Ptok, A. Various Charge-Ordered States in the Extended Hubbard Model with On-Site Attraction in the Zero-Bandwidth Limit. *J. Supercond. Nov. Magn.* **2017**, *30*, 109–115. [CrossRef]
68. Kapcia, K.J.; Barański, J.; Ptok, A. Diversity of Charge Orderings in Correlated Systems. *Phys. Rev. E* **2017**, *96*, 042104. [CrossRef]
69. Lee, S.J.; Lee, J.R.; Kim, B. Patterns of Striped Order in the Classical Lattice Coulomb Gas. *Phys. Rev. Lett.* **2001**, *88*, 025701. [CrossRef]
70. Mancini, F.; Mancini, F.P. One-Dimensional Extended Hubbard Model in the Atomic Limit. *Phys. Rev. E* **2008**, *77*, 061120. [CrossRef]
71. Mancini, F.; Mancini, F.P. Extended Hubbard Model in the Presence of a Magnetic Field. *Eur. Phys. J. B* **2009**, *68*, 341–351. [CrossRef]
72. Mancini, F.; Plekhanov, E.; Sica, G. Exact Solution of the 1D Hubbard Model with NN and NNN Interactions in the Narrow-Band Limit. *Eur. Phys. J. B* **2013**, *86*, 408. [CrossRef]
73. Kaneko, R.; Nonomura, Y.; Kohno, M. Thermal Algebraic-Decay Charge Liquid Driven by Competing Short-Range Coulomb Repulsion. *Phys. Rev. B* **2018**, *97*, 205125. [CrossRef]
74. Kapcia, K.J. Charge Order of Strongly Bounded Electron Pairs on the Triangular Lattice: The Zero-Bandwidth Limit of the Extended Hubbard Model with Strong Onsite Attraction. *J. Supercond. Nov. Magn.* **2019**, *32*, 2751–2757. [CrossRef]
75. Houtappel, R.M.F. Statistics of Two-Dimensional Hexagonal Ferromagnetics with "Ising"-Interaction Between Nearest Neighbours Only. *Physica* **1950**, *16*, 391–392. [CrossRef]
76. Houtappel, R.M.F. Order-Disorder in Hexagonal Lattices. *Physica* **1950**, *16*, 425–455. [CrossRef]
77. Wannier, G.H. Antiferromagnetism. The Triangular Ising Net. *Phys. Rev.* **1950**, *79*, 357–364. [CrossRef]
78. Wannier, G.H. Antiferromagnetism. The Triangular Ising Net (erratum). *Phys. Rev. B* **1973**, *7*, 5017. [CrossRef]

79. Schick, M.; Walker, J.S.; Wortis, M. Phase Diagram of the Triangular Ising Model: Renormalization-Group Calculation with Application to Adsorbed Monolayers. *Phys. Rev. B* **1977**, *16*, 2205–2219. [CrossRef]
80. Metcalf, B.D. Ground State Spin Orderings of the Triangular Ising Model with the Nearest and Next Nearest Neighbor Interaction. *Phys. Lett. A* **1974**, *46*, 325–326. [CrossRef]
81. Oitmaa, J. The Triangular Lattice Ising Model with First and Second Neighbour Interactions. *J. Phys. A Math. Gen.* **1982**, *15*, 573–585. [CrossRef]
82. Saito, Y.; Igeta, K. Antiferromagnetic Ising Model on a Triangular Lattice. *J. Phys. Soc. Jpn.* **1984**, *53*, 3060–3069. [CrossRef]
83. Tanaka, Y.; Uryû, N. Ground State Spin Configurations of the Triangular Ising Net with the First, Second and Third Nearest Neighbor Interactions. *Prog. Theor. Phys.* **1976**, *55*, 1356–1372. [CrossRef]
84. Kudo, T.; Katsura, S. A Method of Determining the Orderings of the Ising Model with Several Neighbor Interactions under the Magnetic Field and Applications to Hexagonal Lattices. *Prog. Theor. Phys.* **1976**, *56*, 435–449. [CrossRef]
85. Arrigoni, E.; Strinati, G.C. Doping-Induced Incommensurate Antiferromagnetism in a Mott-Hubbard Insulator. *Phys. Rev. B* **1991**, *44*, 7455–7465. [CrossRef]
86. Bąk, M. Mixed Phase and Bound States in the Phase Diagram of the Extended Hubbard Model. *Acta Phys. Pol. A* **2004**, *106*, 637–646. [CrossRef]
87. Bursill, R.J.; Thompson, C.J. Variational Bounds for Lattice Fermion Models II. Extended Hubbard Model in the Atomic Limit. *J. Phys. A Math. Gen.* **1993**, *26*, 4497–4511. [CrossRef]
88. Kapcia, K.; Robaszkiewicz, S. Stable and Metastable Phases in the Atomic Limit of the Extended Hubbard Model with Intersite Density-Density Interactions. *Acta. Phys. Pol. A.* **2012**, *121*, 1029–1031. [CrossRef]
89. Becker, C.; Soltan-Panahi, P.; Kronjäger, J.; Dörscher, S.; Bongs, K.; Sengstock, K. Ultracold Quantum Gases in Triangular Optical Lattices. *New J. Phys.* **2010**, *12*, 065025. [CrossRef]
90. Struck, J.; Ölschläger, C.; Le Targat, R.; Soltan-Panahi, P.; Eckardt, A.; Lewenstein, M.; Windpassinger, P.; Sengstock, K. Quantum Simulation of Frustrated Classical Magnetism in Triangular Optical Lattices. *Science* **2011**, *333*, 996–999. [CrossRef] [PubMed]
91. Cao, Y.; Fatemi, V.; Fang, S.; Watanabe, K.; Taniguchi, T.; Kaxiras, E.; Jarillo-Herrero, P. Unconventional Superconductivity in Magic-Angle Graphene Superlattices. *Nature* **2018**, *556*, 43–50. [CrossRef] [PubMed]
92. Cao, Y.; Fatemi, V.; Demir, A.; Fang, S.; Tomarken, S.L.; Luo, J.Y.; Sanchez-Yamagishi, J.D.; Watanabe, K.; Taniguchi, T.; Kaxiras, E.; et al. Correlated Insulator Behaviour at Half-Filling in Magic-Angle Graphene Superlattices. *Nature* **2018**, *556*, 80–84. [CrossRef]
93. Yankowitz, M.; Chen, S.; Polshyn, H.; Zhang, Y.; Watanabe, K.; Taniguchi, T.; Graf, D.; Young, A.F.; Dean, C.R. Tuning Superconductivity in Twisted Bilayer Graphene. *Science* **2019**, *363*, 1059–1064. [CrossRef] [PubMed]
94. Wu, F.; MacDonald, A.H.; Martin, I. Theory of Phonon-Mediated Superconductivity in Twisted Bilayer Graphene. *Phys. Rev. Lett.* **2018**, *121*, 257001. [CrossRef] [PubMed]
95. Xu, C.; Balents, L. Topological Superconductivity in Twisted Multilayer Graphene. *Phys. Rev. Lett.* **2018**, *121*, 087001. [CrossRef]
96. Lian, B.; Wang, Z.; Bernevig, B.A. Twisted Bilayer Graphene: A Phonon-Driven Superconductor. *Phys. Rev. Lett.* **2019**, *122*, 257002. [CrossRef]
97. Wang, G.; Chernikov, A.; Glazov, M.M.; Heinz, T.F.; Marie, X.; Amand, T.; Urbaszek, B. Colloquium: Excitons in Atomically Thin Transition Metal Dichalcogenides. *Rev. Mod. Phys.* **2018**, *90*, 021001. [CrossRef]
98. Xu, Y.; Liu, S.; Rhodes, D.A.; Watanabe, K.; Taniguchi, T.; Hone, J.; Elser, V.; Mak, K.F.; Shan, J. Correlated Insulating States at Fractional Fillings of Moiré Superlattices. *Nature* **2020**, *587*, 214–218. [CrossRef]
99. Kapcia, K. Metastability and phase separation in a simple model of a superconductor with extremely short coherence length. *J. Supercond. Nov. Magn.* **2014**, *27*, 913–917. [CrossRef]

Short Biography of Author

Konrad Jerzy Kapcia is an assistant professor (adjunct) at Condensed Matter Theory Division of Faculty of Physics of Adam Mickiewicz University (AMU) in Poznań (Poland). He graduated in physics (M.Sc.) in 2009 and defended his dissertation (PhD) in theoretical physics in 2014 (both at AMU). Next, he worked in the Institute of Physics of the Polish Academy of Sciences (PAS) in Warsaw and in the Henryk Niewodniczański Institute of Nuclear Physics of PAS in Kraków. He was a visiting scientist at SISSA in Trieste (Italy) and CFEL-DESY in Hamburg (Germany). He is interested in theoretical physics, mainly in the field of strongly correlated systems (including cold gases on optical lattices) and superconductivity as well as in ab initio modeling of real materials (electron and phonon properties). He was the head and principal investigator of several projects financed by National Science Centre (Poland) and a recipient of a few scholarships of Ministry of Science and Higher Education (Poland). He is a member of the Polish Physical Society, the Institute of Physics (IoP, UK), and the American Physical Society.

 *nanomaterials*

MDPI

Article

Energy Stored and Capacitance of a Circular Parallel Plate Nanocapacitor

Orion Ciftja

Department of Physics, Prairie View A&M University, Prairie View, TX 77446, USA; ogciftja@pvamu.edu

Abstract: Nanocapacitors have received a great deal of attention in recent years due to the promises of high energy storage density as device scaling continues unabated in the nanoscale era. High energy storage capacity is a key ingredient for many nanoelectronic applications in which the significant consumption of energy is required. The electric properties of a nanocapacitor can be strongly modified from the expected bulk properties due to finite-size effects which means that there is an increased need for the accurate characterization of its properties. In this work, we considered a theoretical model for a circular parallel plate nanocapacitor and calculated exactly, in closed analytic form, the electrostatic energy stored in the nanocapacitor as a function of the size of the circular plates and inter-plate separation. The exact expression for the energy is used to derive an analytic formula for the geometric capacitance of this nanocapacitor. The results obtained can be readily amended to incorporate the effects of a dielectric thin film filling the space between the circular plates of the nanocapacitor.

Keywords: nanocapacitor; energy; capacitance; circular plate; dielectric thin film

PACS: 07.50.-e; 77.55.-g; 77.55.F-

Citation: Ciftja, O. Energy Stored and Capacitance of a Circular Parallel Plate Nanocapacitor. *Nanomaterials* **2021**, *11*, 1255. https://doi.org/10.3390/nano11051255

Academic Editor: Jung Sang Cho

Received: 30 March 2021
Accepted: 5 May 2021
Published: 11 May 2021

Publisher's Note: MDPI stays neutral with regard to jurisdictional claims in published maps and institutional affiliations.

Copyright: © 2021 by the author. Licensee MDPI, Basel, Switzerland. This article is an open access article distributed under the terms and conditions of the Creative Commons Attribution (CC BY) license (https://creativecommons.org/licenses/by/4.0/).

1. Introduction

Extensive research efforts in nanotechnology during the last two decades have led to great advances in the fabrication of novel materials such as carbon nanotubes, single electron transistors, nanowires, semiconductor nano dots, to mention a few, with length scales in the nanometer range [1–18]. New and improved experimental techniques have enabled the creation of various nanoparticles that can serve as fundamental building blocks in the assemblage of complex structural components in more elaborate nanoscale systems [19–21]. Additionally, the novel properties of artificially made nanostructures can be exploited to tailor device performance as well as enhance reliability. Nanoelectronic technologies are now fully focused on the design of nanoscale transistors that, in principle, must be integrated into nanoscale circuits (nanocircuits). A nanocapacitor is one of the fundamental elements required to design a feasible nanocircuit with the other elements being the nanoinductors and nanoresistors [22]. Therefore, the fabrication and characterization of nanocapacitors are a required step to build prototypes of functional nanocircuits.

Energy storage devices such as supercapacitors and batteries have always drawn much attention for their potential applications [23]. Conventional batteries can store substantial amounts of energy but they have the drawback of their charging time that is much longer than that of a capacitor. As a result, there is a revamped effort to fabricate capacitors with high energy storage capacity. Such capacitors are essentially parallel-plate electrostatic capacitors which can store charge on the surfaces of the two metallic conducting plates. The nanoscale counterpart of such a bulk capacitor, the nanocapacitor, has been shown to have the capability to make use of densely packed interfaces and thin films. As a result, nanocapacitors may potentially serve as the basis of next-generation electronic devices [24]. There are many methods used to fabricate nanocapacitors. However, as miniaturization reaches the nanoscale, all devices fabricated through these approaches start to show size-dependent properties and are becoming extremely difficult to characterize [25].

In this work, we introduce a model for a circular parallel plate nanocapacitor consisting of two identical circular plates placed face-to-face opposite to each other at an arbitrary separation distance. All spatial dimensions of the system are constrained to be in the nanoscale. Each of the plates of the nanocapacitor is assumed to be a perfectly two-dimensional (2D) circular disk with a radius, R, that measures on the nanoscale. The separation distance between the two circular plates is considered arbitrary, however, of the same order of magnitude as the radius. Given this layout, one expects the properties of this nanocapacitor to be very sensitive to the geometry and depend in a non-obvious way on the size of the circular plates and inter-plate separation distance. For such a model, we do not consider any quantum effects and/or material-dependent properties that do affect realistic experimentally manufactured nanocapacitor-related systems [26–29]. The focus of this work was to introduce a model for a nanocapacitor that would allow us to obtain an exact expression for the energy stored and/or its capacitance as a function of size and geometry. The influence of a dielectric material between the plates of the nanocapacitor can be later added to the model through the addition of a phenomenological size/thickness-dependent relative permittivity parameter. It was shown that the model that is currently investigated allows one to obtain an exact expression for the stored energy and the capacitance of such a nanocapacitor as a function of its size and geometry. It was found that the energy storage capability of the nanocapacitor when the plates are charged and its corresponding capacitance depend in a non-obvious way on the separation distance between the two circular plates.

The article is organized as follows. In Section 2, we explain the theory and the model that we use to study the nanocapacitor under consideration. In Section 3, we discuss the results obtained for the energy stored in this nanocapacitor. In Section 4, we provide an exact expression for the corresponding nanocapacitance. In Section 5, we discuss the scope of the research in more detail, elaborate on the rationale of the work and add some possible application examples. In Section 6, we briefly summarize the findings and provide some concluding remarks.

2. Theory and Model

The schematic model for a circular parallel plate nanocapacitor is depicted in Figure 1. The circular plates have an identical radius R and contain, respectively, a charge of $+Q$ and $-Q$. It is assumed that charge is uniformly distributed over each of the two circular plates. The respective uniform surface charge densities are denoted as $\pm\sigma$ where $\sigma = Q/(\pi R^2)$ represents the surface charge density of the positively charged circular plate. The system of coordinates is so chosen that the two circular plates lie parallel, opposite to each other, and perpendicular to the z direction. The positively charged circular plate, denoted 1, lies in the $z = 0$ plane. The negatively charged circular plate, denoted 2, lies at some arbitrary z plane. The origin of the system of coordinates coincides with the center of circular plate 1. The separation distance between the two parallel circular plates is denoted as $d = |z| \geq 0$.

For this choice of the coordinative system, the elementary charges in the circular plate 1 are all confined to the domain, D_1 written as

$$D_1 : \begin{cases} 0 \leq \sqrt{x_1^2 + y_1^2} \leq R \\ \\ z_1 = 0 \, . \end{cases} \tag{1}$$

The elementary charges in the circular plate 2 are all confined in the domain, D_2 given as

$$D_2 : \begin{cases} 0 \leq \sqrt{x_2^2 + y_2^2} \leq R \\ \\ z_2 = z \, . \end{cases} \tag{2}$$

It is assumed that any two arbitrary elementary charges, dq_1 and dq_2, interact with each other via the standard Coulomb interaction potential, $k_e \, dq_1 \, dq_2 / |\vec{r}_1 - \vec{r}_2|$ where k_e is Coulomb's electric constant and $|\vec{r}_1 - \vec{r}_2|$ is the separation distance between the pair of elementary charges.

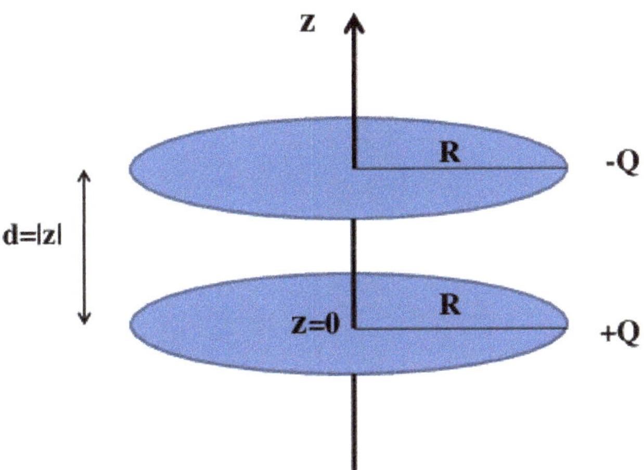

Figure 1. Schematic view of a circular parallel plate nanocapacitor. The two circular plates have a radius R and are at a distance, $d = |z| \geq 0$ apart. The respective $\pm Q$ charge of each of the circular plates is assumed to be uniformly spread over the corresponding surfaces.

The energy stored in the nanocapacitor will depend in a non-trivial way on the geometry of the model since there are no assumptions made about R and d. In turn, this should lead to a finite-size geometric capacitance that should resemble the macroscopic one only for $d \ll R$, or otherwise, will be strikingly different. At this point, we take the opportunity to clarify that we use the term "macroscopic capacitor" as opposed to "nanocapacitor" to describe any device where the standard formula of a parallel plate capacitor, $C_m = \epsilon_0 \, A/d$, is used to describe its capacitance (in free space) where A is the area of the plates and d is their separation distance. In reality, for any capacitor, micro or macro, the simple textbook formula is only a valid approximation when the separation distance of the plates is much smaller than their radius (for a circular plate) or length (for a square plate). To start with, let us write the total electrostatic energy stored in the nanocapacitor as

$$U = U_{11} + U_{22} + U_{12} , \tag{3}$$

where U_{11} (U_{22}) represent, respectively, the Coulomb self-energy stored in circular plate 1 (2) while U_{12} represents the Coulomb electrostatic interaction energy between the two circular plates of the nanocapacitor separated by an arbitrary distance. The positive Coulomb self-energy of each of the two circular plates is identical. Thus, one can write the total electrostatic energy of the circular parallel plate capacitor as

$$U = 2 \, U_{11} + U_{12} . \tag{4}$$

The Coulomb self-energy of a uniformly charged circular plate, namely, a uniformly charged disk is readily available, in the literature [30–32] and one way to write it is:

$$U_{11} = \frac{8}{3\pi} \frac{k_e \, Q^2}{R}. \tag{5}$$

The calculation of the negative interaction energy, U_{12}, is more difficult from a mathematical point of view. For this reason, we explain all the necessary steps in some detail in order to better understand the derivation of the final result. On starts with the following integral expression:

$$U_{12} = -k_e\, \sigma^2 \int_{D_1} d^2\rho_1 \int_{D_2} d^2\rho_2\, \frac{1}{|\vec{r}_1 - \vec{r}_2|}\,, \tag{6}$$

where $\vec{\rho}_i = (x_i, y_i)$ $(i = 1, 2)$ are 2D position vectors while $\vec{r}_1 = (x_1, y_1, z_1 = 0)$ and $\vec{r}_2 = (x_2, y_2, z_2 = z)$. To facilitate the calculation of the energy expression in Equation (6), one writes that quantity as

$$U_{12} = -\sigma \int_{D_2} d^2\rho_2\, V(\vec{r}_2)\,, \tag{7}$$

where:

$$V(\vec{r}_2) = k_e\, \sigma \int_{D_1} d^2\rho_1\, \frac{1}{|\vec{r}_1 - \vec{r}_2|}\,, \tag{8}$$

represents the electrostatic potential due to the circular plate 1 at the location of circular plate 2. Note that: $|\vec{r}_1 - \vec{r}_2|^2 = |\vec{\rho}_1 - \vec{\rho}_2|^2 + |z|^2$. The form of Equation (7) is convenient since a general formula for the electrostatic potential created by a uniformly charged disk at some arbitrary point in space is available in the literature [33]. In fact, one may start from that point and calculate quite generally the electrostatic interaction energy between any two coaxial parallel uniformly charged disks with different radii and different charges [34]. A general expression that applies to such a situation for two uniformly charged disks with respective charges Q and Q' (and radii, R and R') is given in Ref. [34]. For the current case of two circular parallel plates in a nanocapacitor, one has $Q' = -Q$ and $R' = R$. By combining Equations (17) and (19) in Ref. [34] one has:

$$U_{12}(a) = -\left\{ -2a + \frac{2a}{3\pi} \left[(4 - a^2)\, E\left(-\frac{4}{a^2} \right) + (4 + a^2)\, K\left(-\frac{4}{a^2} \right) \right] \right\} \frac{k_e\, Q^2}{R}\,, \tag{9}$$

where the parameter $a = |z|/R \geq 0$ is explicitly shown as an argument of the interaction energy and:

$$K(m) = \int_0^{\pi/2} \frac{d\theta}{\sqrt{1 - m\, \sin^2(\theta)}}\,, \tag{10}$$

and:

$$E(m) = \int_0^{\pi/2} d\theta\, \sqrt{1 - m\, \sin^2(\theta)}\,, \tag{11}$$

are, respectively, the complete elliptic integrals of the first and second kind. The notation adopted above for the complete elliptic integrals of the first and second kind follow that of Ref. [35].

3. Energy

To calculate the total electrostatic energy of the circular parallel plate nanocapacitor, we substitute the results from Equations (5) and (9) into the expression provided by Equation (4). This leads to an exact analytical expression for the total energy stored in a circular parallel plate nanocapacitor with free space between the plates written in compact form as

$$U(a) = F(a)\, \frac{k_e\, Q^2}{R}\,, \tag{12}$$

where $F(a)$ is the following auxiliary function:

$$F(a) = \left\{ \frac{16}{3\pi} + 2a - \frac{2a}{3\pi} \left[(4 - a^2)\, E\left(-\frac{4}{a^2} \right) + (4 + a^2)\, K\left(-\frac{4}{a^2} \right) \right] \right\} \frac{k_e\, Q^2}{R}\,. \tag{13}$$

Note that, from this point on, the parameter $a = |z|/R \geq 0$ is explicitly shown in the argument of the total energy expression displayed in Equation (12). The reader can verify that:

$$U(a = 0) = 0 \quad ; \quad F(a = 0) = 0 \,. \tag{14}$$

Therefore, one can expand the expression for $F(a)$ in Equation (13) in a power series of about $a = 0$ to obtain the dependence of $U(a)$ for a small a. The result to the lowest order (linear order) of a parameter, a, is:

$$U_{linear}(a) = 2\,a\,\frac{k_e\,Q^2}{R} \,, \tag{15}$$

where $a = |z|/R$ and $|z|$ is the separation distance between the two circular plates of the capacitor. One can check that $U_{linear}(a)$ in Equation (15) is identical to the expression for the energy stored in a macroscopic ideal parallel plate capacitor constructed from two very large uniformly charged circular plates of area A in free space:

$$U_m = \frac{Q^2}{2\,C_m} \,, \tag{16}$$

where the capacitance (in free space) is:

$$C_m = \epsilon_0\,\frac{A}{d} = \epsilon_0\,\frac{\pi\,R^2}{|z|} \,. \tag{17}$$

The equivalence of $U_{linear}(a)$ in Equation (15) to U_m in Equation (16) is easy to verify if one starts from Equation (16) and rewrites it in terms of $k_e Q^2/R$ and $a = |z|/R$ without forgetting that the Coulomb's electric constant is $k_e = 1/(4\,\pi\,\epsilon_0)$.

Figure 2 shows the dependence of the energy, $U(a)$, stored in a circular parallel plate nanocapacitor as a function of parameter $a = |z|/R$ (solid circles) in conjunction with $U_{linear}(a)$ (solid line), its counterpart for a macroscopic capacitor. The energies are expressed in units of $k_e\,Q^2/R$. It can be seen that the amount of energy stored in the finite-size nanocapacitor, $U(a)$, is substantially different from that obtained when the formula of a macroscopic capacitor $U_{linear}(a)$ is used. The differences are sizable even when the separation distance between the plates is of the order of $|z|/R \propto 10\,\%$.

In fact, we considered this scenario in more detail, and in order to verify how reasonable the approximation is, we calculated the relative difference between the linearized approximate values (for a macroscopic bulk capacitor) and the exact values of the energy of the nanocapacitor:

$$\left|\frac{U_{linear}(a) - U(a)}{U(a)}\right| \,. \tag{18}$$

We checked that the relative energy difference was slightly more than 14% for a value of $a = |z|/R = 0.1$. This means that the linearized expression of the energy, namely the expression for the energy of a macroscopic capacitor, is only reliable for very small separation distances between the plates (note that, for $a = 0.05$, the relative energy difference is quite sizable at about 8%).

Obviously, in free space, the energy stored in the nanocapacitor is electrostatic in origin and depends only on the geometry of the system. Within the framework of this model, one would account for the presence of a dielectric material between the plates by introducing a phenomenological size/thickness-dependent relative nanopermittivity parameter in the expression for the stored energy. As shown in recent work [26], the nanopermittivity parameter of very thin dielectric films is different from the bulk value and should be determined experimentally as a function of size and/or the thickness of the dielectric film in a case-by-case basis.

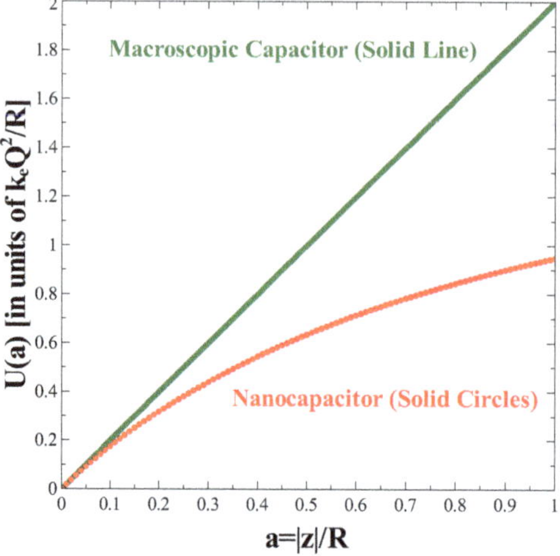

Figure 2. Energy stored in a circular parallel plate nanocapacitor, $U(a)$, in units of $k_e Q^2/R$ as a function of the parameter $a = |z|/R$ (solid circles) where $|z|$ is the separation distance between the two identical circular parallel plates placed opposite to each other and R is their radius. The circular plates contain, respectively, a charge of $\pm Q$. Charge is uniformly spread over each of the two circular plates resulting in uniform surface charge densities, $\pm \sigma$. The exact result, $U(a)$, is compared to the approximate expression, $U_{linear}(a)$ which represents the energy of an ideal macroscopic circular parallel plate capacitor (solid line).

4. Capacitance

The expression for the size-dependent geometric capacitance of the circular parallel plate nanocapacitor is extracted from the formula of the energy written as a function of the charge and capacitance:

$$C(a) = \frac{Q^2}{2\,U(a)}\,, \tag{19}$$

where $U(a)$ is given from Equations (12) and (13). One uses Equations (12) and (13) to write $C(a)$ as:

$$C(a) = \epsilon_0\, R\, \frac{2\,\pi}{F(a)}\,, \tag{20}$$

where $\epsilon_0 = 1/(4\,\pi\,k_e)$ and $F(a)$ is given from Equation (13). The expression in Equation (20) applies to a circular parallel plate nanocapacitor in free space. One can extend the result in Equation (20) to the case when there is a dielectric material between the plates by amending the formula in Equation (20) the following way:

$$C_\epsilon(a) = \epsilon(a)\,\epsilon_0\, R\, \frac{2\,\pi}{F(a)}\,, \tag{21}$$

where $\epsilon(a)$ represents the relative nanopermittivity of a given dielectric film. The value of this parameter is expected to be size/thickness-dependent and different from the bulk, given that properties of a typical material in the nanoscale do not mirror the properties of the same material in the bulk regime. For example, it has been seen experimentally that the relative permittivity of a h-BN thin film is almost twice higher than that of its bulk BN counterpart [26]. In the present model, the role played by a dielectric film in the nanocapacitor is included in a conventional way by adding a phenomenological size/thickness-dependent relative nanopermittivity parameter in the expression of the capacitance. This means that this model may be considered as a genuine candidate to explain experiments where one notices an increase in the capacitance as a result of the decrease in the thickness of the nanocapacitor.

Note that the statement above is self-evident for all cases when the capacitance of a nanocapacitor is dominated by geometric effects. However, it is not the norm when it comes to many nanocapacitors whose properties are dominated by quantum effects. In these cases, the expected behavior is a decrease in the capacitance as a result of the decrease in the thickness of the dielectric film inside the plates. Recent experiments for a nanocapacitor made of graphene square plates/electrodes and h-BN dielectric films showed an increase in capacitance with decrease in thickness [26]. For this reason, this so-called anomalous size-dependent increase in capacitance was quickly attributed to subtle quantum effects exhibited by these materials. Note that this type of behavior is the same as that exhibited by the geometric capacitance of the model in Equation (21) as well as any other variant of such model including the case of a nanocapacitor with two square plates instead of circular plates.

5. Discussions

At this juncture, it is important to remark that the calculation of the electrostatic energy stored and/or capacitance of a parallel-plate capacitor is a long-standing problem in potential theory that has been addressed by many authors [36]. To the best of our knowledge, an exact analytic solution to the problem (the one that makes the system an equipotential) does not seem possible. There have been reports from few authors that have obtained such a solution, but sooner or later, all these attempts have been found faulty. For example, a recently proposed analytic solution by Atkinson et al. [37] was shown to be incorrect by Hughes [38]. With few words, an exact solution to this celebrated problem in potential theory (identifying the solution that makes the system an equipotential) does not seem possible, in the sense that to this date, no explicit analytical solution that have been reported are acceptable. To the best of our knowledge, the closest that one comes to an analytic solution is by using Love's integral expressions [39] for the potential and related quantities, which are then solved numerically. For example, numerical values for the potential in the vicinity of a parallel plate capacitor were calculated using both the Love integral-equation method and a relaxation method [40]. Similarly, the capacitance of the circular parallel plate capacitor has been numerically calculated by several authors using various computational techniques [41–45].

The advantage of the model reported in this work is that it allows us to obtain an exact analytic expression for the electrostatic energy stored and/or capacitance. We are certainly not claiming that energy and/or capacitance have never been computed before. On the contrary, as we already explained, there are several works in this direction, but all of them are numerical calculations [36]. Another advantage of the model that we report is its simplicity. Potential theory problems of this nature are very difficult to solve. Calculating the equilibrium charge distribution on the two plates of the capacitor, namely obtaining the precise analytic form of the surface charge density that leads to an equipotential surface (same potential all over the surface) is an unsolvable analytical problem. A single circular disk/plate represents a rare example for which this problem is analytically solvable. The equilibrium surface charge density of a charged conducting disk is different from that of a uniformly charged disk, but the energy stored in such a circular disk does not differ much from its uniformly charged counterpart. In fact, we found that the relative energy difference between them is about 8% [32]. Based on these general physical considerations, we expected that the results for the electrostatic energy stored and/or capacitance obtained from our model would likely be not very different from this order of accuracy if compared to numerical values found for the same system.

As is known, capacitors play an important role in integrated circuits such as storing electrical energy and blocking direct current while allowing alternating currents to propagate. Nanocapacitors, for instance those based on graphene layers as plates, are becoming critical for advanced electric energy storage and for building nanoelectronic circuits. The model in this work can be directly applied to predict the capacitance of nanocapacitors made of graphene plates and hexagonal boron nitride (h-BN) films (as the dielectric material filling the space between the two plates). Such a system can achieve superior capacitor properties. Furthermore, the thickness can be experimentally tuned up to the thinnest possible value, essentially consisting of only monolayer materials: h-BN dielectric film with graphene electrodes [26]. In fact, we argued in a recent work [46] that the anomalous size-dependent increase in capacitance with a decrease in thickness seen in this family of h-BN/graphene nanocapacitors [26] is consistent with conventional electrostatic principles and can be explained without appealing to quantum effects. The effects of the differences on the geometry of plates, namely a circular plate versus a square plate, are expected to be small as long as the plates have the same area.

Our results for the stored electrostatic energy and/or capacitance of a circular parallel plate nanocapacitor may provide a sound theoretical basis to understand various micro-electromechanical systems (MEMS). Applications of MEMS technology [47,48] is centered around sensors and actuators. Generally speaking, sensors provide a way to monitor some physical variables. On the other hand, actuators take an input control signal (such as voltage or current) and produce a force or torque to generate motion. As a basic example, reconsider the circular parallel plate nanocapacitor model of this work. If the distance between the two plates changes as a result of an external force, the overall capacitance also changes. The change of capacitance can be exactly calculated in this model. The value is expected to be very accurate since it represents a difference of two quantities calculated under same conditions. With other words, although the assumption of uniform charge distribution in each of the plates may introduce some error, the difference of two capacitances for two different separation distances will tend to be very accurate since these errors are systemic and will tend to cancel out. As a result, by calculating $C_\epsilon(a_2) - C_\epsilon(a_1)$, one obtains a very accurate sense of the force. In fact, this principle forms the basis for the electrostatic sensing of position when a parallel plate capacitor is used as an actuator. Assume that the bottom plate is held fixed, while the top plate is suspended by an ideal elastic spring that is free to move. One may calibrate the system so that the spring is initially underformed. This means that a variation of capacitance (in a charge-controlled actuator) provides us information on the elastic force exerted from the spring on the top plate, which in turn, provides a sensing of position. The expected high accuracy of the model used in this application can be very useful to characterize various electrostatic sensors and actuators that are found in a wide array of MEMS-based devices.

6. Conclusions

We introduced a model for a circular parallel plate nanocapacitor consisting of two identical uniformly charged circular plates opposite to each other at an arbitrary separation distance. It is assumed that both the radii of the circular plates and their separation distance are finite and of the same order of magnitude. An analytic result for the total energy stored due to such a nanocapacitor has been found. The final result is expressed in terms of an auxiliary function that depends on a single dimensionless parameter and can be readily evaluated. As an example for the utility of our result, the energy stored in the nanocapacitor was calculated not only generally, but also when the two electrodes are in very close proximity to each other. The result in the latter case was found to converge to the familiar expression for the energy stored in a macroscopic capacitor. As a second example, we graphically displayed the relative energy difference between the exact expression and its macroscopic counterpart showing that the macroscopic formula should be used with extreme caution and only when the spacing between the two circular plates is very small, in the order of less than 5% relative to the radius. The exact analytic formula shown in Equation (13) can help understand how the energy is stored in multi-layered capacitive nanostructures with circular symmetry [49–53].

The expression for the energy stored in a circular parallel plate nanocapacitor was used to derive an analytic formula for the corresponding nanocapacitance of the system. The initial result for the nanocapacitance derived under the assumption of free space between the two circular plates can be extended to a more general case scenario where such a space between the plates is filled with a dielectric film. For the given lengths in the nanoscale, it is expected that the electric properties of the very thin dielectric film will differ from its corresponding bulk counterpart as already seen in experiments with graphene boron nitrade nanocapacitors [26]. For this reason, one must assume a phenomenological size/thickness-dependent relative nanopermittivity for the dielectric film and extract its value experimentally.

At this juncture, we also point out that the results obtained thus far can also be helpful to gauge the accuracy of various theoretical approximations and numerical methods used to study the properties of systems with circular symmetry. This situation may apply to certain 2D compositions formed in semiconductor quantum dots, heterostructures or thin films [54–57] which lead to the creation of a two-dimensional electron gas (2DEG) system. In particular, the physics of quantum Hall effect (QHE) has at its heart the model of a 2D system of electrons confined in a neutralizing background represented by a uniformly charged circular disk lying on the same plane as the layer of electrons [58–63]. A tweak of this model in the context of QHE studies would be to consider the 2D layer of electrons as being space separated at some distance above the depleted positively charged jellium disk region ("the donor charge" region). In this setup, the Hartree term in a Hartree–Fock method [54] would result in a capacitive-like effect similar to the one presently studied .

Funding: This research was funded in part by National Science Foundation (NSF) Grant No. DMR-2001980.

Institutional Review Board Statement: Not applicable.

Informed Consent Statement: Not applicable.

Data Availability Statement: The data presented in this study are available upon request from the author.

Acknowledgments: We acknowledge useful discussions with B. Ciftja.

Conflicts of Interest: The author declares no conflict of interest. The funders had no role in the design of the study; in the collection, analyses, or interpretation of data; in the writing of the manuscript, or in the decision to publish the results.

Abbreviations

The following abbreviations are used in this manuscript:

2D	Two-dimensional
MEMS	Micro-electromechanical systems
2DEG	Two-dimensional electron gas
QHE	Quantum Hall effect

References

1. Theodore, L. *Nanotechnology: Basic Calculations for Engineers and Scientists*; Wiley: Hoboken, NJ, USA, 2006.
2. Ventra, M.D.; Evoy, S.; Heflin, J.R., Jr. (Eds.) *Introduction to Nanoscale Science and Technology*; Series: Nanostructure Science and Technology; Kluwer Academic Publishers: Boston, MA, USA, 2004.
3. Banyai, L.; Koch, S.W. *Semiconductor Quantum Dots*; Series on Atomic, Molecular and Optical Physics; World Scientific: Singapore, 1993; Volume 2.
4. Jacak, L.; Hawrylak, P.; Wojs, A. *Quantum Dots*; Nanoscience and Technology Series; Springer: Berlin/Heidelberg, Germany, 1998.
5. Masumoto, Y.; Takagahara, T. (Eds.) *Semiconductor Quantum Dots: Physics, Spectroscopy and Applications*; Nanoscience and Technology Series; Springer: Berlin/Heidelberg, Germany, 2002.
6. Michler, P. (Ed.) *Single Quantum Dots: Fundamentals, Applications and New Concepts*; Topics in Applied Physics; Springer: Berlin/Heidelberg, Germany, 2003; Volume 90.
7. Bird, J.P. (Ed.) *Electron Transport in Quantum Dots*; Kluwer Academic Publishers: Boston, MA, USA, 2003.
8. Ashoori, R.C.; Stormer, H.L.; Weiner, J.S.; Pfeiffer, L.N.; Baldwin, K.W.; West, K.W. Single-electron capacitance spectroscopy of discrete quantum levels. *Phys. Rev. Lett.* **1992**, *68*, 3088. [CrossRef]
9. Kastner, M.A. Artificial atoms. *Phys. Today* **1993**, *46*, 24. [CrossRef]
10. Tarucha, S.; Austing, D.G.; Honda, T.; Hage, R.J.v.; Kouwenhoven, L.P. Shell filling and spin effects in a few electron quantum dot. *Phys. Rev. Lett.* **1996**, *77*, 3613. [CrossRef]
11. Ciftja, O. Classical behavior of few-electron parabolic quantum dots. *Physica B* **2009**, *404*, 1629. [CrossRef]

12. Ciftja, O. Generalized description of few-electron quantum dots at zero and nonzero magnetic field. *J. Phys. Condens. Matter* **2007**, *19*, 046220. [CrossRef]
13. Tavernier, M.B.; Anisimovas, E.; Peeters, F.M.; Szafran, B.; Adamowski, J.; Bednarek, S. Four-electron quantum dot in a magnetic field. *Phys. Rev. B* **2003**, *68*, 205305. [CrossRef]
14. Ciftja, O. Electric field controlled interference in a system with Rashba spin-orbit coupling. *AIP Adv.* **2016**, *6*, 055217. [CrossRef]
15. Pfannkuche, D.; Gerhardts, R.R.; Maksym, P.A.; Gudmundsson, V. Theory of quantum dot helium. *Physica B* **1993**, *189*, 6. [CrossRef]
16. Kainz, J.; Mikhailov, S.A.; Wensauer, A.; Rössler, U. Quantum dots in high magnetic fields: Calculation of ground-state properties. *Phys. Rev. B* **2002**, *65*, 115305. [CrossRef]
17. Ciftja, O. A Jastrow correlation factor for two-dimensional parabolic quantum dots. *Mod. Phys. Lett. B* **2009**, *23*, 3055. [CrossRef]
18. Merkt, U.; Huser, J.; Wagner, M. Energy spectra of two electrons in a harmonic quantum dot. *Phys. Rev. B* **1991**, *43*, 7320. [CrossRef] [PubMed]
19. Ciftja, O. Quantitative analysis of shape-sensitive interaction of a charged nanoplate and a charged nanowire. *Nano* **2015**, *10*, 1550114. [CrossRef]
20. Hong, N.H.; Raghavender, A.T.; Ciftja, O.; Phan, M.H.; Stojak, K.; Srikanth, H.; Zhang, Y.H. Ferrite nanoparticles for future heart diagnostics. *Appl. Phys. A* **2013**, *112*, 323. [CrossRef]
21. Ciftja, O. Spin dynamics of an ultra-small nanoscale molecular magnet. *Nanoscale Res. Lett.* **2007**, *2*, 168. [CrossRef]
22. Engheta, N.; Salandrino, A.; Alù, A. Circuit elements at optical frequencies: Nanoinductors, nanocapacitors, and nanoresistors. *Phys. Rev. Lett.* **2005**, *95*, 095504. [CrossRef] [PubMed]
23. Saha, S.K.; DaSilva, M.; Hang, Q.; Sands, T.; Janes, D.B. A nanocapacitor with giant dielectric permittivity. *Nanotechnology* **2006**, *17*, 2284. [CrossRef]
24. González, G.; Kolosovas-Machuca, E.S.; López-Luna, E.; Hernxaxndez-Arriaga, H.; Gonzxaxlez, F.J. Design and fabrication of interdigital nanocapacitors coated with HfO2. *Sensors* **2015**, *15*, 1998. [CrossRef] [PubMed]
25. Wang, F.; Clément, N.; Ducatteau, D.; Troadec, D.; Tanbakuchi, H.; Lagrand, B.; Dambrine, G.; Thxexron, D. Quantitative impedance characterization of sub-10 nm scale capacitors and tunnel junctions with an interferometric scanning microwave microscope. *Nanotechnology* **2014**, *25*, 405703. [CrossRef]
26. Shi, G.; Hanlumyuang, Y.; Liu, Z.; Gong, Y.; Gao, W.; Li, B.; Kono, J.; Lou, J.; Vajtai, R.; Sharma, P.; et al. Boron Nitride-Graphene nanocapacitor and the origins of anomalous size-dependent increase of capacitance. *Nano Lett.* **2014**, *14*, 1739. [CrossRef]
27. Kim, Y.; Han, H.; Vrejoiu, I.; Lee, W.; Hesse, D.; Alexe, M. Cross talk by extensive domain wall motion in arrays of ferroelectric nanocapacitors. *Appl. Phys. Lett.* **2011**, *99*, 202901. [CrossRef]
28. Li, Q.; Patel, C.; Ardebili, H. Mitigating the dead-layer effect in nanocapacitors using graded dielectric films. *Int. J. Smart Nano Mater.* **2012**, *3*, 23. [CrossRef]
29. Li, L.-J.; Zhu, B.; Ding, S.-J.; Lu, H.-L.; Sun, Q.-Q.; Jiang, A.; Zhang, D.W.; Zhu, C. Three-dimensional AlZnO/Al2O3/AlZnO nanocapacitor arrays on Si substrate for energy storage. *Nanoscale Res. Lett.* **2012**, *7*, 544. [CrossRef] [PubMed]
30. Ciftja, O. A result for the Coulomb electrostatic energy of a uniformly charged disk. *Results Phys.* **2017**, *7*, 1674. [CrossRef]
31. Ciftja, O. A uniformly charged circular disk with an anisotropic Coulomb interaction potential. *J. Electrostat.* **2020**, *107*, 103472. [CrossRef]
32. Ciftja, O. Results for charged disks with different forms of surface charge density. *Results Phys.* **2020**, *16*, 102962. [CrossRef]
33. Ciftja, O.; Hysi, I. The electrostatic potential of a uniformly charged disk as the source of novel mathematical identities. *Appl. Math. Lett.* **2011**, *24*, 1919. [CrossRef]
34. Ciftja, O. Electrostatic interaction energy between two coaxial parallel uniformly charged disks. *Results Phys.* **2019**, *15*, 102684. [CrossRef]
35. Abramowitz, M.; Stegun, I.A. (Eds.) *Handbook of Mathematical Functions with Formulas, Graphs, and Mathematical Tables*, 9th ed.; Dover: New York, NY, USA, 1972.
36. Norgren, M.; Jonsson, B.L.G. The capacitance of the circular parallel plate capacitor obtained by solving the Love integral equation using an analytic expansion of the kernel. *Prog. Electromagn. Res. PIER* **2009**, *97*, 357. [CrossRef]
37. Atkinson, W.J.; Young, J.H.; Brezovich, I.A. An analytic solution for the potential due to a circular parallel plate capacitor. *J. Phys. A Math. Gen.* **1983**, *16*, 2837. [CrossRef]
38. Hughes, B.D. On the potential due to a circular parallel plate capacitor. *J. Phys. A Math. Gen.* **1984**, *17*, 1385. [CrossRef]
39. Love, E.R. The electrostatic field of two equal circular co-axial conducting disks. *Quart. J. Mech. Appl. Math.* **1949**, *2*, 428. [CrossRef]
40. Bartlett, D.F.; Corle, T.R. The circular parallel plate capacitor: A numerical solution for the potential. *J. Phys. A Math. Gen.* **1985**, *18*, 1337. [CrossRef]
41. Hutson, V. The circular plate condenser at small separations. *Proc. Camb. Phil. Soc.* **1963**, *59*, 211. [CrossRef]
42. Wintle, H.J.; Kurylowicz, S. Edge corrections for strip and disc capacitors. *IEEE Trans. Instr. Meas.* **1985**, *34*, 41. [CrossRef]
43. Carlson, G.T.; Illman, B.L. The circular disk parallel plate capacitor. *Am. J. Phys.* **1994**, *62*, 1099. [CrossRef]
44. Donolato, C. Approximate evaluation of capacitances by means of Green's reciprocal theorem. *Am. J. Phys.* **1996**, *64*, 1049. [CrossRef]

45. Hwang, C.O.; Given, J.A. Last-passage Monte Carlo algorithm for mutual capacitance. *Phys. Rev. E* **2006**, *74*, 027701. [CrossRef] [PubMed]

46. Ciftja, O. Origin of the anomalous size-dependent increase of capacitance in boron nitride-graphene nanocapacitors. *RSC Adv.* **2019**, *9*, 7849. [CrossRef]

47. Petersen, K.E. Silicon as a mechanical material. *Proc. IEEE* **1982**, *70*, 420. [CrossRef]

48. Liu, C. *Foundations of MEMS*, 2nd ed.; Pearson: Upper Saddle River, NJ, USA, 2012.

49. Akbar, S.; Lee, I.-H. Electron-electron interactions in square quantum dots. *Phys. Rev. B* **2001**, *63*, 165301. [CrossRef]

50. Ishizuki, M.; Takemiya, H.; Okunishi, T.; Takeda, K.; Kusakabe, K. Shape of polygonal quantum dots and ground-state instability in the spin polarization. *Phys. Rev. B* **2012**, *85*, 155316. [CrossRef]

51. Califano, M.; Harrison, P. Approximate methods for the solution of quantum wires and dots: Connection rules between pyramidal, cuboidal, and cubic dots. *J. Appl. Phys.* **1999**, *86*, 5054. [CrossRef]

52. Aristone, F.; Sanchez-Dehesa, J.; Marques, G.E. Electronic levels of cubic quantum dots. *J. Korean Phys. Soc.* **2001**, *39*, 493.

53. Welander, E.; Burkard, G. Electric control of the exciton fine structure in nonparabolic quantum dots. *Phys. Rev. B* **2012**, *86*, 165312. [CrossRef]

54. Quang, N.H.; Huong, N.Q.; Dung, T.A.; Tuan, H.A.; Thang, N.T. Strongly confined 2D parabolic quantum dot: Biexciton or quadron? *Physica B* **2021**, *602*, 412591. [CrossRef]

55. Huong, N.Q. Nonlinear optical response of hybrid exciton state in quantum dot-dendrite systems. *J. Phys. Conf. Ser.* **2020**, *1461*, 012117. [CrossRef]

56. Ogloblya, O.V.; Kuznietsova, H.M.; Strzhemechny, Y.M. Effects of coulomb repulsion on conductivity of heterojunction carbon nanotube quantum dots with spin-orbital coupling and interacting leads. *Physica B* **2017**, *504*, 96. [CrossRef]

57. Zvanut, M.E.; Dashdorj, J.; Sunay, U.R.; Leach, J.H.; Udwary, K. Effect of local fields on the Mg acceptor in GaN films and GaN substrates. *J. Appl. Phys.* **2016**, *120*, 135702. [CrossRef]

58. Laughlin, R.B. Anomalous quantum Hall effect: An incompressible quantum fluid with fractionally charged quasiparticles. *Phys. Rev. Lett.* **1983**, *50*, 1395. [CrossRef]

59. Morf, R.; Halperin, B.I. Monte Carlo evaluation of trial wave functions for the fractional quantized Hall effect: Disk geometry. *Phys. Rev. B* **1986**, *33*, 2221. [CrossRef] [PubMed]

60. Ciftja, O.; Wexler, C. Monte Carlo simulation method for Laughlin-like states in a disk geometry. *Phys. Rev. B* **2003**, *67*, 075304. [CrossRef]

61. Qiu, R.-Z.; Haldane, F.D.M.; Wan, X.; Yang, K.; Yi, S. Model anisotropic quantum Hall states. *Phys. Rev. B* **2012**, *85*, 115308. [CrossRef]

62. Ciftja, O. Exact results for a quantum Hall state with broken rotational symmetry. *J. Phys. Chem. Solids* **2019**, *130*, 256. [CrossRef]

63. Ciftja, O. Energy of the Bose Laughlin quantum Hall state of few electrons at half filling of the lowest Landau level. *Ann. Phys.* **2020**, *421*, 168279. [CrossRef]

 *nanomaterials*

Article

Graphene Nanoribbon Gap Waveguides for Dispersionless and Low-Loss Propagation with Deep-Subwavelength Confinement

Zhiyong Wu [1], Lei Zhang [2], Tingyin Ning [3], Hong Su [1], Irene Ling Li [1], Shuangchen Ruan [1], Yu-Jia Zeng [1] and Huawei Liang [1,*]

1   Shenzhen Key Laboratory of Laser Engineering, College of Physics and Optoelectronic Engineering, Shenzhen University, Shenzhen 518060, China; 1800281004@email.szu.edu.cn (Z.W.); hsu@szu.edu.cn (H.S.); liling@szu.edu.cn (I.L.L.); scruan@szu.edu.cn (S.R.); yjzeng@szu.edu.cn (Y.-J.Z.)
2   Key Laboratory for Physical Electronics and Devices of the Ministry of Education and Shanxi Key Lab of Information Photonic Technique, School of Electronic Science and Engineering, Xi'an Jiaotong University, Xi'an 710049, China; eiezhanglei@mail.xjtu.edu.cn
3   Shandong Provincial Engineering and Technical Center of Light Manipulations & Shandong Provincial Key Laboratory of Optics and Photonic Device, School of Physics and Electronics, Shandong Normal University, Jinan 250358, China; ningtingyin@sdnu.edu.cn
*   Correspondence: hwliang@szu.edu.cn; Tel.: +86-755-8652-2081

Abstract: Surface plasmon polaritons (SPPs) have been attracting considerable attention owing to their unique capabilities of manipulating light. However, the intractable dispersion and high loss are two major obstacles for attaining high-performance plasmonic devices. Here, a graphene nanoribbon gap waveguide (GNRGW) is proposed for guiding dispersionless gap SPPs (GSPPs) with deep-subwavelength confinement and low loss. An analytical model is developed to analyze the GSPPs, in which a reflection phase shift is employed to successfully deal with the influence caused by the boundaries of the graphene nanoribbon (GNR). It is demonstrated that a pulse with a 4 μm bandwidth and a 10 nm mode width can propagate in the linear passive system without waveform distortion, which is very robust against the shape change of the GNR. The decrease in the pulse amplitude is only 10% for a propagation distance of 1 μm. Furthermore, an array consisting of several GNRGWs is employed as a multichannel optical switch. When the separation is larger than 40 nm, each channel can be controlled independently by tuning the chemical potential of the corresponding GNR. The proposed GNRGW may raise great interest in studying dispersionless and low-loss nanophotonic devices, with potential applications in the distortionless transmission of nanoscale signals, electro-optic nanocircuits, and high-density on-chip communications.

Keywords: graphene plasmons; dispersionless; deep-subwavelength gap; electro-optic switch; on-chip integration

Citation: Wu, Z.; Zhang, L.; Ning, T.; Su, H.; Li, I.L.; Ruan, S.; Zeng, Y.-J.; Liang, H. Graphene Nanoribbon Gap Waveguides for Dispersionless and Low-Loss Propagation with Deep-Subwavelength Confinement. *Nanomaterials* **2021**, *11*, 1302. https://doi.org/10.3390/nano11051302

Academic Editor: Orion Ciftja

Received: 25 April 2021
Accepted: 13 May 2021
Published: 14 May 2021

Publisher's Note: MDPI stays neutral with regard to jurisdictional claims in published maps and institutional affiliations.

Copyright: © 2021 by the authors. Licensee MDPI, Basel, Switzerland. This article is an open access article distributed under the terms and conditions of the Creative Commons Attribution (CC BY) license (https://creativecommons.org/licenses/by/4.0/).

1. Introduction

A perfect slow-light guiding system, concurrently holding the robust broadband dispersionless transmission, low loss, and deep-subwavelength mode confinement with a large refractive index, is highly desirable in modern nanophotonics [1]. Although both dispersionless and low-loss light-guiding can be simultaneously attained using all-dielectric structures [2,3], the mode sizes are comparatively large due to the diffraction limit. Surface plasmon polaritons (SPPs), propagating along the interface between plasmonic (e.g., metallic) and dielectric materials [4–6], are promising candidates for overcoming the diffraction limit and reducing the mode size to the deep-subwavelength scale [7–11]. The nanoscale confinement and low propagation loss of electromagnetic wave have been simultaneously realized using the hybrid plasmonic waveguide [12–14], metallic V-groove [15,16], metallic wedge [17,18], graphene monolayer [19,20], and so on [21–26]. However, the dispersion of SPPs in these waveguides is relatively high. So far, the dispersionless SPPs have mainly shown strong field confinement along the one-dimensional (1D) direction perpendicular to the interface between plasmonic and dielectric materials, while the mode fields are uniform in the other direction of cross sections [27,28]. Even regardless of the propagation loss, it is still very challenging to acquire broadband dispersionless SPPs with a two-dimensional (2D) deep-subwavelength confinement, owing to the strong frequency dispersion of common plasmonic materials [1]. Although some progress has been made using twisted bilayer plasmonic materials [29], multilayered axially uniform hybrid plasmonic-dielectric systems [1], and nonlinear graphene configurations [30,31], the robust dispersionless and deep-subwavelength guiding systems with low propagation loss are still very underdeveloped both in theory and experiment.

In this study, a graphene nanoribbon gap waveguide (GNRGW), which can be fabricated through a layer-by-layer stacking method [32], is proposed to realize the robust dispersionless light-guiding with a 2D deep-subwavelength confinement and low loss. An analytical model is established for the first time, to our knowledge, to study the gap SPPs (GSPPs) propagating in the GNRGW, in which the influence caused by the boundaries of the graphene nanoribbon (GNR) is dealt with by introducing a reflection phase shift. The analytical results agree well with the numerical simulations obtained using COMSOL Multiphysics. Significantly, the dispersion relation of the fundamental GSPP mode is linear within an ultra-wide spectral range (several microns in the frequency domain), which means that the broadband dispersionless propagation of light can be obtained. The corresponding mode width is only 10 nm, which is 3 orders of magnitude smaller than the free-space wavelength, and simultaneously the propagation length could be as long as 10 μm ($175\lambda_{\mathrm{GSPP}}$, λ_{GSPP} is the wavelength of the GSPPs). Furthermore, the dispersionless low-loss propagation with deep-subwavelength mode confinement has strong robustness against both the width change and the curving of GNRs. When a GNRGW array is employed as a multichannel optical switch, each channel can be controlled independently by tuning the chemical potential of the corresponding GNR via the gate voltage, even if the channel separation is as small as 40 nm. The proposed GNRGW not only possesses robust dispersionless and low-loss light-guiding properties with 2D deep-subwavelength field confinement, but also can be integrated with ultra-compact planar optoelectronic devices.

2. Theoretical Model and Method

The GNRGW is schematically shown in Figure 1a. A GNR is separated from an underlying graphene sheet by a dielectric spacer with a thickness of d and a relative permittivity of ε_3. The asymmetric structure is adopted to avoid the difficulty in alignment between two graphene layers, which facilitates the fabrication. Theoretically, the excellent tunability of graphene stems from the Pauli blocking of inter-band transitions [33,34]. The chemical potentials of the GNR and the graphene sheet can be modified by the applied voltages V [32,35]. The relationship between them can be written as $\mu_c = \mathrm{sgn}(n)\hbar v_F \sqrt{\pi|n|}$, where v_F is the Fermi velocity and $n = C_g(V + V_0)/e$ is the charge density. C_g and V_0 are the gate capacitance and offset voltage, respectively [36]. The entire construction is embedded in the air with a relative permittivity of $\varepsilon_1 = 1$, as shown in Figure 1b. The substrate with a relative permittivity of ε_5 is semi-infinitely thick. To prevent unrealistically sharp edges, both terminations of the GNR are set to be semicircular with a curvature of $\delta/2$, where $\delta = 0.34$ nm is the thickness of graphene [37–39]. The relative permittivity of graphene, ε_g, is calculated using $\varepsilon_g = 1 + i\sigma_g/(\omega\varepsilon_0\delta)$ [40,41], where ε_0 and ω are the permittivity of vacuum and the angular frequency of the incident wave, respectively. The surface conductivity of graphene, σ_g, can be evaluated using the Kubo formula [36,42,43], as shown in Figure S1 in Supplementary Materials A.

To analyze GSPPs guided by the GNRGW, the Helmholtz equation is solved using the method of separation of variables, and their dispersion equation is deduced as (see Supplementary Materials B)

$$\frac{e^{2k_2\delta}(\tilde{\zeta}_1 + \tilde{\zeta}_2)\zeta_A + (\tilde{\zeta}_1 - \tilde{\zeta}_2)\zeta_B}{e^{2k_2\delta}(\tilde{\zeta}_1 + \tilde{\zeta}_2)\zeta_C + (\tilde{\zeta}_1 - \tilde{\zeta}_2)\zeta_D} = e^{2k_4\delta}\frac{\tilde{\zeta}_5 + \tilde{\zeta}_4}{\tilde{\zeta}_5 - \tilde{\zeta}_4}, \tag{1}$$

where

$$\zeta_A = e^{2k_3 d}(\xi_2 + \xi_3)(\xi_3 - \xi_4) + (\xi_2 - \xi_3)(\xi_3 + \xi_4), \tag{2}$$

$$\zeta_B = e^{2k_3 d}(\xi_2 - \xi_3)(\xi_3 - \xi_4) + (\xi_2 + \xi_3)(\xi_3 + \xi_4), \tag{3}$$

$$\zeta_C = e^{2k_3 d}(\xi_2 + \xi_3)(\xi_3 + \xi_4) + (\xi_2 - \xi_3)(\xi_3 - \xi_4), \tag{4}$$

$$\zeta_D = e^{2k_3 d}(\xi_2 - \xi_3)(\xi_3 + \xi_4) + (\xi_2 + \xi_3)(\xi_3 - \xi_4), \tag{5}$$

$\xi_j = k_j/\varepsilon_j$ (j = 1, 2, 3, 4, or 5), $k_j = (\beta^2 - \varepsilon_j k_0^2 + q_x^2)^{1/2}$, and $k_0 = (2\pi)/\lambda$. $\beta = \beta_1 + i\beta_2$ is the complex propagation constant along the z-axis; λ is the free-space wavelength; ε_j is the relative permittivity at the corresponding region. β_2 represents the loss factor of the propagated electromagnetic waves. The square root of the eigenvalue, q_x, is influenced by the reflection on boundaries of the GNR, and the mode field distribution along the x direction, closely related to q_x, is similar to that of a standing wave. To evaluate the influence of the boundaries on the GSPPs, a reflection phase shift, φ, is introduced for the first time, which satisfies the following boundary condition:

$$q_x = \frac{m\pi - \varphi}{L}, \tag{6}$$

where m is a natural number characterizing the mode order of the GSPPs, and L is the width of the GNR. The expression of φ is derived in detail in Supplementary Materials C (see Figure S2). When φ is determined, q_x can be obtained. Then, β can be solved using Equation (1). The value of the reflection phase shift is negative for $m = 0$; thus, Equation (6) can still hold, which means that the fundamental mode can be guided. Significantly, the reflection phase shift can be employed to evaluate the boundary effect not only in GNRGW, but also in some other types of plasmonic waveguides with finite widths, such as the metal nanowire on a metal substrate [44,45] and the metal nanowire on graphene [46,47].

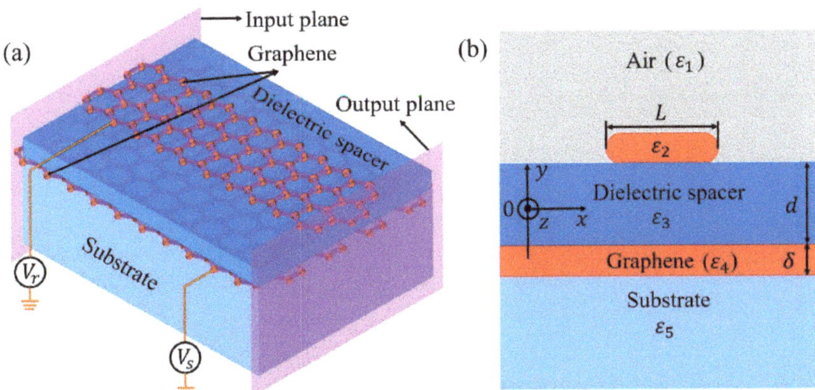

Figure 1. The schematic diagram of the GNRGW: (**a**) the 3D model; (**b**) the cross-sectional view in the x-y plane.

3. Results and Discussion

The field distributions of the four lowest-order GSPPs supported by the GNRGW with L = 120 nm are calculated using the commercially available finite element method package (COMSOL Multiphysics), as shown in Figure 2a. Other parameters are set to be: ε_3 = 11.7 (silicon), ε_5 = 2.25 (silica), d = 5.0 nm, both the chemical potentials of the GNR (μ_{cr}) and graphene sheet (μ_{cs}) are 1.0 eV, and the angular frequency of the incident wave $\omega = 1.78 \times 10^{14}$ rad/s (or the equivalent wavelength λ = 10.6 μm), unless otherwise stated. According to the quantum-corrected model (QCM) unfolded in Supplementary Materials D (see Figures S3 and S4), the quantum tunneling effect is ignorable for a thickness of the dielectric spacer larger than 0.8 nm, which is also applicable to metallic nanocavities [48,49]. It can be seen that electromagnetic energy is mainly squeezed into the deep-subwavelength spacer between the two pieces of graphene. The m-th order mode has m field nodes with zero intensity along the x direction, and E_y is symmetric and anti-symmetric for even-order and odd-order modes, respectively. Such GSPP modes could be excited by the near-field coupling method [50].

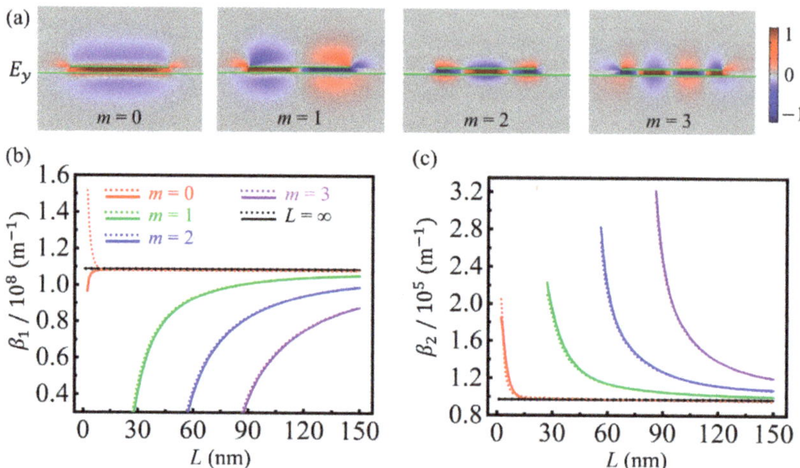

Figure 2. (**a**) Distributions of the y-components of electric fields, E_y, for the four lowest-order GSPPs. The locations of graphene are indicated by the green lines. Dependences of the real (**b**) and imaginary (**c**) parts of β on L for the four modes. The solid and dashed lines correspond to the analytical model and simulation, respectively.

The dependences of real and imaginary parts of β on L are further calculated using both the analytical model and COMSOL Multiphysics, as shown in Figure 2b,c. β_1 and β_2 of GSPPs supported by infinitely extended double-layer graphene sheets (corresponding to $L = \infty$) are also provided for comparison, as shown by the black horizontal lines. The cutoff width of the GNR for each mode, L_m, can be predicted by the analytical model (see Supplementary Materials C), where β_1 and β_2 tend to 0 and ∞, respectively. When $L < 30$ nm, all the high-order GSPPs ($m > 0$) are cut off. For the fundamental mode ($m = 0$), β is independent on L in the range of $L > 8$ nm, which indicates its strong robustness against the width change. When L gets smaller from 8 nm, the fundamental GSPP mode gradually fades, and the edge mode gradually dominates (see Figure S5 in Supplementary Materials E). Thus, there is an obvious deviation between the analytical and simulated results in this range, as shown in Figure 2b, despite the fact that the proposed model is accurate for analyzing the GSPPs. In comparison with the fundamental mode, the complex propagation constants of the higher-order GSPPs are obviously affected by L.

The dispersion characteristics of guided modes are very important for both the fundamental principles and practical device-engineering [24], so the dependences of β_1 and β_2 on the angular frequency are calculated for the fundamental GSPP mode in a GNRGW with $L = 10$ nm. As shown in Figure 3a, the variation of β_2 is very small; thus, the influence of the propagation loss on the dispersion is insignificant. The propagation length $L_p = 1/\beta_2$ is approximately 10 μm ($175\lambda_{GSPP}$ for $\lambda = 10.6$ μm), which is quite long for a guided mode with a 10 nm width. The small deviation between the simulated and analytical results in the high frequency range is caused by the evolution from GSPPs to edge plasmon modes (see Figure S5 in Supplementary Materials E). Notably, β_1 increases linearly with ω, which means that the fundamental GSPP mode is broadband dispersionless. β_1 is two orders of magnitude larger than the wave vector in free space, and the group velocity is

$v_g = d\omega/d\beta_1 = c/180$, where c is the speed of light in vacuum. The analytical relationship between β_1 and ω is further discussed in Supplementary Materials F (see Figure S6).

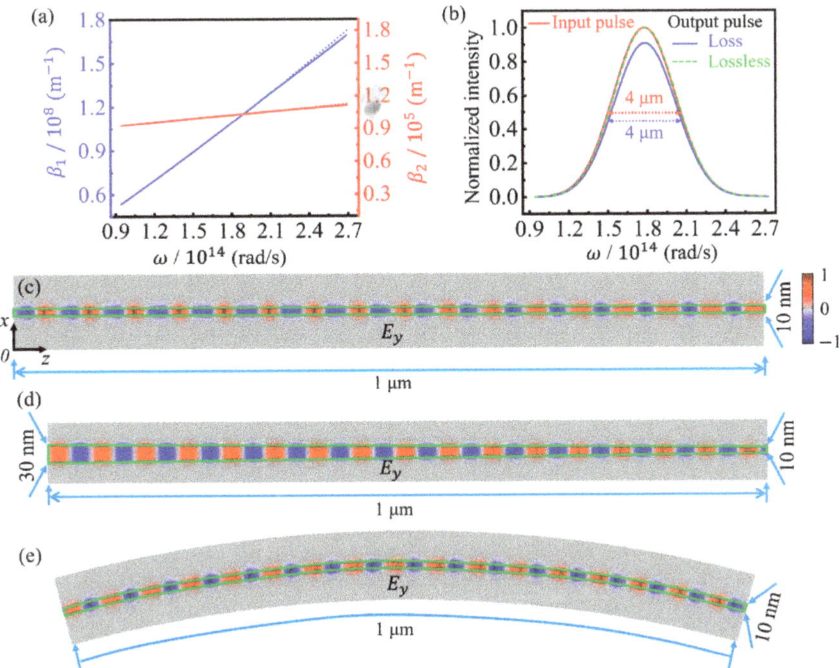

Figure 3. (**a**) Dependences of the real and imaginary parts of β on the angular frequency of incident wave for the fundamental GSPP mode. (**b**) The envelope profiles of pulses with a center wavelength of 10.6 μm and a FWHM of 4 μm in the frequency domain, probed on the input plane (red solid line) and the output plane (blue solid line), respectively. The green dashed line shows the pulse on the output plane without considering the propagation loss. Distributions of E_y on the center planes of dielectric spacers ($y = 0$) in GNRGWs with the straight (**c**), wedge-shaped (**d**), and curved (**e**) GNRs, respectively, for $\lambda = 10.6$ μm. The interior areas of the green boxes represent GNRs. All electric field distributions share the same color legend.

To demonstrate the dispersionless and low-loss propagation of the deep-subwavelength GSPPs, the evolution of a pulse with a Gaussian envelope is simulated for a propagation distance of 1 μm. The normalized envelope profile of the input pulse with a center wavelength of 10.6 μm and a full width at half maximum (FWHM) of 4 μm in the frequency domain is described by the red solid line in Figure 3b. The peak value of the output pulse, as shown by the blue solid line in Figure 3b, is 10% smaller than that of the input pulse, owing to the inevitable propagation loss. However, if the amplitude of the output pulse is multiplied by 1.10 to ignore the propagation loss, its envelope profile, as shown by the green dashed line in Figure 3b, entirely coincides with that of the input pulse, which means a dispersionless propagation. The time-domain envelope profiles of the input and output pulses are also the same, as shown in Figure S7 in Supplementary Materials G. In the propagation process, the electromagnetic energy is well confined in the dielectric spacer covered by the GNR with a width of 10 nm, as shown in Figure 3c, so the mode width is approximately $\lambda/1000$. Therefore, all the propagation characteristics of the broadband dispersionless, low loss, and deep-subwavelength mode confinement can be attained simultaneously using the GNRGW.

To analyze the robustness of the dispersionless propagation with a deep-subwavelength mode width against the shape change of the GNR, the evolution of a broadband pulse propagating in a GNRGW with a wedge-shaped GNR is simulated. The distribution of E_y indicates that the electromagnetic energy is always confined in the dielectric spacer covered by the GNR, and the mode width varies with L, as shown in Figure 3d. The electric field on the output plane is 1.39 times higher than that on the input plane; thus, the wedge-shaped structure can be used as an effective coupler between wide and narrow GNRGWs. To ignore the field enhancement caused by the focusing, the frequency- and time-domain amplitudes of the output pulse are multiplied by 0.72, as shown by the green dashed line in Figure S8a,b in Supplementary Materials H, and then the envelope profiles of the input and output pulses entirely coincide. Therefore, the dispersionless propagation in the GNRGW exhibits excellent robustness against the width change of the GNR. Furthermore, a GNRGW with a curved GNR is employed to guide the broadband pulse. The distribution of E_y signifies that the electromagnetic energy can still be confined in the dielectric spacer covered by the curved GNR, as shown in Figure 3e. The amplitude of the output pulse is only 0.17% lower than that on the output plane of the straight GNRGW with the same length; thus, the bending loss is negligible. If the frequency- and time-domain amplitudes of the output pulse are multiplied by 1.10 to ignore the loss, as shown in Figure S9a,b in Supplementary Materials H, the envelope profiles completely coincide with those of the input pulse. Thus, the dispersionless propagation is also very robust against the curving of the GNR.

The dependences of the real and imaginary parts of β on the chemical potential of the GNR are further investigated, as shown in Figure 4a. Both β_1 and β_2 decreases as the chemical potential increases, so the mode characteristics can be adjusted by modifying the gate voltage. β_1 is much larger than the wave vector of light in vacuum, corresponding to a deep-subwavelength mode concentration, as shown in the inset of Figure 4a. Therefore, the GNRGWs can be employed as highly integrated photonic devices. Even if the separation between two adjacent GNRs is as small as 40 nm, the GSPPs can still propagate in each channel independently (see Figure S10 in Supplementary Materials I). Moreover, the dependences of β on ε_3 and ε_5 (see Figure S11 in Supplementary Materials J) indicate that the GSPPs can be guided by various GNRGWs with different spacers and substrates; thus, they are highly compatible with some other planar photonic devices.

The propagation of GSPPs in a GNRGW array with five GNRs is further simulated. The distributions of E_y on the output plane of the array for GNRs with different chemical potentials are shown in Figure 4b, and the corresponding distributions on the center plane of the dielectric spacer are shown in Figure 4c,d, respectively. The left graphene sheet is used to couple light into the GNRGW array. The interval, l, is set to be 5 nm to separate the GNRs from the left graphene sheet, which ensures that the chemical potential of each GNR can be controlled independently by applying the gate voltage. According to Figure 2b, β_1 of the fundamental GSPP mode is independent on the width of the GNR in the range of $L > 8$ nm; thus, the mode matching condition, β_1, of GSPPs supported by the GNRGW on the right side and by the double-layer graphene sheets on the left side matching each other can be satisfied by applying the same gate voltage on the two parts. At this time, the light can be coupled into the corresponding GNRGW, resulting in an "on" state, as shown by I, III, and V channels in Figure 4c as well as II and IV channels in Figure 4d. When the mode condition is mismatched, the coupling efficiency decreases sharply, exhibiting an "off" state, as depicted by II and IV channels in Figure 4c as well as I, III, and V channels in Figure 4d. Both "on" and "off" states can be controlled independently by each GNRGW via adjusting the gate voltage, which allow the GNRGW array to be employed as high-density electro-optical switches with 20 channels per micron.

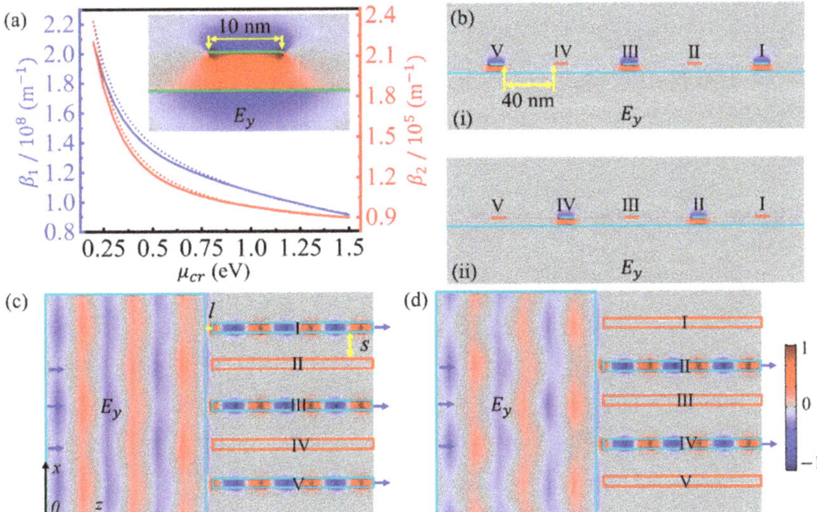

Figure 4. (a) Dependences of the real and imaginary parts of β on the chemical potential of the GNR. The inset in (a) shows the distribution of E_y. (b) Distributions of E_y on the output plane of a multichannel optical switch, parts (i) and (ii) corresponding to (c) and (d), respectively. (c,d) Distributions of E_y on the center plane of the dielectric spacer for GNRs with different chemical potentials. The interior areas of the cyan and red boxes (or lines) represent graphene with chemical potentials of 1.0 and 0.3 eV, respectively. All electric field distributions share the same color legend.

4. Conclusions

In conclusion, a GNRGW is proposed to realize robust dispersionless light-guiding with 2D deep-subwavelength field confinement and low propagation loss. An analytical model, which can successfully deal with the influence caused by the extremely narrow width of the GNR, is presented to analyze the GSPPs supported by the proposed GNRGW. It is demonstrated that when a pulse with a 10 nm mode size and a 4 μm bandwidth propagates over 1 μm in the GNRGW, no waveform distortion is observed, which is very robust against the shape change of the GNR. Furthermore, an electronically controlled multichannel optical switch is achieved by using a GNRGW array. Even if the separation between adjacent GNRGWs is as small as 40 nm, GSPPs in each channel can still be controlled independently by tuning the chemical potential of the corresponding GNR. The proposed GNRGW is expected to stimulate a new pathway to realize the highly integrated, broadband dispersionless photonic devices with deep-subwavelength mode confinement and low propagation loss.

Supplementary Materials: The following are available online at https://www.mdpi.com/article/10.3390/nano11051302/s1, Figure S1: Dependences of surface conductivity of graphene on the angular frequency of incident waves and the chemical potential of graphene, Figure S2: Schematic diagram of the reflection process, Figure S3: Dependences of ε_3 on d obtained using the quantum-corrected model and the classical method, Figure S4: Dependences of β on d obtained using the quantum-corrected model and the classical method, Figure S5: The influence of L on the mode field distribution, Figure S6: Dependences of β_1 and reflection phase shift on the angular frequency of incident waves, Figure S7: Time-domain envelope profiles of the input and output pulses in a GNRGW with a straight GNR, Figure S8: Frequency- and time-domain envelope profiles of the input and output pulses in a GNRGW with a wedge-shaped GNR, Figure S9: Frequency- and time-domain envelope profiles of the input and output pulses in a GNRGW with a curved GNR, Figure S10: Crosstalk analysis on two GNRGWs, Figure S11: Dependences of the complex propagation constant on the relative permittivities of the dielectric spacer and substrate.

Author Contributions: Conceptualization, Z.W. and H.L.; methodology, Z.W.; validation, Z.W.; formal analysis, Z.W. and H.L.; investigation, Z.W.; resources, S.R. and H.L.; writing—original draft preparation, Z.W.; writing—review and editing, H.L., L.Z., T.N., Y.-J.Z., I.L.L. and S.R.; supervision, H.L.; funding acquisition, H.L., H.S., and L.Z. All authors have read and agreed to the published version of the manuscript.

Funding: This work was supported in part by the National Natural Science Foundation of China under Grant No. 11874270, the Shenzhen Science and Technology Project under Grant Nos. JCYJ2020010-9105825504 and JCYJ20190808143409787, and the Fundamental Research Funds for the Central Universities of Xi'an Jiaotong University under Grant No. Z201805196.

Data Availability Statement: The data presented in this study are available on request from the corresponding author.

Conflicts of Interest: The authors declare no conflict of interest.

References

1. Karalis, A.; Joannopoulos, J.D.; Soljacic, M. Plasmonic-Dielectric Systems for High-Order Dispersionless Slow or Stopped Subwavelength Light. *Phys. Rev. Lett.* **2009**, *103*, 043906. [CrossRef] [PubMed]
2. Guo, Y.; Jafari, Z.; Agarwal, A.M.; Kimerling, L.C.; Li, G.; Michel, J.; Zhang, L. Bilayer dispersion-flattened waveguides with four zero-dispersion wavelengths. *Opt. Lett.* **2016**, *41*, 4939–4942. [CrossRef] [PubMed]
3. Guo, Y.H.; Jafari, Z.; Xu, L.J.; Bao, C.J.; Liao, P.C.; Li, G.F.; Agarwal, A.M.; Kimerling, L.C.; Michel, J.; Willner, A.E.; et al. Ultra-flat dispersion in an integrated waveguide with five and six zero-dispersion wavelengths for mid-infrared photonics. *Photonics Res.* **2019**, *7*, 1279–1286. [CrossRef]
4. Pitarke, J.M.; Silkin, V.M.; Chulkov, E.V.; Echenique, P.M. Theory of surface plasmons and surface-plasmon polaritons. *Rep. Prog. Phys.* **2006**, *70*, 1–87. [CrossRef]
5. Teng, D.; Wang, K. Theoretical Analysis of Terahertz Dielectric-Loaded Graphene Waveguide. *Nanomaterials* **2021**, *11*, 210. [CrossRef]
6. Unutmaz, M.A.; Unlu, M. Terahertz Spoof Surface Plasmon Polariton Waveguides: A Comprehensive Model with Experimental Verification. *Sci. Rep.* **2019**, *9*, 8. [CrossRef]
7. Gramotnev, D.K.; Bozhevolnyi, S.I. Plasmonics beyond the diffraction limit. *Nat. Photonics* **2010**, *4*, 83–91. [CrossRef]
8. Chau, Y.F.C.; Chao, C.T.C.; Huang, H.J.; Kooh, M.R.R.; Kumara, N.; Lim, C.M.; Chiang, H.P. Ultrawide Bandgap and High Sensitivity of a Plasmonic Metal-Insulator-Metal Waveguide Filter with Cavity and Baffles. *Nanomaterials* **2020**, *10*, 2030. [CrossRef]
9. Gerislioglu, B.; Dong, L.L.; Ahmadivand, A.; Hu, H.T.; Nordlander, P.; Halas, N.J. Monolithic Metal Dimer-on-Film Structure: New Plasmonic Properties Introduced by the Underlying Metal. *Nano. Lett.* **2020**, *20*, 2087–2093. [CrossRef]
10. Ahmadivand, A.; Gerislioglu, B.; Ramezani, Z. Gated graphene island-enabled tunable charge transfer plasmon terahertz metamodulator. *Nanoscale* **2019**, *11*, 8091–8095. [CrossRef]
11. Butt, M.A.; Khonina, S.N.; Kazanskiy, N.L. Plasmonics: A Necessity in the Field of Sensing-A Review (Invited). *Fiber Integr. Opt.* **2021**, 1–34. [CrossRef]
12. Oulton, R.F.; Sorger, V.J.; Genov, D.A.; Pile, D.F.P.; Zhang, X. A hybrid plasmonic waveguide for subwavelength confinement and long-range propagation. *Nat. Photonics* **2008**, *2*, 496–500. [CrossRef]
13. Zhang, S.; Xu, H. Optimizing Substrate-Mediated Plasmon Coupling toward High-Performance Plasmonic Nanowire Waveguides. *Acs Nano* **2012**, *6*, 8128–8135. [CrossRef] [PubMed]
14. Sun, Y.; Zheng, Z.; Cheng, J.T.; Liu, J.W.; Liu, J.S.; Li, S.N. The un-symmetric hybridization of graphene surface plasmons incorporating graphene sheets and nano-ribbons. *Appl. Phys. Lett.* **2013**, *103*, 3. [CrossRef]
15. Bozhevolnyi, S.I.; Volkov, V.S.; Devaux, E.; Ebbesen, T.W. Channel plasmon-polariton guiding by subwavelength metal grooves. *Phys. Rev. Lett.* **2005**, *95*, 4. [CrossRef] [PubMed]
16. Bozhevolnyi, S.I.; Volkov, V.S.; Devaux, E.; Laluet, J.Y.; Ebbesen, T.W. Channel plasmon subwavelength waveguide components including interferometers and ring resonators. *Nature* **2006**, *440*, 508–511. [CrossRef]
17. Pile, D.F.P.; Ogawa, T.; Gramotnev, D.K.; Okamoto, T.; Haraguchi, M.; Fukui, M.; Matsuo, S. Theoretical and experimental investigation of strongly localized plasmons on triangular metal wedges for subwavelength waveguiding. *Appl. Phys. Lett.* **2005**, *87*, 061106. [CrossRef]
18. Moreno, E.; Rodrigo, S.G.; Bozhevolnyi, S.I.; Martin-Moreno, L.; Garcia-Vidal, F.J. Guiding and focusing of electromagnetic fields with wedge plasmon polaritons. *Phys. Rev. Lett.* **2008**, *100*, 4. [CrossRef] [PubMed]
19. Nikitin, A.Y.; Guinea, F.; Garcia-Vidal, F.J.; Martin-Moreno, L. Edge and waveguide terahertz surface plasmon modes in graphene microribbons. *Phys. Rev. B* **2011**, *84*, 4. [CrossRef]
20. Lu, W.B.; Zhu, W.; Xu, H.J.; Ni, Z.H.; Dong, Z.G.; Cui, T.J. Flexible transformation plasmonics using graphene. *Opt. Express* **2013**, *21*, 10475–10482. [CrossRef]
21. Christensen, J.; Manjavacas, A.; Thongrattanasiri, S.; Koppens, F.H.L.; de Abajo, F.J.G. Graphene Plasmon Waveguiding and Hybridization in Individual and Paired Nanoribbons. *ACS Nano* **2012**, *6*, 431–440. [CrossRef] [PubMed]

22. Liang, H.; Ruan, S.; Zhang, M.; Su, H.; Li, I.L. Modified surface plasmon polaritons for the nanoconcentration and long-range propagation of optical energy. *Sci. Rep.* **2014**, *4*, 1–4. [CrossRef]
23. Goncalves, P.A.D.; Dias, E.J.C.; Xiao, S.S.; Vasilevskiy, M.I.; Mortensen, N.A.; Peres, N.M.R. Graphene Plasmons in Triangular Wedges and Grooves. *ACS Photonics* **2016**, *3*, 2176–2183. [CrossRef]
24. Goncalves, P.A.D.; Bozhevolnyi, S.I.; Mortensen, N.A.; Peres, N.M.R. Universal description of channel plasmons in two-dimensional materials. *Optica* **2017**, *4*, 595–600. [CrossRef]
25. Teng, D.; Wang, K.; Li, Z. Graphene-Coated Nanowire Waveguides and Their Applications. *Nanomaterials* **2020**, *10*, 229. [CrossRef] [PubMed]
26. Huong, N.T.; Vy, N.D.; Trinh, M.T.; Hoang, C.M. Tuning SPP propagation length of hybrid plasmonic waveguide by manipulating evanescent field. *Opt. Commun.* **2020**, *462*, 6. [CrossRef]
27. Tsakmakidis, K.L.; Pickering, T.W.; Hamm, J.M.; Page, A.F.; Hess, O. Completely Stopped and Dispersionless Light in Plasmonic Waveguides. *Phys. Rev. Lett.* **2014**, *112*. [CrossRef] [PubMed]
28. Li, D.; Du, K.; Liang, S.; Zhang, W.; Mei, T. Wide band dispersionless slow light in hetero-MIM plasmonic waveguide. *Opt. Express* **2016**, *24*, 22432–22437. [CrossRef] [PubMed]
29. Hu, G.W.; Ou, Q.D.; Si, G.Y.; Wu, Y.J.; Wu, J.; Dai, Z.G.; Krasnok, A.; Mazor, Y.; Zhang, Q.; Bao, Q.L.; et al. Topological polaritons and photonic magic angles in twisted alpha-MoO(3)bilayers. *Nature* **2020**, *582*, 209–213. [CrossRef]
30. Smirnova, D.A.; Shadrivov, I.V.; Smirnov, A.I.; Kivshar, Y.S. Dissipative plasmon-solitons in multilayer graphene. *Laser Photonics Rev.* **2014**, *8*, 291–296. [CrossRef]
31. Nesterov, M.L.; Bravo-Abad, J.; Nikitin, A.Y.; Garcia-Vidal, F.J.; Martin-Moreno, L. Graphene supports the propagation of subwavelength optical solitons. *Laser Photonics Rev.* **2013**, *7*, L7–L11. [CrossRef]
32. Epstein, I.; Alcaraz, D.; Huang, Z.Q.; Pusapati, V.V.; Hugonin, J.P.; Kumar, A.; Deputy, X.M.; Khodkov, T.; Rappoport, T.G.; Hong, J.Y.; et al. Far-field excitation of single graphene plasmon cavities with ultracompressed mode volumes. *Science* **2020**, *368*, 1219–1223. [CrossRef]
33. Ahmadivand, A.; Gerislioglu, B.; Noe, G.T.; Mishra, Y.K. Gated Graphene Enabled Tunable Charge-Current Configurations in Hybrid Plasmonic Metamaterials. *ACS Appl. Electron. Mater.* **2019**, *1*, 637–641. [CrossRef]
34. Bonaccorso, F.; Sun, Z.; Hasan, T.; Ferrari, A.C. Graphene photonics and optoelectronics. *Nat. Photonics* **2010**, *4*, 611–622. [CrossRef]
35. Jin, Y.; Joo, M.K.; Moon, B.H.; Kim, H.; Lee, S.; Jeong, H.Y.; Lee, Y.H. Coulomb drag transistor using a graphene and MoS2 heterostructure. *Commun. Phys.* **2020**, *3*, 8. [CrossRef]
36. Liang, H.W.; Zhang, L.; Zhang, S.; Cao, T.; Alu, A.; Ruan, S.C.; Qiu, C.W. Gate-Programmable Electro-Optical Addressing Array of Graphene-Coated Nanowires with Sub-10 nm Resolution. *ACS Photonics* **2016**, *3*, 1847–1853. [CrossRef]
37. Eda, G.; Fanchini, G.; Chhowalla, M. Large-area ultrathin films of reduced graphene oxide as a transparent and flexible electronic material. *Nat. Nanotechnol.* **2008**, *3*, 270–274. [CrossRef] [PubMed]
38. Wu, Z.Y.; Ning, T.Y.; Li, J.Q.; Zhang, M.; Su, H.; Li, I.L.; Liang, H.W. Tunable photonic-like modes in graphene-coated nanowires. *Opt. Express* **2019**, *27*, 7. [CrossRef] [PubMed]
39. Wu, Z.Y.; Zhang, L.; Zhang, M.; Li, I.L.; Su, H.; Zhao, H.C.; Ruan, S.C.; Liang, H.W. Graphene Plasmon Resonances for Electrically-Tunable Sub-Femtometer Dimensional Resolution. *Nanomaterials* **2020**, *10*, 1381. [CrossRef] [PubMed]
40. Luo, L.B.; Wang, K.Y.; Ge, C.W.; Guo, K.; Shen, F.; Yin, Z.P.; Guo, Z.Y. Actively controllable terahertz switches with graphene-based nongroove gratings. *Photonics Res.* **2017**, *5*, 604–611. [CrossRef]
41. Chen, T.; Wang, L.L.; Chen, L.J.; Wang, J.; Zhang, H.K.; Xia, W. Tunable terahertz wave difference frequency generation in a graphene/AlGaAs surface plasmon waveguide. *Photonics Res.* **2018**, *6*, 186–192. [CrossRef]
42. Gusynin, V.P.; Sharapov, S.G.; Carbotte, J.P. Magneto-optical conductivity in graphene. *J. Phys. Condes. Matter* **2007**, *19*, 25. [CrossRef]
43. Hanson, G.W. Dyadic Green's functions and guided surface waves for a surface conductivity model of graphene. *J. Appl. Phys.* **2008**, *103*, 8. [CrossRef]
44. He, X.B.; Tang, J.B.; Hu, H.T.; Shi, J.J.; Guan, Z.Q.; Zhang, S.P.; Xu, H.X. Electrically Driven Highly Tunable Cavity Plasmons. *ACS Photonics* **2019**, *6*, 823–829. [CrossRef]
45. Li, Y.; Hu, H.T.; Jiang, W.; Shi, J.J.; Halas, N.J.; Nordlander, P.; Zhang, S.P.; Xu, H.X. Duplicating Plasmonic Hotspots by Matched Nanoantenna Pairs for Remote Nanogap Enhanced Spectroscopy. *Nano Lett.* **2020**, *20*, 3499–3505. [CrossRef]
46. Li, Z.; Pan, Y.; You, Q.; Zhang, L.; Zhang, D.; Fang, Y.; Wang, P. Graphene-coupled nanowire hybrid plasmonic gap mode-driven catalytic reaction revealed by surface-enhanced Raman scattering. *Nanophotonics* **2020**, *9*, 4519–4527. [CrossRef]
47. Rappoport, T.G.; Epstein, I.; Koppens, F.H.L.; Peres, N.M.R. Understanding the Electromagnetic Response of Graphene/Metallic Nanostructures Hybrids of Different Dimensionality. *ACS Photonics* **2020**, *7*, 2302–2308. [CrossRef]
48. Esteban, R.; Aguirregabiria, G.; Borisov, A.G.; Wang, Y.M.M.; Nordlander, P.; Bryant, G.W.; Aizpurua, J. The Morphology of Narrow Gaps Modifies the Plasmonic Response. *ACS Photonics* **2015**, *2*, 295–305. [CrossRef]
49. Yang, D.J.; Zhang, S.P.; Im, S.J.; Wang, Q.Q.; Xu, H.X.; Gao, S.W. Analytical analysis of spectral sensitivity of plasmon resonances in a nanocavity. *Nanoscale* **2019**, *11*, 10977–10983. [CrossRef]
50. Fei, Z.; Rodin, A.S.; Andreev, G.O.; Bao, W.; McLeod, A.S.; Wagner, M.; Zhang, L.M.; Zhao, Z.; Thiemens, M.; Dominguez, G.; et al. Gate-tuning of graphene plasmons revealed by infrared nano-imaging. *Nature* **2012**, *487*, 82–85. [CrossRef] [PubMed]

Article

Enhanced Stability and Mechanical Properties of a Graphene–Protein Nanocomposite Film by a Facile Non-Covalent Self-Assembly Approach

Chunbao Du [1], Ting Du [1], Joey Tianyi Zhou [2], Yanan Zhu [1], Xingang Jia [1] and Yuan Cheng [3,4,*]

1   College of Chemistry and Chemical Engineering, Xi'an Shiyou University, Xi'an 710065, China; duchunbao218@126.com (C.D.); duting102@126.com (T.D.); zhuyanan@xsyu.edu.cn (Y.Z.); jiaxingang76@xsyu.edu.cn (X.J.)
2   Institute of High Performance Computing, A*STAR, Singapore 138632, Singapore; joey_zhou@ihpc.a-star.edu.sg
3   Monash Suzhou Research Institute, Monash University, Suzhou Industrial Park, Suzhou 215000, China
4   Department of Materials Science and Engineering, Monash University, Clayton, VIC 3800, Australia
*   Correspondence: yuan.cheng@monash.edu

Abstract: Graphene-based nanocomposite films (NCFs) are in high demand due to their superior photoelectric and thermal properties, but their stability and mechanical properties form a bottleneck. Herein, a facile approach was used to prepare nacre-mimetic NCFs through the non-covalent self-assembly of graphene oxide (GO) and biocompatible proteins. Various characterization techniques were employed to characterize the as-prepared NCFs and to track the interactions between GO and proteins. The conformational changes of various proteins induced by GO determined the film-forming ability of NCFs, and the binding of bull serum albumin (BSA)/hemoglobin (HB) on GO's surface was beneficial for improving the stability of as-prepared NCFs. Compared with the GO film without any additive, the indentation hardness and equivalent elastic modulus could be improved by 50.0% and 68.6% for GO–BSA NCF; and 100% and 87.5% for GO–HB NCF. Our strategy should be facile and effective for fabricating well-designed bio-nanocomposites for universal functional applications.

Keywords: graphene; nanocomposite film; film-forming ability; stability; mechanical properties

Citation: Du, C.; Du, T.; Zhou, J.T.; Zhu, Y.; Jia, X.; Cheng, Y. Enhanced Stability and Mechanical Properties of a Graphene–Protein Nanocomposite Film by a Facile Non-Covalent Self-Assembly Approach. *Nanomaterials* **2022**, *12*, 1181. https://doi.org/10.3390/nano12071181

Academic Editor: Orietta Monticelli

Received: 4 March 2022
Accepted: 31 March 2022
Published: 1 April 2022

Publisher's Note: MDPI stays neutral with regard to jurisdictional claims in published maps and institutional affiliations.

Copyright: © 2022 by the authors. Licensee MDPI, Basel, Switzerland. This article is an open access article distributed under the terms and conditions of the Creative Commons Attribution (CC BY) license (https://creativecommons.org/licenses/by/4.0/).

1. Introduction

Two-dimensional (2D) nanomaterials have recently opened a new era for flexible devices owing to their exotic electronic and optical properties [1–3]. Graphene is an emerging constituent for 2D nanomaterials, and graphene films hold great potential for meeting various intellectualized functionalities [4–6]. However, the mechanical properties of pure graphene films have significant flaws, such as limited flexibility and stability [7,8]. Reinforcing components are usually added to produce nanocomposite films (NCFs) to improve overall characteristics, which opens new avenues for graphene's use. Synthetic polymers are used in most graphene-based NCFs owing to their superior designability and usefulness [9,10]. However, synthetic polymers do not easily decompose naturally, resulting in considerable solid waste [11,12]. Therefore, the present trend is to develop environmentally friendly graphene-based NCFs to reduce carbon emissions and allow more recycling of materials.

Biomacromolecules (BMMs), which are indispensable for in vivo life, including proteins, polypeptides, enzymes, DNA, RNA, lipids, and polysaccharides, are being used in in vitro applications because of their exceptional functionality and biodegradability [13–17]. When included in NCFs with nanomaterials that have the desired photoelectric and thermal properties, multifarious applications are available, such as biosensors, artificial tissue, information storage, and drug delivery [18–20]. Various studies report that integrating

BMMs and graphene has already been done, forming some novel composites will multiple applications [20–25]. For example, Liu et al. employed a simple method to prepare low-cost graphene and silk-based pressure sensor, which could be used as artificial skin to monitor the pressure of the human body in real-time [26]. In another case, Chu et al. fabricated a hybrid scaffold using graphene oxide (GO) and an acellular dermal matrix, promoting cell proliferation in the wounds of diabetics [27]. Recently, Chang et al. loaded a heat shock protein 90 inhibitor NVP-AUY922 on a GO-based GO/BaHoF5/PEG nanocomposite to perform sensitized photothermal therapy (PTT). The achieved nano-platform, GO/BaHoF5/PEG/NVP-AUY922, had excellent biocompatibility and made tumor cells more sensitive to hyperthermia, which could promote the development of low-laser-hazard PTT [28]. In the work of Zhao et al., photomodule single-layer reduced graphene oxide (rGO) has been organized into a well-defined multilayer stack with the help of amyloid-like protein aggregates [29]. The as-fabricated hybrid film reliably adheres to the plastic substrate with robust interfacial adhesion. The sensitive photothermal effect of rGO in the bilayer film can be initiated with a blue laser from 100 m away, indicating that the combination of GO with BMMs exhibited great potential in remote light control of robots and devices. Our previous work has summarized the bio–nanomaterial interaction mechanisms at the molecular level of some typical 2D nanomaterials and BMMs, including non-covalent and covalent interactions, and proposed the challenges for the future development of 2D materials and biomacromolecules [30]. Despite significant advances, insufficient attention has been devoted to the stability of graphene-based NCFs in applicable environments involving acidity, alkalinity, salt, heat, and so on. Furthermore, the production processes of NCFs with graphene and biomacromolecules at the molecular level, including the species, conformations, film-forming ability, and mechanical characteristics, need to be investigated further. As a result, there are still major opportunities in, and obstacles to, extending more general biomacromolecules, including understanding and controlling molecular pathways.

In this work, after considering the desirability of simplicity, low cost, and reproducibility for BMMs to be used in scaled-up applications, ordinary and commercialized proteins, i.e., bovine serum albumin (BSA) and hemoglobin from bovine-blood (HB), were chosen for assembly with GO to fabricate NCFs (i.e., GO–BSA NCF and GO–HB NCF; Figure 1a). Using lysozyme (Lyz)-formed NCF (i.e., GO–Lyz NCF) for a comparison, the film-forming abilities of the NCFs were investigated comprehensively by tracking the experimental progress and analyzing the microstructures. X-ray photoelectron spectrometry (XPS), scanning electron microscopy (SEM), and circular dichroism (CD) spectroscopy, along with theoretical simulations, were adopted to reveal the binding mechanism. Moreover, the stability, thermostability, and mechanical properties were investigated by dissolution experiments, differential scanning calorimetry (DSC), and nanometer indentation. Although the structures and properties of BSA, HB, and Lyz are different, consistent stability and improved mechanical strength were achieved, which might provide some inspiration for fabricating other stable nanocomposites.

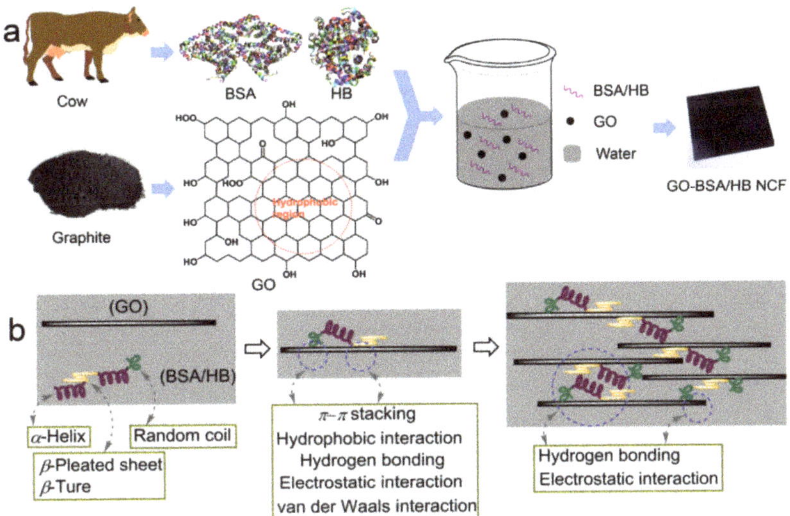

Figure 1. Scheme of self-assembly of GO with BSA/HB for GO–BSA and GO–HB NCF (**a**). Possible binding mechanism of a GO sheet with BSA/HB and the fabrication of a NCF (**b**).

2. Materials and Methods

2.1. Experimental Materials

GO with a thickness of 1.2–5 nm and a lateral size of 50 nm to 3 μm was purchased from Hangzhou Nano-Mall Technology Co., Ltd., Hangzhou, China. BSA, HB, and Lyz were supplied by Macklin. NaCl (AR, >99.5%), HCl (AR, 36.0–38.0%), and NaOH (AR, >99.6%) were products of Sinopharm Chemical Reagent Co., Ltd., Shanghai, China.

2.2. Fabrication of NCF

The fabrication of NCF was conducted by using simple vacuum filtration. Typically, protein (i.e., BSA, HB, or Lyz) was dissolved in the deionized water to form the homogeneous solution with 4 mg mL^{-1}. After that, the aqueous GO solution (4 mg mL^{-1}, 5 mL) was added to the protein solution (5 mL) after the ultrasonic treatment of the aqueous GO solution at 100 W, at 20 °C, for 30 min. Afterward, the mixed solution was stirred at room temperature (26.4 °C) for 16 h to complete the self-assembly process. Then, the mixed solution was poured into the filter flask to remove the water and obtain the wettish NCF. After drying at 30 °C in a vacuum drying oven for 2 h, the dried GO–BSA, GO–HB, and GO–Lyz NCF were kept in a desiccator. The preparation method of the GO film was the same as that of the above NCFs without adding protein. Notably, the structure and characteristics of the composite films varied depending on the proportions of GO and protein [31]. During the experiments, other ratios of GO and protein were also tested, but the films obtained were all poor. When the amount of GO was higher than protein, the film was brittle. In the opposite case, the film was thin and difficult to remove from the substrate.

2.3. Analysis and Characterization

SEM (FEI Talos F200X) was used to observe the morphology and structure of each film at a high voltage of 10.0 kV. Chemical surface characterization was performed by XPS (Shimadzu Kratos, Manchester, United Kingdom) with monochromatic Al Ka radiation (1486.6 eV); the deconvolution method using Gaussian and Lorentz curve fittings was employed to conduct the semiquantitative analysis of the elements. The thermal properties of films were characterized by DSC (Nestal DSC214, Selb, Germany). The CD spectroscopy experiments were carried out using a CD spectrometer (Applied Photophysics Ltd. Chirascan, Leatherhead, UK); the GO–BMM compound solution after self-assembly was diluted by a factor of 27; CD spectra were obtained by scanning the diluted GO–BMM compound solution and deducting the background of the GO solution. Secondary structures of BSA and HB were determined by fitting the far-UV CD data using CDNN algorithms. An optical microscope (Olympus BX51, Tokyo, Japan) was used to observe the morphology of GO and GO–BMM solutions at room temperature. The mechanical properties of NCFs were characterized by a nanometer indentation instrument (UNHT) at room temperature (28.3 °C) with five random positions (RP) and the compression rate of 1 mm min^{-1}; F_m (mN) and h (nm) were confirmed in the obtained curved; H_{IT} (GPa) and E^* (GPa) were calculated by the supporting analysis software. The stability of films was measured in separate NaCl, HCl, and NaOH aqueous solutions, each with a concentration of 0.1 M.

3. Results and Discussion

3.1. Structural Property of NCFs

The fabrication of NCFs was conducted using simple vacuum filtration (Figure 1a). Protein (i.e., BSA, HB, and Lyz; see Figure S1a–c) was dissolved in the deionized water to form the homogeneous solution and then mixed with the ultrasonically treated GO aqueous solution. After stirring, the mixed solution was poured into a filter flask to remove the water and obtain the wettish film. Finally, the GO–BSA, GO–HB, and GO–Lyz NCFs were obtained after drying. The most spread out BSA, HB, and Lyz were in the three dimensions of the crystalline state was about $14 \times 8 \times 5$ nm^3, which was far less than the size of GO sheet (50 nm–3 µm) [32]. BSA, HB, and Lyz were more likely to be bound on the surface of a GO sheet rather than the edge. The interactions between the GO sheet and BSA/HB/Lyz were dominated by the multiple non-covalent interactions (Figure 1b). Notably, because protein molecules could not be trapped by the filter membrane, the pure protein film could only be obtained by solvent evaporation and could not be obtained using vacuum filtration. GO is a 2D material with a high aspect ratio that is useful for adsorbing protein molecules as a skeleton when creating films, and its outstanding mechanical strength is very advantageous [33,34]. To unravel the protein–GO interaction mechanism, specific experiments were designed, and the results obtained are discussed step by step as follows.

The sizes (layers, transverse, and longitudinal direction) and properties (functional groups and groups density) of GO were detected first. The GO used in this work was prepared through a typical Hummers sonication method [35]. The thickness and lateral size of the GO sheet were 1.2–5 nm and 50 nm to 3 µm, respectively. Therefore, this ensured that the GO sheet was stretching rather than crimping into spherical particles [36]. XPS data (Figure S1d) showed the main elements of GO were C and O, and there was very little N. The functional groups of the GO sheet included the following percentages of the total carbon: carbonyl (C=O), 2.0%; hydroxyl (-OH), 46.5%; carboxyl (-C=(O)-OH), 2.5% (Figures 2a and S1e). Those percentages imply that the graphene structure in GO sheet largely remained intact. The typical elements in BSA, HB, and Lyz other than C and O were N and S. Figure 2b,c shows the N 1s and S 2p XPS survey spectra of GO–BSA, GO–HB, and GO–Lyz NCF, further indicating the existence of BSA, HB, and Lyz. Moreover, the different atomic concentrations of each element in these films could also support the formation of these NCFs (Figure 2d). The polar binding sites of GO sheets usually existed on the edges and in the defects on the surface. Therefore, BSA/HB/Lyz was more likely to be bound on the surface of the GO sheet rather than the edge due to the huge difference between the size of any of these proteins and that of GO sheets.

The insets in Figures S2a and 3a–c show the megashapes of GO film, GO–BSA NCF, GO–HB NCF, and GO–Lyz NCF. GO–BSA and GO–HB NCFs had complete structures and exhibited bendability, whereas the GO film and GO–Lyz NCV were easily broken quickly. The reasons for this phenomenon were associated with the properties of these proteins and their binding situations with GO sheet. To further investigate the binding situations between the GO sheet and these proteins, SEM was employed to discover the surface microtopography. For the GO film, due to the polar sites on the surface and corner of the GO sheet, it was not easy to accomplish the π–π tight stacking in the GO sheets. That caused the GO film to have an uneven surface with great roughness (Figure S2a). On the contrary, GO sheet–BSA and GO sheet–HB compounds were achieved by adequate self-assembly, and their surfaces showed relatively good uniformity (Figure 3a,b). Thereinto, BSA and HB played the role of plasticizer to adjust the interfacial compatibility of GO. However, for the GO–Lyz NCF, its microscopic surface was the same as that of the GO film (Figure 3c), which meant the effect of this approach was negligible. These differences were also reflected in the GO sheet–protein compound solutions macroscopically. For both GO sheet–BSA and GO sheet–HB, the compound solutions, after a 16 h self-assembly process before suction filtration, were stable suspensions without precipitation or aggregation (Figure S3b,c). Most interestingly, both GO sheet–BSA and GO sheet–HB compound solutions were stable in polar aqueous solutions, which were more stable than the GO solution (Figure S3a), indicating that the external surfaces of GO sheet–BSA and GO sheet–HB compounds are also polar. Nevertheless, the external surface of GO sheet–Lyz compound was hydrophobic, and aggregation of the GO sheet–Lyz compound occurred through the spontaneous hydrophobic interactions, causing the apparent precipitation phenomenon of GO sheet–Lyz in the aqueous phase (Figure S3d). The stabilities of these compounds in aqueous solutions were closely related to the properties of the obtained NCFs.

SEM images of the internal cross-section were more valid evidence to confirm the above analysis. As shown in Figure 3d,e, both GO–BSA NCF and GO–HB NCF displayed a prominent layered hierarchical structure of natural nacre. Moreover, their compactness and smoothness in section micromorphology were better than those of GO and GO–Lyz NCFs when compared with the inter-layer gaps of the latter films (Figures 3f and S2b). Although GO sheets were stable in the aqueous phase due to their polar sites, there were inevitably irregular gaps between GO sheets among the π–π tight stacking. By contrast, BSA and HB had better adhesion to GO sheets to facilitate the mutual attraction and fill the gaps for the dense structures. For Lyz, the structural change induced by the GO sheet was not beneficial for the tight and homogeneous binding of GO sheet–Lyz, and the instability of GO sheet–Lyz in an aqueous solution also caused inhomogeneity of GO–Lyz NCF. The excellent interfacial compatibility of GO sheet–protein compounds contributed to enhancing the mechanical properties of NCFs. Thereinto, the conformational changes of proteins induced by GO sheet were essential. Considering film-forming ability, GO–BSA NCF and GO–HB NCF are better candidates than GO–Lyz NCF for practical applications. Furthermore, using the SEM images of interior cross-sections, the thicknesses of GO–BSA and GO–HB NCF were determined to be around 2.1 and 2.2 µm, respectively.

The mechanical properties of GO–BSA and GO–HB NCF could also be reflected by the conformational changes of BSA and HB. To prove the secondary structure changes in BSA/HB induced by GO sheets, CD spectra of BSA and HB aqueous solutions with and without the addition of GO were obtained. The secondary structures (i.e., α-helix, β-pleated sheet, β-turn, and random coil) of proteins were confirmed by the positions of α-helixes (222 and 208 nm positive peaks, 192 nm negative peak), β-pleated sheets (217–218 nm positive peaks, 195–198 nm negative peak), β-turns (220–230 nm weak positive peaks, 180–190 nm strong positive peaks, 205 nm negative peak), and random coils (198 nm positive peaks, 220 nm negative peak) [37]. Figure 2e showed that there were apparent changes in the secondary structures of BSA and HB before and after binding of GO sheets, indicating that GO had the apparent effect on the structures of BSA and HB. After analyzing the data of contents of the secondary structures (Table 1), the changes in the secondary structures in BSA and HB showed the same pattern, which was a decrease in α-helixes and increases in β-pleated sheets (antiparallel and parallel), β-turns, and random coils. Before introducing GO sheets, α-helixes predominated in BSA and HB with the contents of 50.7 and 46.0%, respectively. After interactions with GO sheets, the α-helix contents of BSA and HB decreased to 14.6 and 16.4%, respectively. Our previous work has revealed the binding mechanism of the α-helix fragments of BSA with graphene by using molecular dynamics simulations [38]. The adsorption of an α-helix on the surface of graphene induces a transition from to the 3_{10}-helix structure, which was reflected in the substantial increase in random coils from 24.1 to 42.2% for BSA in this work. This induction mode might also work for HB because the content of the random coils of HB increased from 26.5 to 41.4%. For the β-pleated sheets (antiparallel and parallel), the tiled state on GO sheets was more stable due to the interactions of more binding sites, which was in accordance with the molecular dynamics simulations of our previous work that showed graphene was advantageous to the stability of β-pleated sheets [39]. The increase in β-turns always accompanied the increase in β-pleated sheets. Therefore, the increases in contents of the β-pleated sheets (antiparallel and parallel) and random coils in BSA and HB were beneficial for the self-assembly of GO sheet with BSA and HB.

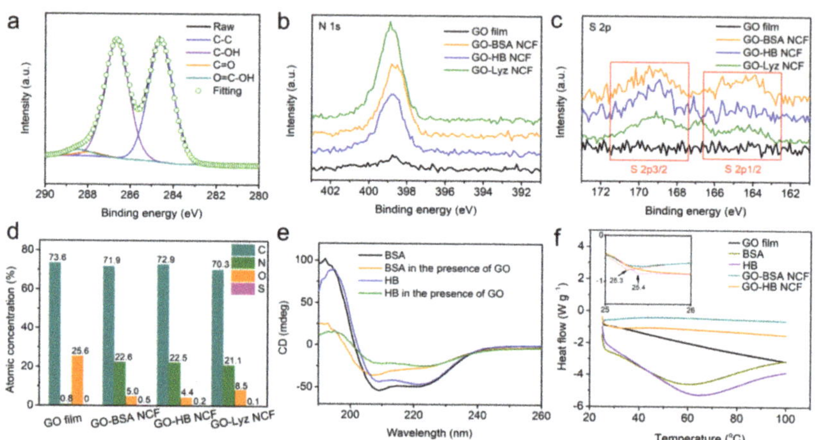

Figure 2. C 1s high magnification of films (**a**). N 1s high magnification of GO film (**b**). S 2p high magnification of films (**c**). The atomic compositions of films (**d**). CD spectra of BSA and HB with and without GO induction (**e**). DSC curves of GO film, BSA, HB, GO–BSA NCF, and GO–HB NCF with a heating rate of 5 °C min^{-1} in N_2 flow from 25 to 100 °C (**f**).

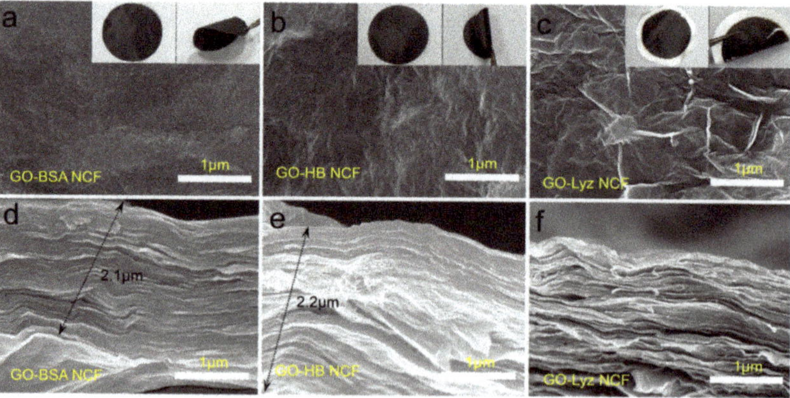

Figure 3. SEM images of surface (**a–c**); photographs with a uniform diameter of about 5 cm (insets of (**a**,**c**,**e**)). Internal cross-sections (**d–f**) of GO–BSA NCF, GO–HB NCF, and GO–Lyz NCF.

Table 1. The contents of the secondary structure elements of BSA and HB with and without GO.

Secondary Structure		BSA (%)	BSA in the Presence of GO Sheet (%)	HB (%)	HB in the Presence of GO Sheet (%)
α-Helix		50.7	14.6	46.0	16.4
β-Pleated sheet	Antiparallel	5.6	11.6	5.8	11.1
	Parallel	5.2	14.3	6.5	13.7
β-Turn		14.4	17.3	15.2	17.4
Random coil		24.1	42.2	26.5	41.4

3.2. Stability of NCFs

The thermostability of films is very important to determine their applications at different temperatures. The thermostability of GO film, GO–BSA NCF, and GO–HB NCF was characterized by DSC. As shown in Figure 2f, heat release of the GO film proceeded the increase in temperature, indicating that the GO sheet was very sensitive to heat. After binding with BSA and HB, only transient heat release occurred in GO–BSA and GO–HB NCFs, and then the heat flows were maintained within a stable range, implying that GO–BSA and GO–HB NCF were thermostable. It is not difficult to see that BSA and HB also exhibited continuous heat release and absorption around 60–70 °C due to their conformational changes with the temperature change. The glass transition is the transition of amorphous material from a glassy state to a high elastic state. As shown in the inset of Figure 2f, the glass-transition temperatures of GO–BSA and GO–HB NCFs were confirmed to be 25.3 and 25.4 °C, respectively, which belong to the scope of room temperature. This indicates that the GO–BSA and GO–HB NCF could be kept in high elastic states at room temperature and remain stable. It was advantageous for GO–BSA and GO–HB NCFs to fully utilize their flexibility for stability. Before the self-assembly process of GO with BSA or HB, the conformational changes of BSA or HB were completed, so the film-forming process would not induce a conformational change in BSA or HB. That is to say, the film-forming process would only involve the self-assembly of "GO sheet-BSA/HB compounds." Combined with the results of CD spectra, it could be concluded that the introduction of BSA or HB on the surface of a GO sheet was helpful to improving the thermostability, which is attributed to the increased contents of the β-pleated sheets (antiparallel and parallel) and random coils.

The stability of films in various complex environments is also essential for their actual applications. It has been confirmed that the functional groups of the GO sheet were OH and –COOH. Even so, there were many hydrophobic areas on the surface of the GO sheet. The formation of the GO film was caused by the polar and non-polar interactions of GO sheets. The polar interactions included hydrogen bonding of –OH with –OH, –OH with –COOH, and –COOH with –COOH; non-polar interactions were π–π stacking. The GO film was barely stable in the aqueous phase and was dissolved partly after 7 days (Figure S4a). After ultrasonication, the mutual attraction of GO sheets could not conquer the destructive effect from outside that dissolved GO film easily. $-COOH \leftrightarrow -COO^- + H^+$ was a dynamic equilibrium process and water could break the hydrogen bonding to a certain degree. GO films were stable in acidic, alkaline, and saline environments for a standing time of 7 days (Figure S4b–d). However, they were unstable under ultrasonication in alkaline and saline environments because the films were dissolved easily, which we attribute to different dissolution mechanisms. The increase in $-COO^-$ groups facilitated the electrostatic repulsion in an alkaline environment, and the saline ions destroyed the electrostatic attraction in a saline environment. Both cases were not beneficial for the stability the GO films. By contrast, GO films were still stable in an acidic environment, even under ultrasonication, because the increase in –COOH groups could tremendously enhance the mutual attraction of GO sheets. After determining the stability and instability mechanisms of GO films in the above environments, the assembly mechanisms of GO–BSA and GO–HB NCFs were also analyzed. The main functional groups of BSA and HB are

–NH$_3$, –COOH, hydrophobic chains (benzene ring and alkyl chain), and other polar groups (–OH, –C(=O)–NH–). The formation of GO–BSA and GO–HB NCFs meant self-assembly of GO sheet–BSA and GO sheet–HB compounds, respectively, which still involved multiple non-covalent interactions, as in the formation of GO films. The process was markedly different for BSA and HB, in types of interactions and binding strengths, owing to their uniqueness. Therefore, the stability of GO–BSA and GO–HB NCFs differed significantly in different environments. GO–BSA NCF was very stable in aqueous, acidic, alkaline, and saline environments with or without ultrasonic treatment, indicating that the favorable interactions for film-forming were far stronger than the adverse interactions (Figure 4a–d). There was no denying that the confirmation of interactions is very complex and needs precision instruments and testing [40,41]. GO–HB NCF was very stable in acidic and saline environments regardless of ultrasonic treatment (Figure 4f,h), but it can dissolve easily in aqueous and alkaline environments with ultrasonic treatment (Figure 4e,g). The stability of these films will determine their ranges of application, which is the case for all BMMs with a unique characteristics.

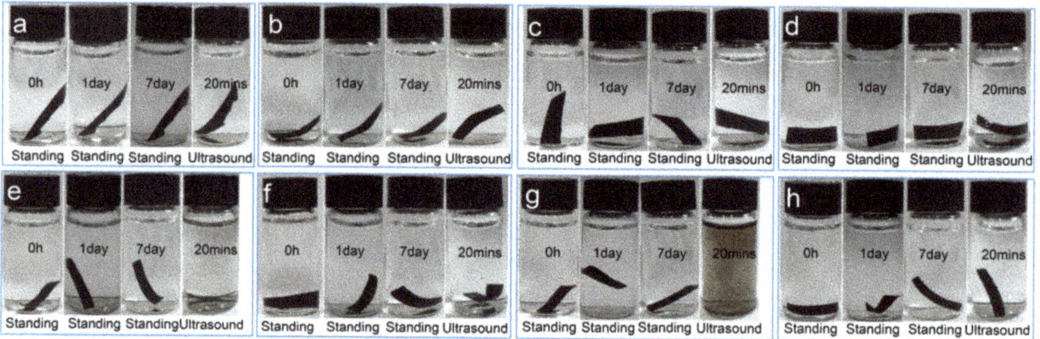

Figure 4. The stability of the GO–BSA NCF in aqueous (**a**), HCl (**b**), NaOH (**c**), and NaCl (**d**) solutions at room temperature. Stability of GO–HB NCF in aqueous (**e**), HCl (**f**), NaOH (**g**) and NaCl solution (**h**) with different treatments at room temperature.

3.3. Mechanical Properties of NCFs

The mechanical properties of GO film, GO–BSA NCF and GO–HB NCF were characterized by nanoindentation with five random positions (RP) under the maximum applied load (F_m, mN) of 5 mN. Generally, there are three stages, namely, loading, holding, and unloading (Figure 5a), which are very apparent in H_{IT}-h curves. There were significant differences in the tracks of five H_{IT}-h curves of the GO films, indicating that the uniformity of GO films was relatively poor (Figure 5b), which is in accordance with the results of SEM in Figure S2a: the surfaces of GO films were uneven with large roughness. However, the tracks for GO–BSA and GO–HB NCFs (Figure 5c,d) were nearly identical, indicating that the uniformity of GO–BSA NCF and GO–HB NCF is much better than that of GO film. In addition, the corresponding indentation hardness (H_{IT}, GPa) and equivalent elastic modulus (E^*, GPa) of each H_{IT}-h curve can be calculated. Figure 5e,f shows the H_{IT} and E^* of each RP of a GO film, GO–BSA NCF, and GO–HB NCF, and there are no abrupt values, indicating these films had good structural homogeneity and no flaws. Numerous studies have strived to improve the mechanical properties of GO-based films because their limited mechanical properties have hindered practical applications [42–44]. Here, the average H_{IT} and E^* of GO–BSA NCF were 0.12 and 2.7 GPa, respectively; the average H_{IT} and E^* of GO–HB NCF were 0.16 and 3.0 GPa, respectively. Compared with average the H_{IT} (0.08 GPa) and E^* (1.6 GPa) of GO film, the mechanical properties of GO–BSA and GO–HB NCF were very much improved. Under F_m of 5 mN, the depths of indentation of GO–BSA and GO–HB NCF were around 1300 and 1200 nm, which were lower than that of GO film (1600–2000 nm), indicating that the existence of BSA and HB in interlayers of films could store the stress. BSA and HB served as filling agents and plasticizers to complement the stretchability of GO. In the work of Li et al., the functional groups on GO sheets had a significant influence on the mechanical properties of a GO–silk-based nanocomposite, and the oxygen-containing groups of GO could form hydrogen bonding with silk fibroins at the interface to improve the adhesive force [45]. A similar principle applies for BSA and HB, because BSA or HB could bind to the surface of a GO sheet strongly through hydrogen bonding, in addition the spontaneous hydrophobic interactions, which would shield the sheet from water molecules surrounding GO–BSA NFC or GO–HB NCF, thereby stabilizing their structures [45]. Therefore, the resistance to instantaneous and continuous external forces in GO–BSA and GO–HB NCF is significantly improved over that of GO films, meaning that GO–BSA and GO–HB NCFs exhibit enhanced applicability. In the previous work of Shao and Fan et al., bacterial cellulose and chitosan were fabricated with GO to improve the mechanical properties of GO–bacterial cellulose and GO–chitosan NCF, respectively, and the assembly process only involved hydrogen bonding and electrostatic interactions [46,47]. Compared with Shao's work with an E^* of 0.5 GPa, GO–BSA NCF and GO–HB NCF had significant advantages in terms of mechanical properties [46]. In addition, although the H_{IT} (0.40 GPa) and E^* (6.5 GPa) of GO–chitosan NCF in Fan's work were much higher than those of GO–BSA and GO–HB NCF, GO–chitosan NCF was extremely unstable in an acidic environment, which will handicap its applicability [47]. Therefore, combining GO and well matched BMMs is an effective and fantastic strategy to construct nacre-like NCFs with good stability and enhanced mechanical properties.

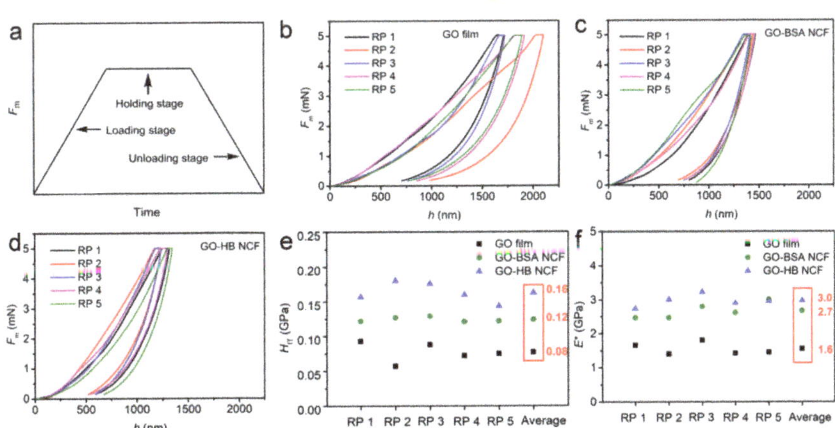

Figure 5. Typical scheme of nanoindentation load–displacement curve (**a**). Representative nanoindentation load–displacement curves for GO film (**b**), GO–BSA NCF (**c**), GO–HB NCF (**d**) with five RP. The corresponding H_{IT} (**e**) and E^* (**f**) of the GO film, GO–BSA NCF, and GO–HB NCF.

4. Conclusions

In summary, GO–BMM-based NCFs with nacre-mimetic structures were fabricated with GO and proteins through a green and straightforward non-covalent self-assembly process. The sequences and conformational features of BSA, HB, and Lyz determined their film-forming ability with GO sheets, implying that GO–BSA and GO–HB, which are stable compounds in an aqueous solution, are outstanding candidates for fabricating stable NCFs. The GO sheet induced increases in the presence of β-pleated sheets, β-turns, and random coils in BSA and HB, along with a decrease in the presence α-helixes, which was more beneficial for the NCFs with dense and uniform microstructures. Compared with the GO film, GO–BSA and GO–HB NCF exhibited good thermostability below 100 °C, and remained remarkably stable in acidic and saline environments. GO–BSA NCF could be kept stable in an alkaline environment, which endows it with broader application potential. Moreover, GO–BSA and GO–HB NCF exhibited significant advantages through appreciable

HIT enhancements of 50.0% and 100%; and enhancements in E^* by 68.6% and 87.5%, respectively. The binding of BMMs into interlayers of 2D nanomaterials could synergistically provide enhancements while maintaining the films' respective characteristics, making them promising for flexible devices.

Supplementary Materials: The following supporting information can be downloaded at: https://www.mdpi.com/article/10.3390/nano12071181/s1. Figure S1: The crystal structures of BSA (a, PDB code 4F5S), HB (b, PDB code 1FN3), and Lyz (c, PDB code 253L) with the style of newcartoon; wide scan of XPS survey spectra of films (d); atomic concentration of C in GO film (e). Figure S2: SEM images of surfaces (a); photographs with uniform diameter of about 5 cm (insets of a). Internal cross section (b) of GO film. Figure S3: Optical microscope photographs of the GO solution (a), GO–BSA compound solution (b), GO–HB compound solution (c) and GO–Lyz compound solution (d); Figure S4: The stability of GO film in aqueous (a), HCl (b), NaOH (c), and NaCl (d) solutions at room temperature.

Author Contributions: Conceptualization, C.D. and Y.C.; methodology, C.D.; formal analysis, C.D. and Y.C.; investigation, C.D. and T.D.; resources, C.D. and Y.C.; writing—original draft preparation, C.D. and Y.C.; writing—review and editing, J.T.Z., Y.Z. and X.J.; project administration, C.D. and Y.C. All authors have read and agreed to the published version of the manuscript.

Funding: This research was funded by the National Natural Science Foundation of China (22002117), the Natural Science Foundation of Shaanxi Province, China (2021JQ-585), and the Scientific Research Program of Shaanxi Provincial Education Department (20JK0839). J.T.Z. appreciates the financial support from the RIE2020 AME Programmatic Grant A18A1b0045 funded by A*STAR-SERC, Singapore.

Data Availability Statement: The data presented in this study are available on request from the corresponding author.

Acknowledgments: J.T.Z. is grateful for the support from the Agency for Science, Technology, and Research (A*STAR), A*STAR Computational Resource Centre, Singapore (ACRC); and the National Supercomputing Centre, Singapore (NSCC). C.D. thanks the Modern Analysis and Testing Center of Xi'an Shiyou University and Shiyanjia Lab (www.shiyanjia.com (accessed on 1 November 2021)) for characterizations.

Conflicts of Interest: Authors declare no conflict of interest.

References

1. Kim, S.J.; Choi, K.; Lee, B.; Kim, Y.; Hong, B.H. Materials for flexible, stretchable electronics: Graphene and 2D materials. *Annu. Rev. Mater. Res.* **2015**, *45*, 63–84. [CrossRef]
2. Cai, Y.; Chen, S.; Gao, J.; Zhang, G.; Zhang, Y.W. Evolution of intrinsic vacancies and prolonged lifetimes of vacancy clusters in black phosphorene. *Nanoscale* **2019**, *11*, 20987–20995. [CrossRef] [PubMed]
3. Cheng, Y.; Zhang, G.; Zhang, Y.; Chang, T.; Pei, Q.-X.; Cai, Y.; Zhang, Y.-W. Large diffusion anisotropy and orientation sorting of phosphorene nanoflakes under a temperature gradient. *Nanoscale* **2018**, *10*, 1660–1666. [CrossRef]
4. Zhao, Y.; Zhang, G.; Nai, M.H.; Ding, G.; Li, D.; Liu, Y.; Hippalgaonkar, K.; Lim, C.T.; Chi, D.; Li, B.; et al. Probing the physical origin of anisotropic thermal transport in black phosphorus nanoribbons. *Adv. Mater.* **2018**, *30*, 1804928. [CrossRef] [PubMed]
5. Du, C.; Han, B. Disaggregation of amyloid-β fibrils on graphene. *Acta Phys. Chim. Sin.* **2019**, *35*, 1045–1046. [CrossRef]
6. Guo, S.; Chen, J.; Zhang, Y.; Liu, J. Graphene-based films: Fabrication, interfacial modification, and applications. *Nanomaterials* **2021**, *11*, 2539. [CrossRef]
7. Chen, J.; Chen, L.; Zhang, Z.; Li, J.; Wang, L.; Jiang, W. Graphene layers produced from carbon nanotubes by friction. *Carbon* **2012**, *50*, 1934–1941. [CrossRef]
8. Zhao, G.; Feng, C.; Cheng, H.; Li, Y.; Wang, Z.S. In situ thermal conversion of graphene oxide films to reduced graphene oxide films for efficient dye-sensitized solar cells. *Mater. Res. Bull.* **2019**, *120*, 110609. [CrossRef]
9. Wang, Y.; Wang, Y.; Yang, Y. Graphene-polymer nanocomposite-based redox-induced electricity for flexible self-powered strain sensors. *Adv. Energy Mater.* **2018**, *8*, 1800961. [CrossRef]
10. He, G.; Huang, S.; Villalobos, L.F.; Zhao, J.; Mensi, M.; Oveisi, E.; Rezaei, M.; Agrawal, K.V. High-permeance polymer-functionalized single-layer graphene membranes that surpass the postcombustion carbon capture target. *Energy Environ. Sci.* **2019**, *12*, 3305–3312. [CrossRef]
11. Gong, J.; Chen, X.; Tang, T. Recent progress in controlled carbonization of (waste) polymers. *Prog. Polym. Sci.* **2019**, *94*, 1–22. [CrossRef]

12. Chen, S.; Liu, Z.; Jiang, S.; Hou, H. Carbonization: A feasible route for reutilization of plastic wastes. *Sci. Total Environ.* **2020**, *710*, 136250. [CrossRef] [PubMed]

13. Knowles, T.P.; Mezzenga, R. Amyloid fibrils as building blocks for natural and artificial functional materials. *Adv. Mater.* **2016**, *28*, 6546–6561. [CrossRef] [PubMed]

14. Huang, J.; Xu, Z.; Qiu, W.; Chen, F.; Meng, Z.; Hou, C.; Guo, W.; Liu, X.Y. Stretchable and heat-resistant protein-based electronic skin for human thermoregulation. *Adv. Funct. Mater.* **2020**, *30*, 1910547. [CrossRef]

15. Ding, S.; Khan, A.I.; Cai, X.; Song, Y.; Lyu, Z.; Du, D.; Dutta, P.; Lin, Y. Overcoming blood-brain barrier transport: Advances in nanoparticle-based drug delivery strategies. *Mater. Today* **2020**, *37*, 112–125. [CrossRef] [PubMed]

16. Sweedan, A.; Cohen, Y.; Yaron, S.; Bashouti, M.Y. Binding capabilities of different genetically engineered pVIII proteins of the filamentous M13/Fd virus and single-walled carbon nanotubes. *Nanomaterials* **2022**, *12*, 398. [CrossRef] [PubMed]

17. Ananikov, V.P. Organic–inorganic hybrid nanomaterials. *Nanomaterials* **2019**, *9*, 1197. [CrossRef]

18. López Barreiro, D.; Martin-Moldes, Z.; Yeo, J.; Shen, S.; Hawker, M.J.; Martin-Martinez, F.J.; Kaplan, D.L.; Buehler, M.J. Conductive silk-based composites using biobased carbon materials. *Adv. Mater.* **2019**, *31*, 1904720. [CrossRef]

19. Li, C.; Adamcik, J.; Mezzenga, R. Biodegradable nanocomposites of amyloid fibrils and graphene with shape-memory and enzyme-sensing properties. *Nat. Nanotechnol.* **2012**, *7*, 421–427. [CrossRef]

20. Pan, L.; Wang, F.; Cheng, Y.; Leow, W.R.; Zhang, Y.W.; Wang, M.; Cai, P.; Ji, B.; Li, D.; Chen, X. A supertough electro-tendon based on spider silk composites. *Nat. Commun.* **2020**, *11*, 1332. [CrossRef]

21. Xu, X.; Guan, C.; Xu, L.; Tan, Y.H.; Zhang, D.; Wang, Y.; Zhang, H.; Blackwood, D.J.; Wang, J.; Li, M.; et al. Three dimensionally free-formable graphene foam with designed structures for energy and environmental applications. *ACS Nano* **2020**, *14*, 937–947. [CrossRef] [PubMed]

22. Cherusseri, J.; Sambath Kumar, K.; Pandey, D.; Barrios, E.; Thomas, J. Vertically aligned grapheme-carbon fiber hybrid electrodes with superlong cycling stability for flexible supercapacitors. *Small* **2019**, *15*, 1902606. [CrossRef] [PubMed]

23. You, R.; Liu, Y.Q.; Hao, Y.L.; Han, D.D.; Zhang, Y.L.; You, Z. Laser fabrication of graphene-based flexible electronics. *Adv. Mater.* **2020**, *32*, 1901981. [CrossRef] [PubMed]

24. Wang, Z.; Zhu, W.; Qiu, Y.; Yi, X.; von dem Bussche, A.; Kane, A.; Gao, H.; Koski, K.; Hurt, R. Biological and environmental interactions of emerging two-dimensional nanomaterials. *Chem. Soc. Rev.* **2016**, *45*, 1750–1780. [CrossRef] [PubMed]

25. Zhang, N.; Hu, X.; Guan, P.; Zeng, K.; Cheng, Y. Adsorption mechanism of amyloid fibrils to graphene nanosheets and their structural destruction. *J. Phys. Chem. C* **2018**, *123*, 897–906. [CrossRef]

26. Liu, Y.; Tao, L.-Q.; Wang, D.-Y.; Zhang, T.-Y.; Yang, Y.; Ren, T.-L. Flexible, highly sensitive pressure sensor with a wide range based on graphene-silk network structure. *Appl. Phys. Lett.* **2017**, *110*, 123508. [CrossRef]

27. Chu, J.; Shi, P.; Yan, W.; Fu, J.; Yang, Z.; He, C.; Deng, X.; Liu, H. PEGylated graphene oxide-mediated quercetin-modified collagen hybrid scaffold for enhancement of MSCs differentiation potential and diabetic wound healing. *Nanoscale* **2018**, *10*, 9547. [CrossRef]

28. Chang, X.; Zhang, M.; Wang, C.; Zhang, J.; Wu, H.; Yang, S. Graphene oxide/BaHoF5/PEG nanocomposite for dual-modal imaging and heat shock protein inhibitor-sensitized tumor photothermal therapy. *Carbon* **2020**, *158*, 372–385. [CrossRef]

29. Zhao, J.; Li, Q.; Miao, B.; Pi, H.; Yang, P. Controlling long-distance photoactuation with protein additives. *Small* **2020**, *16*, 2000043. [CrossRef]

30. Du, C.; Hu, X.; Zhang, G.; Cheng, Y. 2D materials meet biomacromolecules: Opportunities and challenges. *Acta Phys. Chim. Sin.* **2019**, *35*, 1078–1089. [CrossRef]

31. Hampitak, P.; Melendrez, D.; Iliut, M.; Fresquet, M.; Parsons, N.; Spencer, B.; AJowitt, T.; Vijayaraghavan, A. Protein interactions and conformations on graphene-based materials mapped using a quartz-crystal microbalance with dissipation monitoring (QCM-D). *Carbon* **2020**, *165*, 317–327. [CrossRef]

32. Mantina, M.; Chamberlin, A.C.; Valero, R.; Cramer, C.J.; Truhlar, D.G. Consistent van der Waals radii for the whole main group. *J. Phys. Chem. A* **2009**, *113*, 5806–5812. [CrossRef] [PubMed]

33. Demirel, M.C.; Vural, M.; Terrones, M. Composites of proteins and 2D nanomaterials. *Adv. Funct. Mater.* **2017**, *28*, 1704990. [CrossRef]

34. Kim, E.K.; Qin, X.; Qiao, J.B.; Zeng, Q.; Fortner, J.D.; Zhang, F. Graphene oxide/mussel foot protein composites for high-strength and ultra-tough thin films. *Sci. Rep.* **2020**, *10*, 19082. [CrossRef]

35. Su, C.-Y.; Xu, Y.; Zhang, W.; Zhao, J.; Liu, A.; Tang, X.; Tsai, C.H.; Huang, Y.; Li, L.J. Highly efficient restoration of graphitic structure in graphene oxide using alcohol vapors. *ACS Nano* **2010**, *4*, 5285–5292. [CrossRef]

36. Guan, G.; Zhang, S.; Liu, S.; Cai, Y.; Low, M.; Teng, C.P.; Phang, I.Y.; Cheng, Y.; Duei, K.L.; Srinivasan, B.M.; et al. Protein induces layer-by-layer exfoliation of transition metal dichalcogenides. *J. Am. Chem. Soc.* **2015**, *137*, 6152–6155. [CrossRef]

37. Greenfield, N.J. Using circular dichroism spectra to estimate protein secondary structure. *Nat. Protoc.* **2006**, *1*, 2876–2890. [CrossRef]

38. Yeo, J.; Cheng, Y.; Han, Y.T.; Zhang, Y.; Guan, G.; Zhang, Y.-W. Adsorption and conformational evolution of alpha-helical BSA segments on graphene: A Molecular Dynamics Study. *Int. J. Appl. Mech.* **2016**, *8*, 1650021. [CrossRef]

39. Cheng, Y.; Koh, L.D.; Li, D.; Ji, B.; Zhang, Y.; Yeo, J.; Guan, G.; Han, M.Y.; Zhang, Y.W. Peptide-graphene interactions enhance the mechanical properties of silk fibroin. *ACS Appl. Mater. Inter.* **2015**, *7*, 21787–21796. [CrossRef]

40. Shang, L.; Nienhaus, G.U. In situ characterization of protein adsorption onto nanoparticles by fluorescence correlation spectroscopy. *Acc. Chem. Res.* **2017**, *50*, 387–395. [CrossRef]
41. Gu, Z.; Song, W.; Chen, S.H.; Li, B.; Li, W.; Zhou, R. Defect-assisted protein HP35 denaturation on grapheme. *Nanoscale* **2019**, *11*, 19362–19369. [CrossRef] [PubMed]
42. Zhang, X.; Gong, C.; Akakuru, O.U.; Su, Z.; Wu, A.; Wei, G. The design and biomedical applications of selfassembled two-dimensional organic biomaterials. *Chem. Soc. Rev.* **2019**, *48*, 5564–5595. [CrossRef] [PubMed]
43. El Abbassi, M.; Sangtarash, S.; Liu, X.; Perrin, M.L.; Braun, O.; Lambert, C.; van der Zant, H.S.J.; Yitzchaik, S.; Decurtins, S.; Liu, S.X.; et al. Robust graphene-based molecular devices. *Nat. Nanotechnol.* **2019**, *14*, 957–961. [CrossRef]
44. Wang, B.; Facchetti, A. Mechanically flexible conductors for stretchable and wearable e-skin and e-textile devices. *Adv. Mater.* **2019**, *31*, 1901408. [CrossRef] [PubMed]
45. Zhou, X.; Li, D.; Wan, S.; Cheng, Q.; Ji, B. In silicon testing of the mechanical properties of graphene oxide-silk nanocomposites. *Acta. Mech.* **2017**, *230*, 1413–1425. [CrossRef]
46. Shao, W.; Wang, S.; Liu, H.; Wu, J.; Zhang, R.; Min, H.; Huang, M. Preparation of bacterial cellulose/graphene nanosheets composite films with enhanced mechanical performances. *Carbohydr. Polym.* **2016**, *138*, 166–171. [CrossRef]
47. Fan, H.; Wang, L.; Zhao, K.; Li, N.; Shi, Z.; Ge, Z.; Jin, Z. Fabrication, mechanical properties, and biocompatibility of graphene-reinforced chitosan composites. *Biomacromolecules* **2010**, *11*, 2345–2351. [CrossRef]

nanomaterials

Article

Influence of Ink Properties on the Morphology of Long-Wave Infrared HgSe Quantum Dot Films

Suhui Wang, Xu Zhang, Yi Wang , Tengxiao Guo * and Shuya Cao *

State Key Laboratory of NBC Protection for Civilian, Beijing 102205, China; wangsuhui1995@163.com (S.W.); m4a1hitman@sina.com (X.Z.); wangyi102205@sina.com (Y.W.)
* Correspondence: guotengxiao@sklnbcpc.cn (T.G.); caoshuya@sklnbcpc.cn (S.C.)

Abstract: As the core device of the miniature quantum dot (QD) spectrometer, the morphology control of the filter film array cannot be ignored. We eliminated strong interference from additives on the spectrum of a long-wave infrared (LWIR) QD filter film by selecting volatile additives. This work is significant for detecting targets by spectroscopic methods. In this work, a filter film with characteristic spectral bands located in the LWIR was obtained by the natural evaporation of QD ink, which was prepared by mixing various volatile organic solvents with HgSe QD–toluene solution. The factors affecting the morphology of HgSe LWIR films, including ink surface tension, particle size, and solute volume fraction, were the main focus of the analysis. The experimental results suggested that the film slipped in the evaporation process, and the multilayer annular deposition formed when the surface tension of the ink was no more than 24.86 mN/m. The "coffee ring" and the multilayer annular deposition essentially disappeared when the solute particles were larger than 188.11 nm. QDs in the film were accumulated, and a "gully" morphology appeared when the solute volume fraction was greater than 0.1. In addition, both the increase rate of the film height and the decrease rate of the transmission slowed down. The relationship between film height and transmission was obtained by fitting, and the curve conformed to the Lambert–Beer law. Therefore, a uniform and flat film without "coffee rings" can be prepared by adjusting the surface tension, particle size, and volume fraction. This method could provide an empirical method for the preparation of LWIR QD filter film arrays.

Keywords: HgSe QD; long-wave infrared; evaporated film; morphology

Citation: Wang, S.; Zhang, X.; Wang, Y.; Guo, T.; Cao, S. Influence of Ink Properties on the Morphology of Long-Wave Infrared HgSe Quantum Dot Films. *Nanomaterials* **2022**, *12*, 2180. https://doi.org/10.3390/nano12132180

Academic Editor: Orion Ciftja

Received: 1 June 2022
Accepted: 21 June 2022
Published: 24 June 2022

Publisher's Note: MDPI stays neutral with regard to jurisdictional claims in published maps and institutional affiliations.

Copyright: © 2022 by the authors. Licensee MDPI, Basel, Switzerland. This article is an open access article distributed under the terms and conditions of the Creative Commons Attribution (CC BY) license (https://creativecommons.org/licenses/by/4.0/).

1. Introduction

Nanomaterial inkjet printing technology is a cutting-edge technology for the micro-distribution and precise printing of ink droplets by controlling the nozzle voltage, air pressure, platform temperature, and motion trajectory, which can achieve the high-precision patterned deposition of nanomaterials. This technology has attracted extensive attention in the fields of display panel printing [1–4], microelectronic component fabrication [5–7], and flexible printing [8–10] in recent years due to the advantages of rapidity, convenience, and low cost.

As is known to all, semiconductor QDs are synthetic nanomaterials. When the size of a semiconductor quantum dot is smaller than or comparable to the exciton Bohr radius in all three directions, the electron motion is confined and forms a split energy level [11–21]. Therefore, it has many unique optoelectronic properties different from those of bulk materials, such as broad absorption spectra, narrow symmetrical emission spectra, and large Stokes shifts. Its spectrum can be tuned by adjusting various parameters, such as the synthesis time, material ratio, and core–shell structure [22,23]. In addition to the QD's unique optical properties, it also has the advantages of low cost, multiple types, and easy integration. Therefore, people have used QDs as filter materials to prepare visible-light

(Vis) and near-infrared (NIR) filter film arrays and QD micro-spectrometers [24–29]. The existing filter films and their working bands are shown in Table 1.

Table 1. Existing filter films and their working bands.

Type of QD Filter Array	Operating Range	Reference
CdS, CdSe QD filter array	Vis: 390–690 nm	[13,14]
CdS_xSe_{1-x} nanowire filter array	Vis: 500–630 nm	[15]
Perovskite QD filter array	Vis–NIR: 250–1000 nm	[16]
PbS, PbSe QD filter array	NIR: 900–1700 nm	[17,18]

Using a QD filter film array prepared by inkjet printing technology as the spectroscopic element of a micro-spectrometer is an effective method to miniaturize the spectrometer. The uniformity of the filter film directly determines the error when the detector detects the light intensity and affects the performance of the spectrometer. Therefore, it is necessary to control the film morphology. There are many factors that affect the morphology of thin films, and they are usually divided into external environmental factors and internal characteristic factors. External environmental factors mainly include ambient temperature, substrate temperature, substrate roughness, droplet size, etc., while internal characteristic factors refer to the ink characteristics, including ink surface tension, solute particle size, concentration, viscosity, etc. In terms of ink conditioning, surfactants or adhesives are usually added to regulate the surface tension of the ink solvent. The effect of capillary flow on particles can be overcome by triggering a tension gradient (Marangoni flow) [30] in droplets [31–35], so a uniform and flat film without "coffee rings" can be obtained [36]. Sometimes, increasing the particle size or changing the particle shape can reduce the effect of capillary force and weaken the "coffee ring" effect [37–39]. The solute volume fraction can also be varied to adjust the degree of sparsity or density on deposition patterns [40,41]. As mentioned above, optimizing the film morphology by adjusting the properties of the solvent has excellent effects in printing display panels and preparing visible-light or near-infrared filter films.

The characteristic spectral peak of HgSe QDs used in this study was located in the long-wave infrared at 12.5 µm [42]. The ink cannot be modified simply by adding active agents or polymers when preparing a long-wave infrared filter film. Because most surfactants or adhesives are difficult to evaporate and have strong absorption in the long-wave infrared band, the specific filtering function will not be achieved. Therefore, volatile organic solvents were used to modify the ink in this study. The effects of ink surface tension, particle agglomeration, and the solute volume fraction on the morphology of nanomaterials were investigated.

Eight kinds of evaporable organic solvents (isopropanol, n-octane, ethanol, ethyl acetate, butyl acetate, acetone, chloroform, and toluene) were used as surface tension modifiers. QD inks with different surface tensions were prepared by mixing the organic with toluene–QD solution (toluene was used to ensure the dispersion of QDs). Due to the large difference in polarity between n-octane and toluene, the agglomeration degree of QDs can be regulated by adding n-octane to the QD solution. QD inks with different agglomerated particle sizes could be obtained by mixing different proportions of n-octane and toluene–QD solution. QD inks with different solute volume fractions were prepared by mixing different proportions of toluene and toluene–QD solution. Then, 0.5 µL of the QD solution was dropped on a glass slide with a pipette, simulating the situation of ink droplets on the substrate in inkjet printing. The effects of the surface tension, particle size, and solute volume fraction of the ink solvent on the film morphology were analyzed. The fitting curve of the relationship between the solute volume fraction and transmittance was obtained. The results can provide a reference for the preparation of long-wave infrared QD filter films with specific transmittance and good morphology.

2. Materials and Methods

The materials and devices used in our experiments are as follows.

The characteristic absorption peak of the QD that we used is 12.5 µm. The QD solution was prepared by dissolving 50 mg of HgSe QDs in 1 mL of toluene. A 50 mg/mL HgSe QD–toluene solution was used as the solute, and isopropanol, n-octane, ethanol, ethyl acetate, butyl acetate, acetone, chloroform, and toluene were used as different solvents. Eight kinds of QD inks with different solvents were obtained by mixing the solute with solvents in a 1:1 volume ratio. Then, the inks were put into an ultrasonic instrument (Kunshan KQ-50B, Beijing, China) and shaken for 10 min. The surface tension of the inks was measured by an automatic surface tensiometer (Zhongchen POWEREACH, Shanghai, China) with the platinum plate method at 11.2 °C. Then, 0.5 µL of QD inks were dropped by pipette on glass slides, and the films were observed by using an optical microscope (Mingmei ML31, Guangzhou, China) after the solvent evaporated naturally.

The HgSe QD solution was used as the solute, and the solvent was prepared by mixing toluene and n-octane in volume ratios of 5:5, 4:6, 3:7, 2:8, and 1:9, followed by ultrasonic vibration for 10 min. After that, the inks were obtained by mixing the solvents and the solute with ultrasonic vibration for 10 min. The sampling amount is shown in Table 2.

Table 2. Sampling amount.

QD Solution (µL)	Toluene (µL)	N-Octane (µL)	Ratio of Toluene to N-Octane
10	40	50	5:5
10	30	60	4:6
10	20	70	3:7
10	10	80	2:8
10	0	90	1:9

The surface tension of the inks was measured by an automatic surface tensiometer. Then, 0.5 µL of QD inks was dropped by pipette on glass slides. The films were observed using an optical microscope after the solvent evaporated naturally. The area and number of the particles and the "coffee ring" width were counted and measured using the measurement mode.

The HgSe QD solution was used as the solute, and toluene was used as the solvent. Eight kinds of QD inks with volume fractions φ_μ of 0.01, 0.025, 0.05, 0.075, 0.1, 0.25, 0.5, and 0.75 were obtained by mixing different volumes of toluene with the solute (the effect of QD volume on solute volume was ignored in the calculation), followed by ultrasonic vibration for 10 min, where $\varphi_\mu = V_a/(V_a + V_b)$, V_a is the volume of QD solution, and V_b is the volume of toluene. Then, 0.5 µL of QD inks was dropped by pipette on glass slides, and the films were observed using an optical microscope and atomic force microscope after the solvent evaporated naturally. Next, 0.5 µL of QD ink was dropped on the ZnSe window using a pipette. The infrared absorption spectrum of the film was measured using a Fourier transform infrared spectrometer (Thermo iS50 FT-IR, Beijing, China) after the solvent evaporated naturally.

All substrates in the experiments were washed three times with acetone, ethanol, and distilled water sequentially and dried in a vacuum desiccator (DZF-6050, Beijing, China).

3. Results

3.1. Influence of Ink Surface Tension on FILM Morphology

The morphologies of films prepared with eight different solvent inks as observed under an optical microscope are shown in Figure 1.

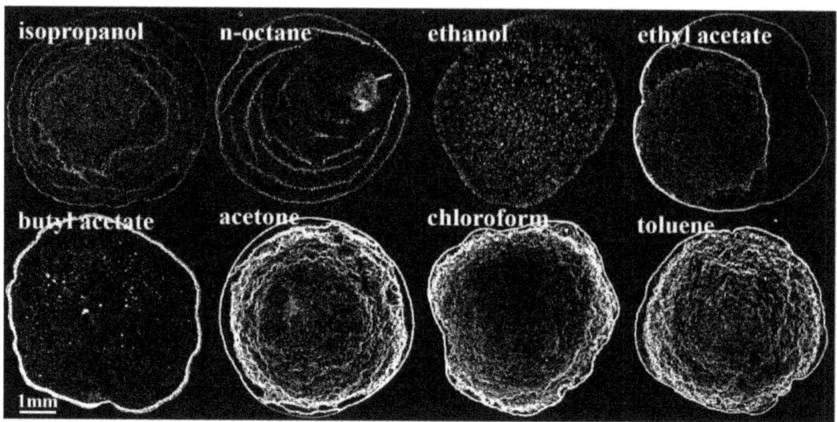

Figure 1. Morphologies of films prepared with different QD inks.

It can be seen in Figure 1 that the morphologies of the films prepared with QD inks with different solvents are different, but they can be roughly divided into three types: the first type is multilayer annular deposition (isopropanol and n-octane as solvents); the second type is nonspecific deposition (ethanol, ethyl acetate, and butyl acetate as solvents); and the third type is "gully" deposition (acetone, chloroform, and toluene as solvents). Then, the ink properties were measured and calculated to explore the reasons for film formation.

The surface tension γ_{g-l} of the ink and the droplet radius R were measured, and the contact angle θ of the droplet, the work of adhesion W_a, the work of immersion W_i, and the spreading coefficient S were calculated, as shown in Table 3.

Table 3. Calculated values of droplet property parameters.

Solvent	Isopropanol	N-Octane	Ethanol	Ethyl Acetate	Butyl Acetate	Acetone	Chloroform	Toluene
Surface tension γ_{g-l} (mN/m)	24.10	24.86	25.63	26.53	26.73	27.29	28.02	28.83
Radius R (mm)	2.33	2.17	2.16	2.11	2.07	1.98	1.79	1.62
Contact angle θ (°)	2.89	3.55	3.63	3.88	4.11	4.67	6.38	8.50
W_a (J/m^2)	48.16	49.68	51.22	52.99	53.39	54.50	55.87	57.35
W_i (J/m^2)	−24.06	−24.82	−25.58	−26.46	−26.66	−27.20	−27.85	−28.52
S	0.03	0.05	0.05	0.06	0.07	0.09	0.17	0.32

Due to the small contact angle of the droplet, it cannot be measured by a contact angle meter. However, the droplet can be regarded as a spherical cap with a volume of 0.5 µL, and then the contact angle can be calculated by measuring the droplet radius R (Figure 2a). The calculation formula is

$$V = \frac{\pi h^2}{3}(3r - h) = \frac{\pi h}{6}\left(3R^2 + h^2\right) = 0.5 \tag{1}$$

$$\theta = \arccos\frac{R^2 - h^2}{R^2 + h^2} \tag{2}$$

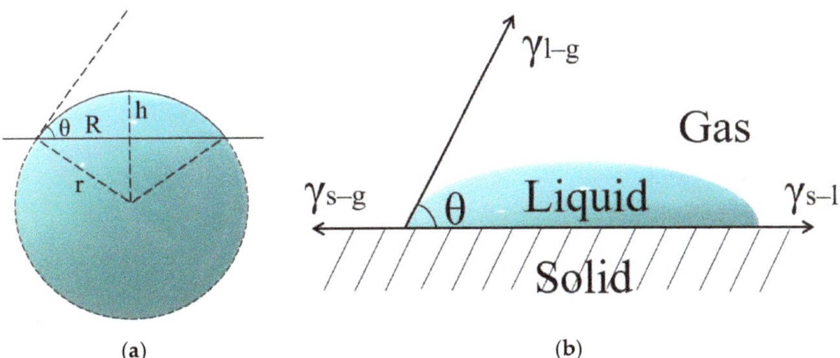

Figure 2. (**a**) Computational model of spherical cap droplet; (**b**) schematic side view of the contact between the droplet and substrate.

The liquid–solid interface wetting of droplets on the substrate can be described by Young's wetting equation [43] (Figure 2b):

$$\gamma_{s-g} = \gamma_{l-s} + \gamma_{l-g} \cos\theta \tag{3}$$

Therefore, W_a, W_i, and S can be calculated from the Gibbs free energy change values during the contact transition of the three solid–liquid–gas interfaces when the droplet is in contact with the substrate [44] ($\gamma_{g-l} = -\gamma_{l-g}$):

$$W_a = \Delta G = \gamma_{l-s} - \gamma_{l-g} - \gamma_{s-g} = -\gamma_{g-l}(1 + \cos\theta) \tag{4}$$

$$W_i = \Delta G = \gamma_{l-s} - \gamma_{g-s} = -\gamma_{g-l}\cos\theta \tag{5}$$

$$S = -\Delta G = \gamma_{g-s} - \gamma_{g-l} - \gamma_{l-s} = \gamma_{g-l}(\cos\theta - 1) \tag{6}$$

As can be seen in Table 3, the relationship between solvents for the parameters γ_{g-l}, θ, W_a, and S was isopropanol < n-octane < ethanol < ethyl acetate < butyl acetate < acetone < chloroform < toluene. The relationship between solvents for the parameters R and W_i was isopropanol > n-octane > ethanol > ethyl acetate > butyl acetate > acetone > chloroform > toluene.

The adhesion of droplets to the substrate increased with the increase in $|W_a|$. The droplets can wet the substrate when $W_i \leq 0$, and the wetting ability decreased with the increase in $|W_i|$. The liquid can spread automatically on the substrate when $S \geq 0$, and the spreading ability decreased with the increase in S.

Among the above inks, the θ of inks with isopropanol and n-octane as solvents was no more than 24.86 mN/m. The droplets had weak adhesion and strong wetting and spreading abilities on the substrate due to the small W_a, and the phenomenon of multilayer ring deposition was more likely to occur. In contrast, the θ and W_a of the inks with toluene and chloroform as solvents were higher. Thus, the droplets had stronger adhesion and weaker wetting and spreading abilities on the substrate with no multilayer annular deposition. Therefore, the film-forming property and uniformity can be improved by appropriately increasing the surface tension of the ink.

3.2. Effect of Particle Size on Film Morphology

It was found that the content of the organic solvent in the ink can affect the agglomeration degree of QDs, which in turn affects the morphology of the film. Due to the large difference in polarity, it will cause obvious agglomeration with the addition of n-octane to the QD–toluene solution. Different agglomerated particles can be obtained by adjusting the ratio between n-octane and toluene. Therefore, inks with volume ratios of toluene to n-octane of 5:5, 4:6, 3:7, 2:8, and 1:9 were prepared. The surface tension, particle size, and "coffee ring" width of the films were measured. The film morphology under the microscope is shown in Figure 3. The particle size distribution in the film is shown in Figure 4. The ink surface tension, film particle size, and "coffee ring" width are shown in Table 4.

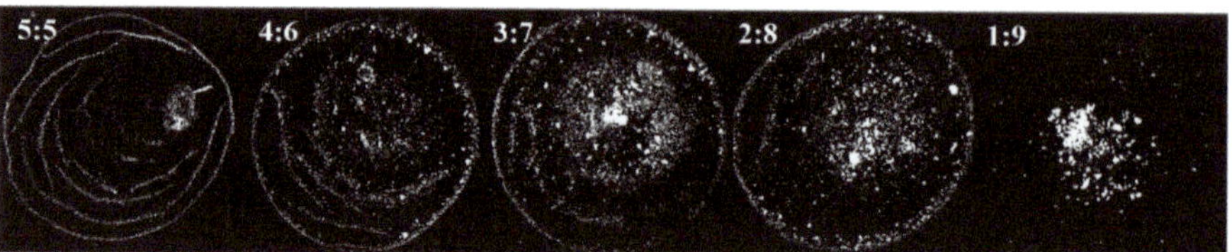

Figure 3. The morphologies of films prepared with different QD inks. The volume ratios of toluene to n-octane in the inks are 5:5, 4:6, 3:7, 2:8, and 1:9.

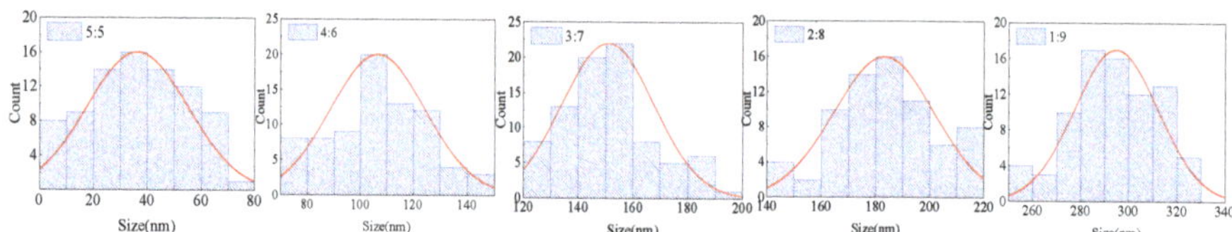

Figure 4. Particle size distribution diagram. The volume ratios of toluene and n-octane in the inks are 5:5, 4:6, 3:7, 2:8, and 1:9.

Table 4. Ink surface tension, particle size, and "coffee ring" width.

Volume Ratio of Toluene to N-Octane	5:5	4:6	3:7	2:8	1:9
Surface tension γ (mN/m)	24.86	24.49	24.36	23.15	22.85
Size (nm)	38.79	106.31	151.78	183.27	294.89
Coffee ring width (µm)	31.41	38.76	50.42	53.23	-

It can be seen in Table 4 that the surface tension of the ink and the dispersion ability of QDs decreased with the increase in n-octane content. The particle size of the QDs increased, and the "coffee ring" became wider due to agglomeration. It can be seen from the discussion in Section 3.1 that the smaller the surface tension of the ink, the more likely the film has the morphology of multilayer ring deposition. However, the number of "coffee rings" in Figure 3 decreases as the surface tension decreases. This was because the dispersion ability of QDs decreased as the n-octane content increased. It was difficult for the capillary flow in the droplet to push the large particles toward the contact line due to agglomeration. The liquid film evaporated to dryness before the large particles reached the contact line, so the "coffee ring" widened. When the agglomerated particle size of QDs was equal to 188.11 nm, the "coffee ring" and multilayer ring deposition essentially disappeared. When the particle size was equal to 303.89 nm, the large-size particles were primarily concentrated in the center of the film. The "coffee ring" and multilayer annular deposition disappeared completely. Therefore, the film-forming property and uniformity can be improved by appropriately increasing the size of the particle.

3.3. Effect of Solute Volume Fraction on Film Morphology

It was found that toluene had the best dispersing effect on QDs. Therefore, toluene was used as the solvent, and QD inks with solute volume fractions of 0.025, 0.05, 0.075, 0.1, 0.25, 0.5, and 0.75 were prepared. The thin films were obtained by evaporation. The morphologies of QD films were observed with an optical microscope, as shown in Figure 5a.

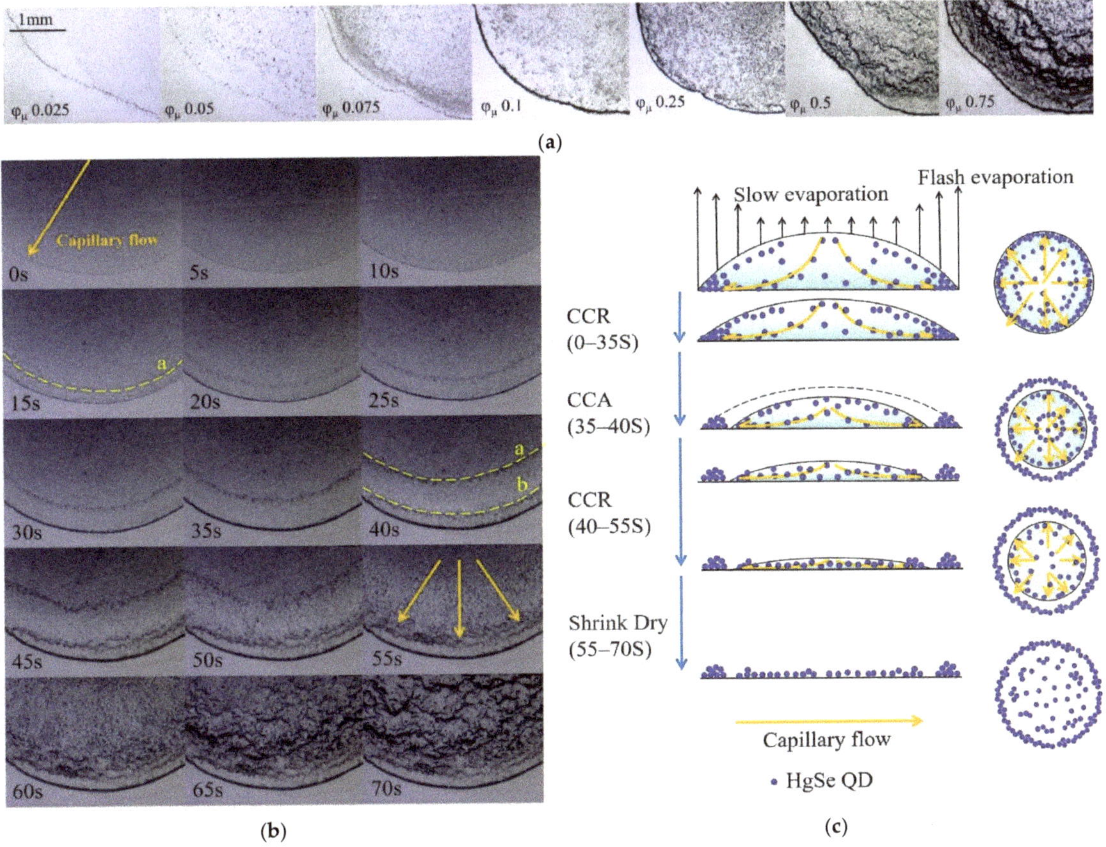

Figure 5. (**a**) The morphologies of films prepared by inks with solute volume fractions of 0.025, 0.05, 0.075, 0.1, 0.25, 0.5, and 0.75 under the microscope. (**b**) The evaporation process when the solute volume fraction was 0.5. (**c**) The schematic side and top views of the liquid film evaporation process when the solute volume fraction was 0.5.

The experimental results show that the solute volume fraction was not the factor that determined the formation of the "coffee ring". This was only determined by the characteristic of the solvent. Since the solute cannot unpin the contact line and redirect the flow, the evaporation rate of the edge of the liquid film was greater than that of the center when the solvent was constant. In order to keep the contact line pinned, there must be a continuous, radially outward capillary flow from the center to the contact line to compensate for the evaporative removal of the liquid, eventually forming a "coffee ring".

When the solute volume fraction was less than 0.1, some QDs were deposited on the glass slide before they moved to the edge due to the flash evaporation rate of the solvent and finally formed a uniform QD film. When the solute volume fraction was greater than 0.1 (for example, φ_μ of 0.5), the film evaporation process was more complicated and formed a gully-like morphology, as shown in Figure 5b. The schematic side and top views of the liquid film evaporation process are shown in Figure 5c.

As can be seen in Figure 5b, when the ink contacts the substrate, it forms a liquid film. Then, the three-phase contact line is pinned at once. The QDs in the liquid film began to move to the three-phase contact line. At this time, the evaporation mode was the constant contact radius model (CCR). When the evaporation proceeded for 15 s, the reverse "coffee ring" 'a' appeared, gradually widened, and moved toward the center of the liquid film. At the same time, some QDs in ring 'a' diffused toward the "coffee ring" under the action of capillary force. The "coffee ring" continued to widen. When the evaporation proceeded for 40 s, the liquid film was released from the pinned "coffee ring", and the short-term constant contact angle (CCA model) evaporation mode occurred. The pinned ring 'b' was the new three-phase contact line, and the film continued to evaporate in CCR mode. As the liquid film gradually became thinner, the temperature difference between the edge and center of the liquid film became smaller, and the moving speed of ring 'a' to the center slowed down. When the evaporation progressed to around 70% (at 50 s), ring 'a' was fixed and flushed out within 5 s. At the same time, the stably distributed QDs in the center of the liquid film also began to move rapidly to the edge. The liquid film was too thin to be fixed after 15 s, so it shrunk rapidly toward the center and evaporated to dryness. QDs piled up on the edges, eventually forming a "gully" morphology of varying depths. Therefore, in order to avoid the appearance of the "gully" and obtain a more uniform and flat QD film, the volume fraction of the ink solute should not be greater than 0.1.

3.4. Infrared Transmittance Analysis of Thin Films

In order to analyze the relationship between film morphology and film transmittance and to provide an empirical method for the subsequent preparation of long-wave infrared QD films, the film morphology was characterized by atomic force microscopy. The 3D surface topography was recorded using a Nanosurf Flex-Axiom atomic force microscope (Nanosurf, AG) in soft tapping mode with a scan speed of 6.25 μm/s to obtain 104 $\times$ 104-pixel images. The experiments were carried out at room temperature (297 $\pm$ 1 K) using cantilevers with the following nominal properties for force–distance curve measurements: a length of 125 μm, a width of 25 μm, a thickness of 2.1 μm, a tip radius of 10 nm, a force constant of 5 N/m, and a resonance frequency of 150 kHz, as shown in Figure 6.

It can be seen in Figure 6a,b that the QDs are distributed in islands on the substrate. The QDs became denser and higher with the increase in the solute volume fraction. It can be seen in Figure 6c that the shape of the QDs appears broader, and the cross-sectional diameter became larger with the further increase in the solute volume fraction.

The arithmetic mean heights (Sa) of films with solute volume fractions of 0.025, 0.05, 0.075, 0.1, 0.25, 0.5, and 0.75 were 53.90 nm, 55.25 nm, 59.23 nm, 61.83 nm, 66.13 nm, 66.82 nm, ands 72.25 nm, respectively, as shown in Figure 7a. The transmissions of films with solute volume fractions of 0.025, 0.05, 0.075, 0.1, 0.25, 0.5, and 0.75 were 88.65%, 81.27%, 61.91%, 59.14%, 45.36%, 37.38%, and 33.85%, respectively, as shown in Figure 7b. The fitting curve of the film height and transmission is shown in Figure 7c.

It can be seen in Figure 7c that the increase rate of the height of the film and the decrease rate of the transmission at the characteristic peak became slower when the solute volume fraction was 0.1. There was a linear relationship between the height and transmission, which conformed to the Lambert–Beer law. This result can provide an important reference for the preparation of thin films with specific transmission.

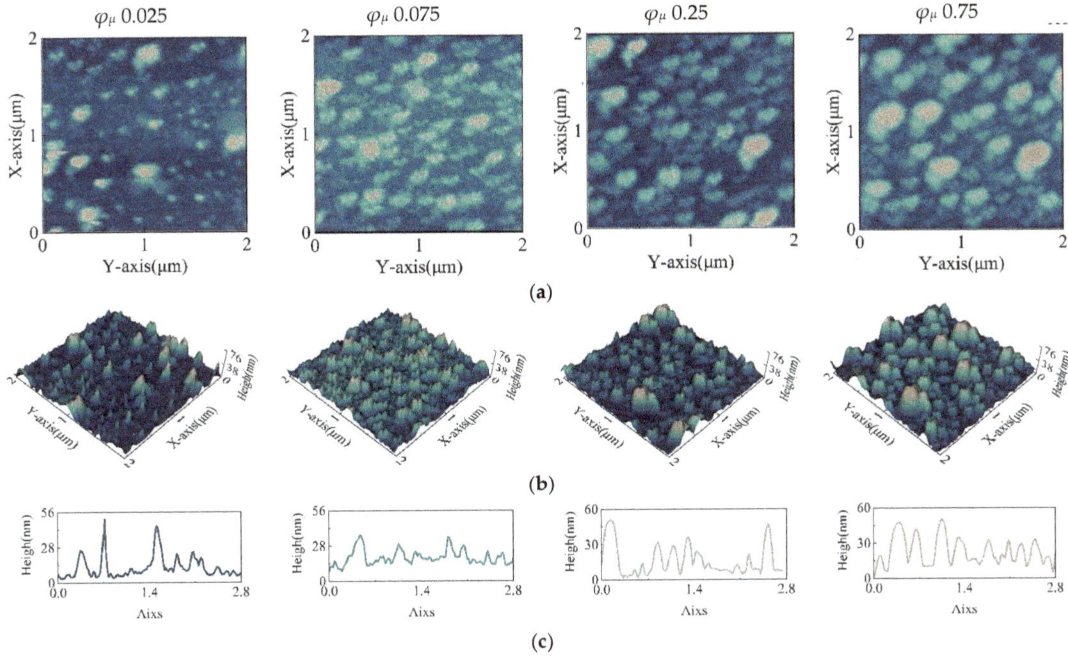

Figure 6. (**a**) Contour map of the film. (**b**) Three-dimensional images of the film under AFM. (**c**) Sectional view of the film's diagonal. (The volume fractions of ink solute are 0.025, 0.075, 0.25, and 0.75.).

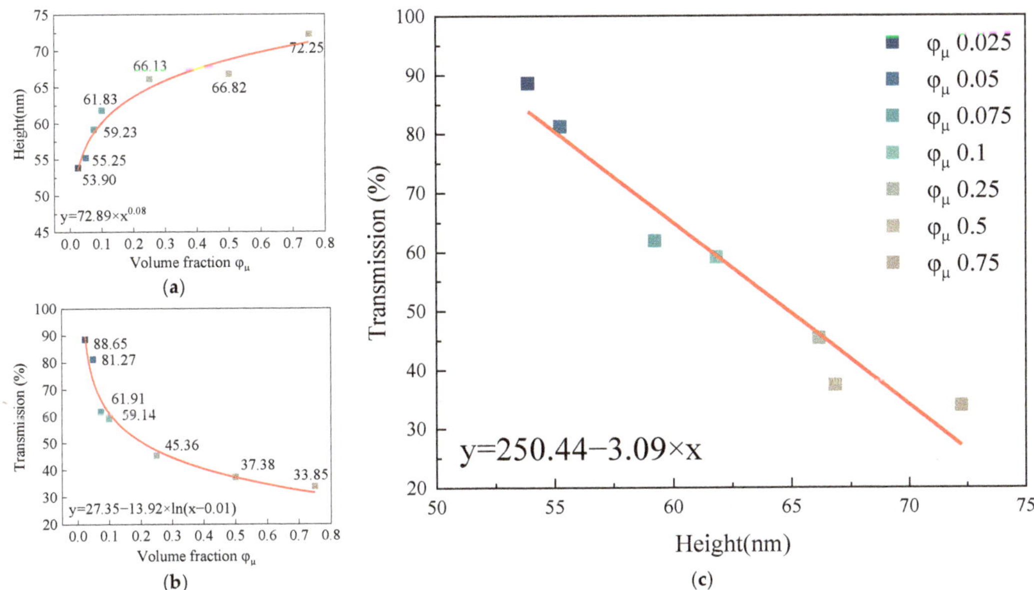

Figure 7. (**a**) Transmittance of films prepared from QD inks with different solute volume fractions. (**b**) Height of films prepared from QD inks with different solute volume fractions. (**c**) The fitting function of solute volume fraction and film transmittance.

4. Conclusions

HgSe QD inks with characteristic spectral bands located in the long-wave infrared were prepared by mixing various organic solvents with a toluene solution of QDs. Among them, QD inks with different tensions were first obtained by mixing eight kinds of organic solvents (isopropanol, n-octane, ethanol, ethyl acetate, butyl acetate, acetone, chloroform, and toluene) with the QD–toluene solution. Secondly, due to the large difference in polarity, QD inks with different agglomerated particle sizes were obtained by mixing different proportions of n-octane and QD–toluene solution. QD inks with different solute volume fractions were then prepared by mixing different proportions of toluene and toluene–QD solution. Finally, films with different morphologies were obtained by naturally evaporating QD ink droplets on the substrate. The effects of the surface tension, particle size, and volume fraction on the film morphology were emphasized in the analysis. After that, the infrared transmission spectra of the films were measured. The experimental results suggest that the film slipped in the evaporation process, and the multilayer annular deposition formed when the surface tension of the ink was no more than 24.86 mN/m. The "coffee ring" and the multilayer annular deposition essentially disappeared when the solute particles were larger than 188.11 nm. When the solute volume fraction was greater than 0.1, the QDs in the film were accumulated, and a "gully" morphology appeared. In addition, the increase rate of the film height and the decrease rate of transmission slowed down. The relationship between the film height and transmission was fitted, and the curve conformed to the Lambert–Beer law. Therefore, the morphology of the film can be improved by adjusting the surface tension of the film, the particle size of the solute, and the volume fraction of the solute. Therefore, a uniform and flat film without "coffee rings" can be prepared by adjusting the surface tension, particle size, and volume fraction. This approach could provide an empirical method for the preparation of LWIR QD filter film arrays. It was also found that the evaporation rate, temperature or type of substrate, and shape of solute particles also affected the film morphology in the experiment. The above factors can be discussed and analyzed in detail in subsequent research. In addition, agglomeration easily occurs due to the large specific surface area of nanoparticles. Therefore, it is also important to modify and passivate the surface to avoid agglomeration when synthesizing nanomaterials, and QD filter films with good morphology can be prepared by improving the ink uniformity.

Author Contributions: S.W.: methodology, validation, investigation, and writing—original draft; X.Z.: methodology; Y.W.: AFM test and writing—review; T.G.: writing—review and editing; S.C.: supervision and writing—review and editing. All authors have read and agreed to the published version of the manuscript.

Funding: This work is funded by the National Key R&D Program of China (2021YFC3330201).

Data Availability Statement: Not applicable.

Acknowledgments: Thanks to Yi Wang for technical support in AFM testing. Thanks to Yong Pan for the theoretical support for calculations.

Conflicts of Interest: The authors declare no conflict of interest.

References

1. Han, J.; Ko, D.; Park, M.; Roh, J.; Jung, H.; Lee, Y.; Kwon, Y.; Sohn, J.; Bae, W.K.; Chin, B.D.; et al. Toward high-resolution, inkjet-printed, quantum dot light-emitting diodes for next-generation displays. *J. Soc. Inf. Disp.* **2016**, *24*, 545–551. [CrossRef]
2. Haverinen, H.M.; Myllyla, R.A.; Jabbour, G.E. Inkjet printed RGB quantum dot-hybrid LED. *J. Disp. Technol.* **2010**, *6*, 87–89. [CrossRef]
3. Hu, Z.; Yin, Y.; Ali, M.U.; Peng, W.; Zhang, S.; Li, D.; Zou, T.; Li, Y.; Jiao, S.; Chen, S.-J.; et al. Inkjet printed uniform quantum dots as color conversion layers for full-color OLED displays. *Nanoscale* **2020**, *12*, 2103–2110. [CrossRef] [PubMed]
4. Roh, H.; Ko, D.; Shin, D.Y.; Chang, J.H.; Hahm, D.; Baet, W.K.; Lee, C.; Kim, J.Y.; Kwalk, J. Enhanced performance of pixelated quantum dot light-emitting diodes by inkjet printing of quantum dot–polymer composites. *Adv. Opt. Mater.* **2021**, *9*. [CrossRef]
5. Burns, S.E.; Cain, P.; Mills, J.; Wang, J.; Sirringhaus, H. Inkjet printing of polymer thin-film transistor circuits. *MRS Bull.* **2011**, *28*, 829–834. [CrossRef]

6.   Godard, N.; Glinsek, S.; Defay, E. Inkjet-printed silver as alternative top electrode for lead zirconate titanate thin films. *J. Alloy Compd.* **2019**, *783*, 801–805. [CrossRef]

7.   Rahumi, O.; Sobolev, A.; Rath, M.K.; Borodianskiy, K. Nanostructured engineering of nickel cermet anode for solid oxide fuel cell using inkjet printing. *J. Eur. Ceram. Soc.* **2021**, *41*, 4528–4536. [CrossRef]

8.   Secor, E.B.; Prabhumirashi, P.L.; Puntambekar, K.; Geier, M.L.; Hersam, M.C. Inkjet printing of high conductivity, flexible graphene patterns. *J. Phys. Chem. Lett.* **2013**, *4*, 1347–1351. [CrossRef]

9.   Wang, C.T.; Huang, K.Y.; Lin, D.T.; Liao, W.C.; Lin, H.W.; Hu, Y.C. A flexible proximity sensor fully fabricated by inkjet printing. *Sensors* **2010**, *10*, 5054–5062. [CrossRef]

10.   Yin, Z.; Huang, Y.; Bu, N.; Wang, X.; Xiong, Y. Inkjet printing for flexible electronics: Materials, processes and equipments. *Chin. Sci. Bull.* **2010**, *55*, 3383–3407. [CrossRef]

11.   Bleuel, M.; Carpenter, J.M.; Micklich, B.J.; Geltenbort, P.; Mishima, K.; Shimizu, H.M.; Iwashita, Y.; Hirota, K.; Hino, M.; Kennedy, S.J.; et al. A small angle neutron scattering (SANS) experiment using very cold neutrons (VCN). *Phys. B Condens. Matter* **2009**, *404*, 2629–2632. [CrossRef]

12.   Ciftja, O. Generalized description of few-electron quantum dots at zero and nonzero magnetic fields. *J. Phys. Condens. Matter* **2007**, *19*, 046220. [CrossRef]

13.   Ciftja, O.; Faruk, M.G. Two-dimensional quantum-dot helium in a magnetic field: Variational theory. *Phys. Rev. B* **2005**, *72*, 1–10. [CrossRef]

14.   Ciftja, O. Understanding electronic systems in semiconductor quantum dots. *Phys. Scr.* **2013**, *88*, 8302. [CrossRef]

15.   Ezaki, T.; Mori, N.; Hamaguchi, C. Electronic structures in circular, elliptic, and triangular quantum dots. *Phys. Rev. B* **1997**, *56*, 6428–6431. [CrossRef]

16.   Harju, A.; Sverdlov, V.; Barbiellini, B.; Nieminen, R.M. Variational wave function for a two-electron quantum dot. *Phys. B Condens. Matter* **1998**, *255*, 145–149. [CrossRef]

17.   Wagner, M.; Merkt, U.; Chaplik, A.V. Spin-singlet–spin-triplet oscillations in quantum dots. *Phys. Rev. B Condens. Matter* **1992**, *45*, 1951–1954. [CrossRef]

18.   Merkt, U.; Huser, J.; Wagner, M.J. Energy spectra of two electrons in a harmonic quantum dot. *Phys. Rev. B* **1991**, *43*, 7320–7323. [CrossRef]

19.   Pfannkuche, D.; Gerhardts, R.R. Quantum-dot helium: Effects of deviations from a parabolic confinement potentia. *Phys. Rev. B* **1991**, *44*, 13132–13135. [CrossRef]

20.   Pfannkuche, D.; Gerhardts, R.R.; Maksym, P.A.; Gudmundsson, V. Theory of quantum dot helium. *Phys. B Condens. Matter* **1993**, *189*, 6–15. [CrossRef]

21.   Pfannkuche, D.; Gerhardts, R.R.; Maksym, P.A. Comparison of a Hartree, a Hartree-Fock, and an exact treatment of quantum-dot helium. *Phys. Rev. B Condens. Matter* **1993**, *47*, 2244–2250. [CrossRef] [PubMed]

22.   Cheng, C.; Cheng, X. *Quantum Dot Nanophotonics and Applications*; Science Press: Beijing, China, 2018.

23.   Kang, Z.; Liu, Y.; Mao, B. *Synthesis and Application of Quantum Dots*; Science Press: Beijing, China, 2018.

24.   Bao, J.; Bawendi, M.G. A colloidal quantum dot spectrometer. *Nature* **2015**, *523*, 67–70. [CrossRef] [PubMed]

25.   Khan, S.A.; Ellerbee Bowden, A.K. Colloidal quantum dots for cost-effective, miniaturized, and simple spectrometers. *Clin. Chem.* **2016**, *62*, 548–550. [CrossRef] [PubMed]

26.   Yang, Z.; Albrow-Owen, T.; Cui, H.; Alexander-Webber, J.; Gu, F.; Wang, X.; Wu, T.-C.; Zhuge, M.; Williams, C.; Wang, P.; et al. Single-nanowire spectrometers. *Science* **2019**, *365*, 1017–1020. [CrossRef]

27.   Zhu, X.; Bian, L.; Fu, H.; Wang, L.; Zou, B.; Dai, Q.; Zhang, J.; Zhong, H. Broadband perovskite quantum dot spectrometer beyond human visual resolution. *Light Sci. Appl.* **2020**, *9*, 73. [CrossRef]

28.   Li, H.; Bian, L.; Gu, K.; Fu, H.; Yang, G.; Zhong, H.; Zhang, J. A Near-infrared miniature quantum dot spectrometer. *Adv. Opt. Mater.* **2021**, *9*, 1–8. [CrossRef]

29.   Wang, S.; Zhang, X.; Sun, Z.; Yang, J.; Guo, T.; Ding, X. Methods of detection multiple chemical substances based on near-infrared colloidal quantum dot arry and spectral restruction algorithm. *Spectrosc. Spectr. Anal.* **2021**, *41*, 3370–3376.

30.   Harris, D.J.; Lewis, J.A. Marangoni effects on evaporative lithographic patterning of colloidal films. *Langmuir* **2008**, *24*, 3681–3685. [CrossRef]

31.   Bhardwaj, R.; Fang, X.; Attinger, D. Pattern formation during the evaporation of a colloidal nanoliter drop: A numerical and experimental study. *New J. Phys.* **2009**, *11*, 075020. [CrossRef]

32.   de Gans, B.J.; Schubert, U.S. Inkjet printing of well-defined polymer dots and arrays. *Langmuir* **2004**, *20*, 7789–7793. [CrossRef]

33.   Pesach, D.; Marmur, A. Marangoni effects in the spreading of liquid mixtures on a solid. *Langmuir* **1986**, *3*, 514–519. [CrossRef]

34.   Park, J.; Moon, J. Control of colloidal particle deposit patterns within picoliter droplets ejected by ink-jet printing. *Langmuir* **2006**, *22*, 3506–3513. [CrossRef] [PubMed]

35.   Still, T.; Yunker, P.J.; Yodh, A.G. Surfactant-induced marangoni eddies alter the coffee-rings of evaporating colloidal drops. *Langmuir* **2012**, *28*, 4984–4988. [CrossRef]

36.   Deegan, R.D.; Bakajin, O.; Dupont, T.F.; Huber, G.; Nagel, S.R.; Witten, T.A. Capillary flow as the cause of ring stains fromdried liquid drops. *Nature* **1997**, *389*, 827–829. [CrossRef]

37.   Bansal, L.; Seth, P.; Murugappan, B.; Basu, S. Suppression of coffee ring: (Particle) size matters. *Appl. Phys. Lett.* **2018**, *112*, 211605. [CrossRef]

38. Kim, D.O.; Pack, M.; Hu, H.; Kim, H.; Sun, Y. Deposition of colloidal drops containing ellipsoidal particles: Competition between capillary and hydrodynamic forces. *Langmuir* **2016**, *32*, 11899–11906. [CrossRef]
39. Shen, X.; Ho, C.M.; Wong, T.S. Minimal size of coffee ring structure. *J. Phys. Chem. B* **2010**, *114*, 5269–5274. [CrossRef]
40. Saroj, S.K.; Panigrahi Kumar, P. Drying pattern and evaporation dynamics of sessile ferrofluid droplet on a PDMS substrate. *Colloids Surf. A Physicochem. Eng. Asp.* **2019**, *580*, 123672. [CrossRef]
41. Ahmad, I.; Hussain, A.; Ali, A.; Jan, R. Energetically and sterically favorable assembly of identical gold nanorods with varying concentrations of nanospheres. *Surf. Rev. Lett.* **2021**, *28*, 2150028. [CrossRef]
42. Lhuillier, E.; Scarafagio, M.; Hease, P.; Nadal, B.; Aubin, H.; Xu, X.Z.; Lequeux, N.; Patriarche, G.; Ithurria, S.; Dubertret, B. Infrared photodetection based on colloidal quantum-dot films with high mobility and optical absorption up to THz. *Nano Lett.* **2016**, *16*, 1282–1286. [CrossRef]
43. Yang, T. An essay on the cohesion of fluids. *Philos. Trans. R. Soc.* **1805**, *95*, 65.
44. Fu, X.; Shen, W.; Yao, T.; Hou, W. *Physical Chemistry*; Higher Education Press: Beijing, China, 2006.

 *nanomaterials*

MDPI

Article

Effect of TiO₂ Film Thickness on the Stability of Au₉ Clusters with a CrOₓ Layer

Abdulrahman S. Alotabi [1,2,3], Yanting Yin [1,3], Ahmad Redaa [4,5], Siriluck Tesana [6,7], Gregory F. Metha [8,*,†] and Gunther G. Andersson [1,3,*,†]

1   Flinders Institute for Nanoscale Science and Technology, Flinders University, Adelaide, SA 5042, Australia
2   Department of Physics, Faculty of Science and Arts in Baljurashi, Albaha University, Baljurashi 65655, Saudi Arabia
3   Flinders Microscopy and Microanalysis, College of Science and Engineering, Flinders University, Adelaide, SA 5042, Australia
4   Department of Earth Sciences, University of Adelaide, Adelaide, SA 5005, Australia
5   Faculty of Earth Sciences, King Abdulaziz University, Jeddah 21589, Saudi Arabia
6   The MacDiarmid Institute for Advanced Materials and Nanotechnology, School of Physical and Chemical Sciences, University of Canterbury, Christchurch 8041, New Zealand
7   National Isotope Centre, GNS Science, Lower Hutt 5010, New Zealand
8   Department of Chemistry, University of Adelaide, Adelaide, SA 5005, Australia
*   Correspondence: greg.metha@adelaide.edu.au (G.F.M.); gunther.andersson@flinders.edu.au (G.G.A.)
†   Current address: Physical Sciences Building, (2111) GPO Box 2100, Adelaide, SA 5001, Australia.

Citation: Alotabi, A.S.; Yin, Y.; Redaa, A.; Tesana, S.; Metha, G.F.; Andersson, G.G. Effect of TiO₂ Film Thickness on the Stability of Au₉ Clusters with a CrOₓ Layer. *Nanomaterials* **2022**, *12*, 3218. https://doi.org/10.3390/nano12183218

Academic Editor: Orion Ciftja

Received: 17 August 2022
Accepted: 9 September 2022
Published: 16 September 2022

Publisher's Note: MDPI stays neutral with regard to jurisdictional claims in published maps and institutional affiliations.

Copyright: © 2022 by the authors. Licensee MDPI, Basel, Switzerland. This article is an open access article distributed under the terms and conditions of the Creative Commons Attribution (CC BY) license (https://creativecommons.org/licenses/by/4.0/).

Abstract: Radio frequency (RF) magnetron sputtering allows the fabrication of TiO₂ films with high purity, reliable control of film thickness, and uniform morphology. In the present study, the change in surface roughness upon heating two different thicknesses of RF sputter-deposited TiO₂ films was investigated. As a measure of the process of the change in surface morphology, chemically -synthesised phosphine-protected Au₉ clusters covered by a photodeposited CrOₓ layer were used as a probe. Subsequent to the deposition of the Au₉ clusters and the CrOₓ layer, samples were heated to 200 °C to remove the triphenylphosphine ligands from the Au₉ cluster. After heating, the thick TiO₂ film was found to be mobile, in contrast to the thin TiO₂ film. The influence of the mobility of the TiO₂ films on the Au₉ clusters was investigated with X-ray photoelectron spectroscopy. It was found that the high mobility of the thick TiO₂ film after heating leads to a significant agglomeration of the Au₉ clusters, even when protected by the CrOₓ layer. The thin TiO₂ film has a much lower mobility when being heated, resulting in only minor agglomeration of the Au₉ clusters covered with the CrOₓ layer.

Keywords: RF magnetron sputtering; TiO₂ film; morphology; triphenylphosphine; Au₉; gold clusters; photodeposition; CrOₓ; Cr(OH)₃; Cr₂O₃ layer

1. Introduction

Titanium dioxide (TiO₂) is a semiconductor widely used for a large range of photocatalytic applications and is also an ideal model system for various types of studies [1,2]. There are various techniques to prepare TiO₂ films, such as sol-gel [3], evaporation [4], chemical vapour deposition [5], atomic layer deposition [6] and radio frequency (RF) magnetron sputtering [7]. Each of these methods has advantages and disadvantages in regard to fabrication costs, uniformity of the film morphology, thermal stability, purity and preparation time. Therefore, the best method of choice for TiO₂ film preparation depends on which application the film will be used in.

Amongst the above-named methods, RF magnetron sputtering is known to produce high-purity TiO₂ films with uniform thickness, ease of use and strong film adhesion to the substrate [8]. The properties of these films are significantly impacted by the sputtering conditions, such as RF power, gas pressure, substrate type, substrate temperature and target-to-substrate distance [9–14]. For instance, it has been reported that control of TiO_2 film thickness is possible by modulating the deposition time and the gas sputtering pressure [15].

TiO_2 films prepared with the RF magnetron sputtering method can be amorphous or have a rutile, anatase, or brookite crystal structure. It is well known that the physical properties of TiO_2 films depend highly on the post-deposition treatment, including heat treatment conditions [16–18]. Çörekçi et al. reported that a correlation between heating treatment and surface morphology with different TiO_2 film thicknesses. It was observed that an increase in surface roughness and grain sizes occurred during heating depending on TiO_2 film thicknesses, which also increased with film thickness. This is because increasing temperatures transform TiO_2 from amorphous to anatase and then to rutile [17], and these phase transitions affect the surface morphology of the TiO_2 film, which includes the roughness and crystallinity of the surface [19].

The aim of this study is to investigate the influence of heat treatment on the surface morphology of RF sputter-deposited TiO_2 films with two different thicknesses, and the effect this has on size-specific Au clusters deposited on the surface. TiO_2 films have attracted interest as substrates for investigating the role of Au clusters as cocatalysts in photocatalysis [20,21]. In these studies, TiO_2 films had been heated as part of the sample preparation procedure. The change in morphology, including surface mobility, upon heating, can lead to agglomeration of the Au clusters. Understanding the change in surface morphology upon heating, thus, is important when using TiO_2 as a substrate for investigating the cocatalyst properties. In the present work, phosphine-protected Au_9 clusters covered by a photodeposited CrO_x layer were used as probes for the TiO_2 mobility during the change of morphology upon heating. Scanning electron microscopy (SEM), X-ray diffraction (XRD), laser scanning confocal microscope (LSCM) and X-ray photoelectron spectroscopy (XPS), have been applied to characterise the thickness, crystal structure, surface morphology and chemical composition and size of the Au cluster. The importance of the present work is to show that morphology changes in RF sputter-deposited TiO_2 depend on the thickness of the TiO_2 layer, and that Au_9 clusters can be used to probe morphology changes in the surface.

2. Experimental Methods and Techniques

2.1. Material and Sample Preparation

2.1.1. Preparation of TiO_2 Films

The RF magnetron sputtering method was used to prepare TiO_2 films on a silicon wafer under high vacuum conditions (HHV/Edwards TF500 sputter coater) [22]. Before the deposition, the silicon wafer was cleaned with ethanol and acetone and then dried in a stream of dry nitrogen. The TiO_2 film was deposited onto a p-type silicon wafer substrate by sputtering a 99.9% pure TiO_2 ceramic target with 500 W sputtering power using Ar^+ (flow rate of 5 sccm) for 50 min. The sputter coating chamber was held under vacuum at 2×10^{-5} mbar. This process resulted in TiO_2 films formed on the silicon wafer with a native oxide layer of TiO_2.

TiO_2 films with two different thickness were fabricated applying the above-described procedure. The TiO_2 films had different colours based on light interference [23]: a TiO_2/Si wafer with a purple colour and a TiO_2/Si wafer with a gold-like colour (see Figure S1). The difference in colour of the films is related to the difference in light interference patterns within the films due to their difference in film thickness [24]. The thickness of TiO_2P is ~400 nm, while TiO_2G is ~1100 nm (*vide infra*). The TiO_2 wafers were cut into 1 cm × 1 cm pieces and used without further treatment. The two TiO_2 wafers are hereafter referred to as (i) TiO_2P and (ii) TiO_2G.

2.1.2. Deposition of Au_9 Clusters

The deposition procedure of $Au_9(PPh_3)_8(NO_3)_3$ (Au_9) was identical for both the TiO_2P and TiO_2G samples. Phosphine-protected Au_9 clusters were synthesised as reported previously [25]. A UV-Vis spectrum of the Au_9 cluster is shown in Figure S2. The TiO_2 films were immersed in Au_9 methanol solutions (2 mL) for 30 min at concentrations of 0.006, 0.06 and 0.6 mM. The TiO_2 samples were rinsed by quickly dipping them into pure methanol and then dried in a stream of dry nitrogen. These samples are hereafter referred to as (i) TiO_2P-Au_9 and (ii) TiO_2G-Au_9.

2.1.3. Photodeposition of CrO_x Layer

Photodeposition of the CrO_x layer was the same for both TiO_2-Au_9 samples (TiO_2P-Au_9 and TiO_2G-Au_9). A 0.5 mM potassium chromate solution was prepared by dissolving K_2CrO_4 ($\geq$99%, Sigma-Aldrich) in deionised water. The TiO_2-Au_9 samples were immersed into the K_2CrO_4 solution (1 mL) and irradiated for 1 h using a UV LED (Vishay, VLMU3510-365-130) with ~1 cm between the sample and the irradiation source. The UV LED had a radiant power of 690 mW at 365 nm wavelength. After photodeposition, the samples were washed by dipping them into deionised water and dried in a stream of dry nitrogen [26]. These samples are hereafter referred to as (i) TiO_2P-Au_9-CrO_x and (ii) TiO_2G-Au_9-CrO_x.

2.1.4. Heat Treatment

To remove the phosphine ligands from Au_9 clusters, all samples were treated with heating at 200 °C for 10 min under ultra-high vacuum (1×10^{-8} mbar) in the XPS chamber.

2.2. Characterization Methods

2.2.1. Scanning Electron Microscopy and Energy Dispersive X-ray Spectroscopy (SEM-EDAX)

The thickness of TiO_2 films (TiO_2P and TiO_2G) was determined by combining SEM imaging and SEM-EDAX (FEI Inspect F50 microscope) scans on cross-sections of the TiO_2 samples. Cross-sectional images were recorded at a magnification of up to 100 k with 15 keV electron energy.

2.2.2. X-ray Diffraction (XRD)

The crystal and phase structure of the TiO_2 films (TiO_2P and TiO_2G) before and after heating were analysed using XRD. A Bruker D8 Advance apparatus was used to record the XRD patterns with an irradiation source of Co-Kα (λ = 1.789 Å) operating at 35 kV and 28 mA.

2.2.3. Laser Scanning Confocal Microscope (LSCM)

The surface morphology of TiO_2 films (TiO_2P and TiO_2G) was measured using a LSCM (Olympus LEXT OLS5000-SAF 3D LSCM) with 100x/0.80NA and 50x/0.60NA LEXT objective lenses. The Olympus Data Analysis software was used to calculate the arithmetic mean deviation (Ra) and root mean square deviation (Rq) values.

2.2.4. X-ray Photoelectron Spectroscopy (XPS)

XPS analysis was performed using an X-ray source with Mg Kα line (hv = 1253.6 eV). A detailed description of the equipment has been given previously [27]. Survey spectrum scans were performed with a pass energy of 40 eV using a SPECS PHOIBOS-HSA 3500 hemispherical analyser. High-resolution XPS spectra were recorded for C, O, P, Si, Ti, Cr and Au with a pass energy of 10 eV. All XPS binding energy scales were normalised using the C 1 s peak at 285 eV. The peaks were fitted to calculate relative intensities considering atomic sensitivity factors. XPS was recorded immediately after sample preparation and heating, thus, reducing the number of atmospheric exposures.

3. Results and Discussion

3.1. Influence of the Thickness of the TiO_2 Films

The influence of the thickness of the RF sputter-deposited TiO_2 on the change in film morphology upon heating is investigated. First, we will determine the thickness of the TiO_2 films for TiO_2P and TiO_2G and describe the crystallinity and morphology of both samples before and after heating. Then, the XPS results will be reported for both TiO_2P and

TiO$_2$G. Subsequently, the agglomeration of Au$_9$ clusters beneath a Cr$_2$O$_3$ overlayer upon heating of the two films is determined and discussed.

3.2. Determination of the TiO$_2$ Film Thickness

Figure 1 shows cross-section SEM images of TiO$_2$P and TiO$_2$G with line measurements of the thickness of the TiO$_2$ films. These SEM images clearly show that the thickness of the film for the TiO$_2$P and TiO$_2$G samples is ~400 nm and ~1100 nm, respectively; the film thickness of TiO$_2$G is more than two times greater than for TiO$_2$P. To confirm the film thickness, EDAX was further processed at the same image spots as SEM. Cross-section SEM-EDAX elemental maps of Ti, O and Si of TiO$_2$P and TiO$_2$G are shown in Figure S3. The EDAX elemental maps confirm that the thickness of the TiO$_2$ film for TiO$_2$G is larger than for TiO$_2$P.

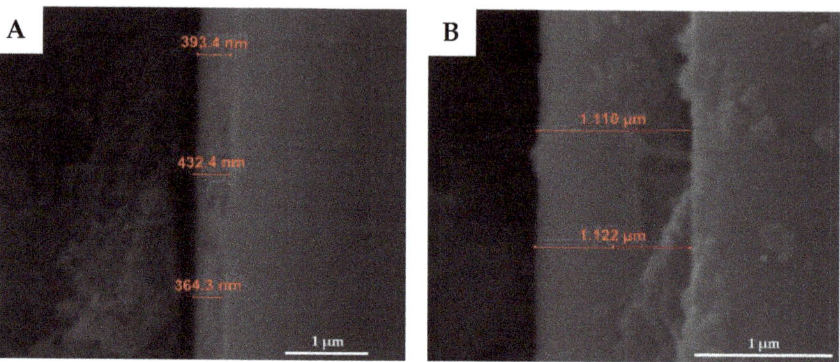

Figure 1. Cross-section SEM images of the (**A**) TiO$_2$P [28] and (**B**) TiO$_2$G layer.

3.3. Crystal Structure of the TiO$_2$P and TiO$_2$G before and after Heating

To assess the crystal structure of the TiO$_2$ film for TiO$_2$P and TiO$_2$G, XRD was conducted (Figure 2). There are no observable anatase, rutile or brookite crystal phase peaks [29], indicating that the films have an amorphous crystal structure. The crystallographic state of the TiO$_2$ is known to be transformed upon heating. The XRD patterns of TiO$_2$ films (TiO$_2$P and TiO$_2$G) after heating at 200 °C for 10 min are shown in Figure 2. Both spectra show an anatase peak at 29.8°, which confirms that the crystal structure of TiO$_2$P and TiO$_2$G has changed to the anatase phase after heating. The intensity of the anatase diffraction peak for TiO$_2$G is more than two times higher than for TiO$_2$P, which is due to the difference in the total amount of TiO$_2$ in each film. The TiO$_2$G layer is more than two times thicker than TiO$_2$P, so we also expect that there is more than twice as much anatase in the TiO$_2$G film. Thus, the percentage change in crystal structure in the films is comparable. The formation of the anatase phase strongly suggests the TiO$_2$ film could be mobile during the heating process, which could influence the morphology of the TiO$_2$ films, as will be discussed below.

3.4. Morphology of the TiO$_2$P and TiO$_2$G Layer before and after Heating

LSCM was conducted on both TiO$_2$ films before and after heating to compare their morphology. Figure 3 shows the surface morphology of TiO$_2$P and TiO$_2$G before and after heating over an area of 16 × 16 µm and the determined Ra and Rq values. The 3D profiles of the same spots are displayed in Figure S4. Before heating, the Ra (and Rq) values of the TiO$_2$P and TiO$_2$G are 0.6 nm (0.8 nm) and 1.0 nm (1.3 nm), respectively. However, after heating, the values become 1.0 nm (1.2 nm) and 12.7 nm (14.7 nm), respectively. The change in Ra (and Rq) for TiO$_2$P is small after heating, especially in comparison to TiO$_2$G, which is 12 times higher after heating. The Ra (and Rq) values were also calculated over a much larger area of 595 × 595 µm and show a similar change (Figure S5). The change in the Ra (and Rq) values indicates that both the TiO_2P and TiO_2G increase in surface roughness after heating. The XRD results show that the TiO_2G and TiO_2P have the same fraction of anatase after heating, so the total amount of anatase in TiO_2G is larger compared to TiO_2P (vide supra). Çörekçi et al. noted a similar finding in their study of different thicknesses of TiO_2 films heated at different temperatures [19]. The authors reported that the surface roughness of the thicker TiO_2 film (300 nm) increased more compared to thinner films (220 and 260 nm) upon heating. In our study, a large change in the surface roughness was observed clearly with the thicker film (more than two times thicker) by a factor of six. Çörekçi et al. assumed that the increase in surface roughness was due to increases in the grain sizes with increasing film thickness and the recrystallization in the TiO_2 films during heating. A number of studies have reported comparable findings that the surface morphology of the TiO_2 films changes upon heating [17,30]. Thus, we conclude that the thicker TiO_2G film is more mobile during heating in comparison to the thinner film in the TiO_2P sample.

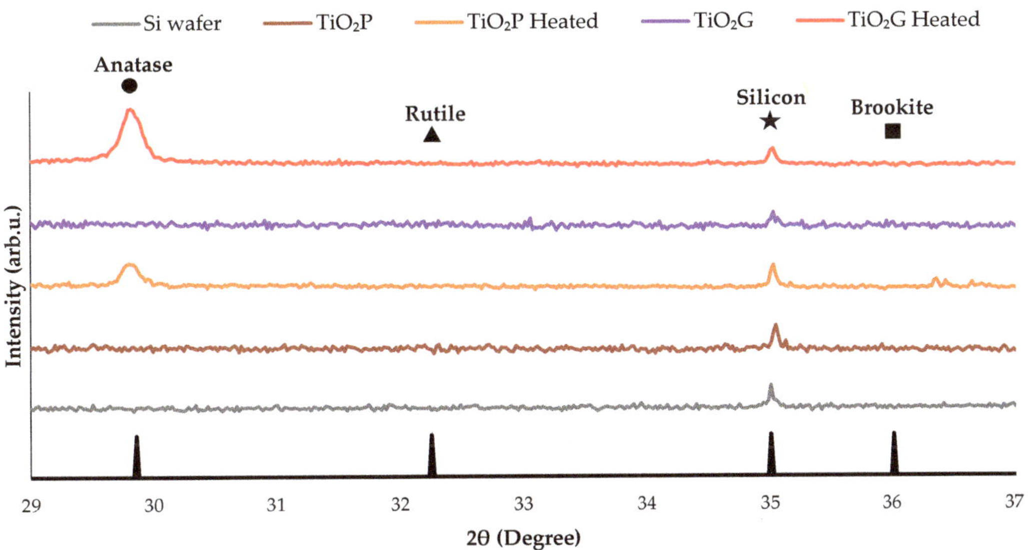

Figure 2. XRD patterns of the Si wafer, TiO_2P and TiO_2P after heating, TiO_2G and TiO_2G after heating to 200 °C. The positions of the diffraction peaks for anatase, rutile and brookite, as well as Si, are indicated using the standard XRD patterns (anatase PDF 01-075-1537, rutile PDF 01-071-4809, brookite PDF 04-007-0758 and Si PDF 00-013-0542).

3.5. Au_9 Clusters on TiO_2P and TiO_2G; a Probe for Mobility during Heating

In order to provide insight into the mobility of the TiO_2 during the recrystallisation process, Au_9 clusters were deposited onto the TiO_2 films and analysed with XPS. XPS was used to investigate the size of phosphine-protected Au_9 clusters deposited onto TiO_2P and TiO_2G. In addition, the effect of the CrO_x overlayer on the Au_9 clusters was investigated, also with XPS. Figures 4 and 5 show the peak positions and relative intensities of Au $4f_{7/2}$ peaks in the XP spectra of three different concentrations (0.006, 0.06 and 0.6 mM) of TiO_2P-Au_9, TiO_2G-Au_9, TiO_2P-Au_9-CrO_x and TiO_2G-Au_9-CrO_x before and after heating. Tables S1 and S2 show a summary of all the Au $4f_{7/2}$ peak positions and full-width-half-maximum (FWHM). Note that all the Au 4f spectra for both substrates (TiO_2P and TiO_2G) are shown in Figures S6 and S7. The TiO_2P XPS results will be first presented, followed by the TiO_2G results.

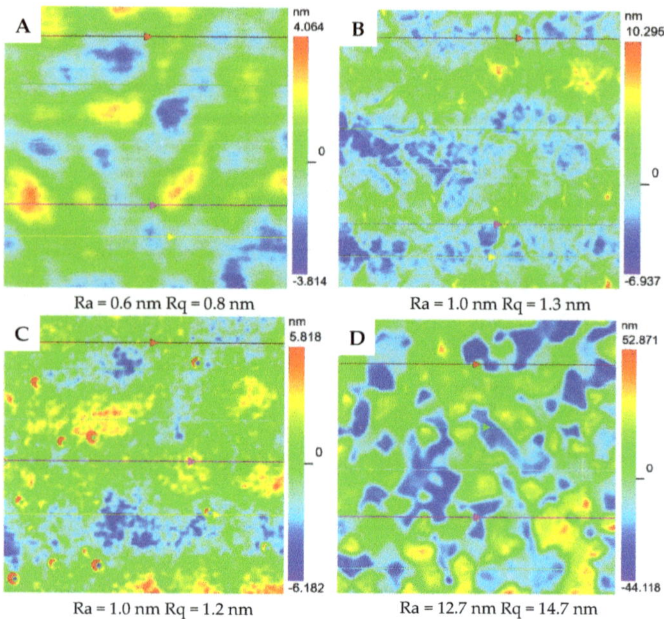

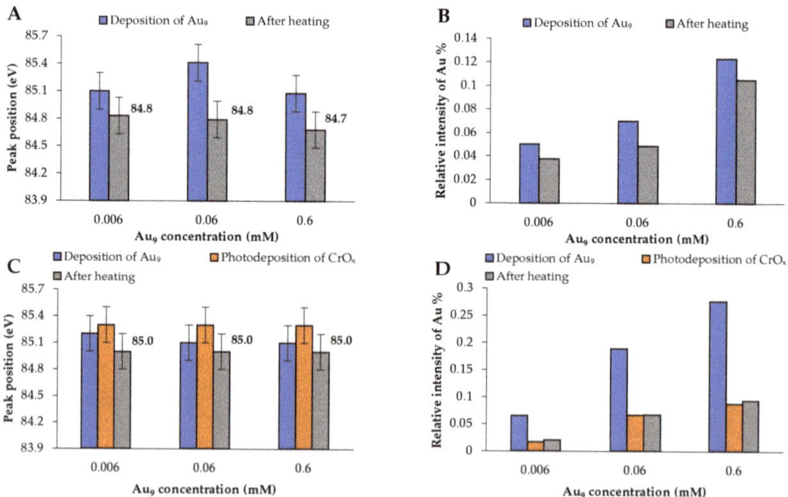

Figure 3. Surface morphology with the Ra and Rq values of (**A**) TiO_2P before heating and (**B**) TiO_2P after heating, (**C**) TiO_2G before heating and (**D**) TiO_2G after heating (area 16×16 μm). Note that the scale bars are different.

Figure 4. XPS results of TiO_2P-Au_9 for three different Au_9 concentrations: (**A**) position of Au $4f_{7/2}$ and (**B**) relative intensity of Au before and after heating. TiO_2P-Au_9-CrO_x (**C**) position of Au $4f_{7/2}$ and (**D**) relative intensity of Au before and after photodeposition of the CrO_x layer and after heating. Note that the vertical scales of (**B**,**D**) are different and that the samples in (**A**,**C**) are different but are prepared in the same manner.

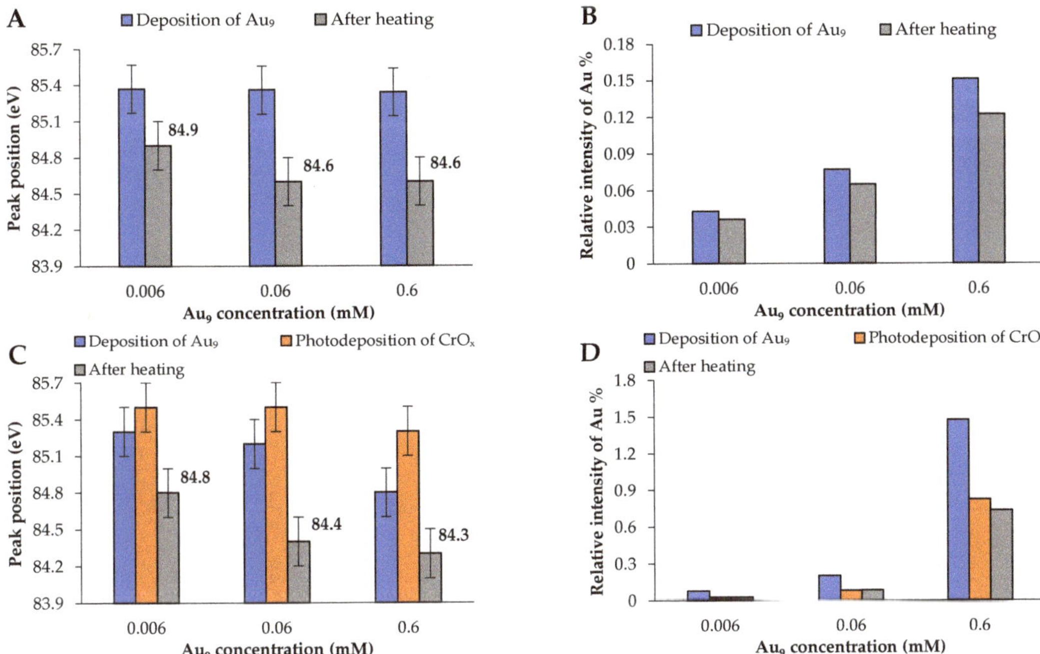

Figure 5. XPS results of Au_9 deposited on TiO_2G for three different Au_9 concentrations: (**A**) position of Au $4f_{7/2}$ and (**B**) relative intensity of Au before and after heating. TiO_2G-Au_9 with the CrO_x layer: (**C**) position of Au $4f_{7/2}$ and (**D**) relative intensity of Au before and after photodeposition of the CrO_x layer and after heating. Note that the vertical scales of (**B,D**) are different and that the samples in (**A,C**) are different but are prepared in the same manner.

3.6. XPS of TiO_2P Sample

Without the CrO_x layer and before heating, the Au $4f_{7/2}$ peaks appeared at 85.1–85.4 eV with an FWHM of 1.7–1.8 eV (Figure 4A), whereas after heating, the Au $4f_{7/2}$ peaks shifted to slightly lower binding energies (84.7–84.8 eV) and FWHM (1.5–1.6 eV), and also showed a decrease in relative Au intensity across all Au_9 concentrations (Figure 4B). The results of the samples covered with a CrO_x layer are shown in Figure 4C,D. The Au $4f_{7/2}$ peak positions of TiO_2P-Au_9 after CrO_x deposition but before heating were observed at 85.3 eV and an FWHM of 1.6 eV for all three concentrations. Note that the Au relative intensities decrease after the photodeposition of the CrO_x layer, confirming the coverage of Au clusters with the CrO_x layer (Figure 4D). After heating, the XPS peak position decreases slightly to 85.0 eV with no significant change in FWHM. The relative Au intensities also remained unchanged upon heating. XPS has been shown previously to be a reliable indicator of the size of phosphine-protected Au_9 clusters through the final state effect [21,28,31–36]. Generally, non-agglomerated Au_9 clusters on TiO_2 appear at a high binding peak (HBP) between 85.0–85.4 eV with an FWHM of 1.7 ± 0.2 eV, and agglomerated Au_9 clusters shift toward a low binding peak (LBP) at 84 eV with a decreasing FWHM that corresponds to bulk Au [28,31–35]. This XPS interpretation has been confirmed by correlating the XPS results with other techniques, such as HRTEM [33,34]. Here, the Au $4f_{7/2}$ peak positions of TiO_2P-Au_9 without the CrO_x layer after heating indicate a small degree of agglomeration of the Au_9 clusters for all concentrations. This is further confirmed by a small decrease in Au intensity after heating, indicating that some of the gold is attenuated due to some larger, agglomerated particles. Electrons emitted from the part of the clusters facing toward the substrate are attenuated when leaving the sample, which decreases the overall Au intensity [31,32]. Therefore, the same total amount of gold deposited on the surface will have a lower intensity for large gold particles than that of small gold clusters. In contrast to the CrO_x layer of the Au $4f_{7/2}$ peaks, positions are unchanged after heating and there is no further decrease in the Au relative intensities, indicating that Au clusters remain non-agglomerated clusters with CrO_x coverage (see Scheme 1A). It is important to note that there is a decrease in Au intensity after photodeposition of the CrO_x layer due to the coverage of Au_9 clusters (Figure 4D). These results are in agreement with our previous report showing that CrO_x overlayers inhibit the agglomeration of Au clusters [28].

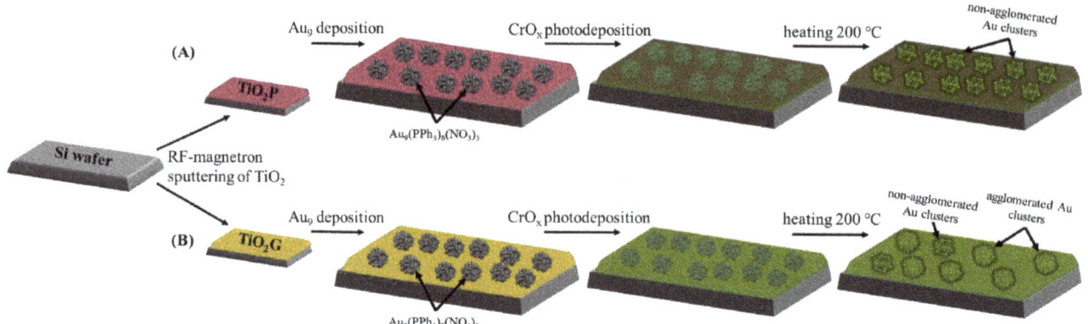

Scheme 1. Schematic illustration of the experimental procedure for preparing (**A**) TiO_2P-Au_9-CrO_x and (**B**) TiO_2G-Au_9-CrO_x.

The P 2p spectra of TiO_2P-Au_9 without and with the CrO_x layer before and after heating are shown in Figure S8 and the peak positions are discussed in the Supplementary Section. The Cr 2p spectra for TiO_2P-Au_9-CrO_x before and after heating at the three different concentrations are shown in Figure S9. A summary of all the Cr $2p_{3/2}$ peak positions is shown in Table S3 and the peak positions are discussed in the Supplementary Section.

3.7. XPS of TiO$_2$G Sample

For the thicker film, TiO_2G-Au_9, the Au $4f_{7/2}$ peak positions before heating for all three different concentrations appeared at the HBP at 85.3 ± 0.1 eV (Figure 5A) and an FWHM of 1.8 ± 0.2 eV, corresponding to non-agglomerated Au clusters. However, after heating, the Au $4f_{7/2}$ shifted toward lower energy (84.6–84.9 eV) and an FWHM of 1.5–1.7 eV with a decrease in Au intensity (Figure 5B), indicating that Au clusters are partially agglomerated. With the CrO_x layer deposited before heating, the Au $4f_{7/2}$ peak positions are observed at the HBP position at 85.3–85.5 eV (Figure 5C), with a decrease in Au $4f_{7/2}$ intensity due to the coverage of the CrO_x layer on Au_9 clusters (Figure 5D). There is a slight increase in the binding energy of the Au 4f peak after the photodeposition of CrO_x, and we do not know if this is a significant change or not. However, the position found can be used as an indication of the presence of non-agglomerated Au clusters. With the CrO_x layer after heating, the Au $4f_{7/2}$ peak positions have further shifted to lower energy (84.3–84.8 eV) positions and an FWHM of 1.3–1.8 eV with a decrease in Au intensity, which is attributed to further agglomeration of the Au clusters based on the final state effect (see Scheme 1B). The degree of agglomeration increases with increasing Au_9 concentration for both cases (without and with the CrO_x layer). Note here the difference; Au clusters on the surface of TiO_2G undergo increased agglomeration after heating, even in the presence of the CrO_x layer. This is different to the TiO_2P, where Au clusters are less likely to agglomerate under the CrO_x layer after heating. This difference will be further discussed below.

The chemical state of the phosphorous ligands of TiO_2G-Au_9 without and with the CrO_x layer, both before and after heating, was determined using the P 2p region (see Figure S10 for more information and accompanying text). Figure S11 shows the Cr 2p spectra for TiO_2G-Au_9-CrO_x before and after heating of the three different concentrations. All the Cr

$2p_{3/2}$ peak positions are given in Table S4 and the peak positions are discussed in the Supplementary Section.

3.8. Effect of the TiO$_2$ Film Thickness

The protective effect of the CrO$_x$ layer on the agglomeration of Au$_9$ clusters is not the same for both the TiO$_2$P and TiO$_2$G substrates. The agglomeration of Au$_9$ clusters is inhibited on TiO$_2$P with the CrO$_x$ overlayer but not on TiO$_2$G, which shows a higher degree of agglomeration. The coverage of the CrO$_x$ layer on Au$_9$ clusters for both substrates is demonstrated by the decrease in the Au-XPS intensities. After heating, it is observed that the relative amount of CrO$_x$ decreases for both films (Table S5). Our previous studies on a similar system revealed that the CrO$_x$ layer diffuses into a TiO$_2$ film after heating to 600 °C due to the differences in surface energy between TiO$_2$ and CrO$_x$ [26]. In this study, both films were heated to only 200 °C, however, CrO$_x$ on TiO$_2$G experienced more diffusion of CrO$_x$ into the film compared to TiO$_2$P. One possibility for the higher degree of Au$_9$ agglomeration and CrO$_x$ diffusion is the mobility of the TiO$_2$ film. Cluster agglomeration can be due to either (i) growth of the clusters over the surface or (ii) mobility of the substrate. In the case of (i), the cluster growth and agglomeration on a substrate can be ascribed to either Smoluchowski ripening or Ostwald ripening mechanisms. For Smoluchowski ripening, the agglomeration of clusters is caused by the collision and coalescence of entire clusters to larger particles [37]. For Ostwald ripening, the growth of larger particles takes place by the detachment of single atoms, which diffuse onto a nearby cluster or nanoparticle [38]. In the case of (ii), a section of the substrate to which a cluster is adsorbed moves closer to another section of the substrate, which has another adsorbed cluster. The significant change in the surface morphology of TiO$_2$G after heating (Ra: 11.7 nm and Rq: 13.4 nm) compared to TiO$_2$P (Ra: 0.4 nm and Rq: 0.5 nm) strongly suggests that the agglomeration of the Au$_9$ clusters with different concentrations on TiO$_2$G after heating is due to the high distortion of the surface upon heating. A higher mobility of the TiO$_2$ substrate during heating means that the local surface beneath an Au cluster moves larger distances compared to a substrate which exhibits lower mobility during heating (see Scheme 2). The high mobility of the thick film is assumed to be due to the recrystallisation during heating, which is in agreement with previous studies [17,19,30]. With increasing mobility, the likelihood of close contact between two or more Au clusters increases, and thus the likelihood of agglomeration is also increased. Furthermore, the degree of agglomeration of the Au clusters is larger for the thicker TiO$_2$G substrate compared to the thinner TiO$_2$P substrate.

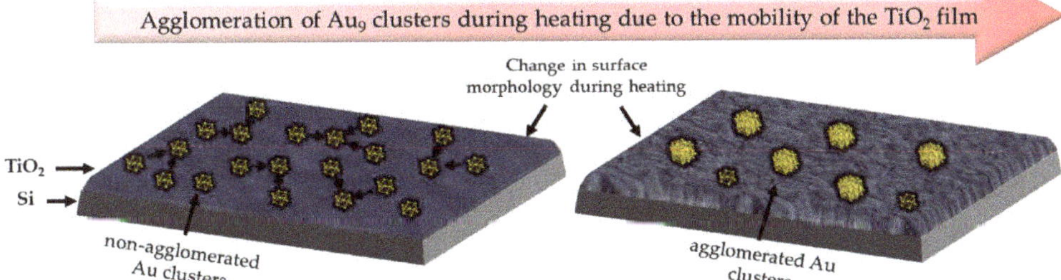

Scheme 2. Schematic illustration showing the agglomeration mechanism of Au$_9$ clusters on the TiO$_2$G film during heating.

4. Conclusions

In summary, the change in surface morphology of two different film thicknesses of RF sputter-deposited TiO$_2$ (~400 nm and ~1100 nm) was examined and compared upon heating. After heating, the thick TiO$_2$ film showed a larger change in surface morphology, which is associated with higher mobility during heating compared to the thin TiO$_2$ film. The difference in mobility is attributed to the differences in the total amount of amorphous

TiO_2 transformed to anatase in each of the films, which then results in differences in the morphology of the surface upon heating. Au_9 clusters were used as a probe for TiO_2 mobility. Au_9 clusters were deposited onto the two different TiO_2 films, followed by photodeposition of the CrO_x layer. After heating, the Au clusters on the thicker film showed a larger degree of agglomeration compared to the thinner film. The higher mobility of the thick film during heating increased the probability of close encounters of Au clusters, which resulted in agglomeration of the Au_9 clusters even in the presence of a CrO_x overlayer. In contrast, the lower mobility of the thin film resulted in less agglomeration of the Au_9 clusters after heating.

Supplementary Materials: The following supporting information can be downloaded at: https://www.mdpi.com/article/10.3390/nano12183218/s1. The supporting information shows the EDAX-SEM elemental mapping of the TiO_2P and TiO_2G cross-section images, the details of the XP spectra, their fitting, and quantification. Figure S1: A photo of the TiO_2P (lift) and TiO_2G (right) films. Figure S2: UV-Vis spectrum of $Au_9(PPh_3)_8(NO_3)_3$ in Methanol. Figure S3: Cross-section SEM-EDAX elemental maps of Ti, O and Si of TiO_2P and TiO_2G. Note that the scale bars are different. Figure S4: 3D Profile of (A) TiO_2P before heating, (B) TiO_2P after heating, (C) before heating, TiO_2G and (D) TiO_2G after heating (area 16 × 16 μm). Figure S5: Surface morphology with the average of Ra and Rq values of (A) TiO_2P before heating, (B) TiO_2P after heating, (C) before heating, TiO_2G and (D) TiO_2G after heating. (area 595 × 595 μm). It is important to know that the scale bars are different. Figure S6: XP spectra of Au 4f of (A) TiO_2P-Au_9: after Au_9 deposition (blue) and after heating (grey) (B) TiO_2P-Au_9-CrO_x: after Au_9 deposition (blue), after CrO_x layer photodeposited (orange) and after heating (grey). Figure S7: XP spectra of Au 4f of (A) TiO_2G-Au_9: after Au_9 deposition (blue) and after heating (grey) (B) TiO_2G-Au_9-CrO_x: after Au_9 deposition (blue), after CrO_x layer photodeposited (orange) and after heating (grey). Figure S8: XP spectra of P 2p of (A) TiO_2P-Au_9: after Au_9 deposition (blue) and after heating (grey) (B) TiO_2P-Au_9-CrO_x: after Au_9 deposition (blue), after CrO_x layer photodeposited (orange), and after heating (grey). Figure S9: XP spectra of Cr 2p of the TiO_2P-Au_9-CrO_x sample of (A) 0.006mM sample, (B) 0.06mM sample and (C) 0.6mM sample: after CrO_x layer photodeposited (orange) and after heating (grey). Figure S10: XP spectra of P 2p of (A) TiO_2G-Au_9: after Au_9 deposition (blue) and after heating (grey) (B) TiO_2G-Au_9-CrO_x: after Au_9 deposition (blue), after CrO_x layer photodeposited (orange), and after heating (grey). Figure S11: XP spectra of Cr 2p of the TiO_2G-Au_9-CrO_x sample of (A) 0.006mM sample, (B) 0.06mM sample and (C) 0.6mM sample: after CrO_x layer photodeposited (orange) and after heating (grey). Table S1: XPS Au $4f_{7/2}$ peak positions and FWHM of TiO_2P-Au_9 and TiO_2P-Au_9-CrO_x. Table S2: XPS Au $4f_{7/2}$ peak positions and FWHM of TiO_2G-Au_9 and TiO_2G-Au_9-CrO_x. Table S3: XPS Cr $2p_{3/2}$ peak positions and FWHM of TiO_2P-Au_9-CrO_x. Table S4: XPS Cr $2p_{3/2}$ peak positions and FWHM of TiO_2G-Au_9-CrO_x. Table S5: XPS relative amount of Cr $2p_{3/2}$ to Ti $2p_{3/2}$ of TiO_2P-Au_9-CrO_x and TiO_2G-Au_9-CrO_x. References [39–44] are cited in the supplementary materials.

Author Contributions: Conceptualization, A.S.A., G.F.M. and G.G.A.; Data curation, A.S.A. and G.G.A.; Formal analysis, A.S.A., Y.Y. and A.R.; Funding acquisition, G.F.M. and G.G.A.; Investigation, A.S.A., Y.Y. and A.R.; Methodology, A.S.A., G.F.M. and G.G.A.; Resources, S.T. and G.G.A.; Supervision, G.G.A.; Visualization, A.S.A.; Writing—original draft, A.S.A.; Writing—review & editing, A.S.A., Y.Y., A.R., S.T., G.F.M. and G.G.A. All authors have read and agreed to the published version of the manuscript.

Funding: Australian Synchrotron, Victoria, Australia (AS1/SXR/15819); US Army project FA5209-16-R-0017; Australian Solar Thermal Research Institute (ASTRI), a project supported by the Australian Government, through the Australian Renewable Energy Agency (ARENA).

Institutional Review Board Statement: Not applicable.

Informed Consent Statement: Not applicable.

Data Availability Statement: The data that support the findings of this study are available from the corresponding author upon reasonable request.

Acknowledgments: Part of this research was undertaken on the soft X-ray spectroscopy beamline at the Australian Synchrotron, Victoria, Australia (AS1/SXR/15819). The work was supported by the US Army project FA5209-16-R-0017. This research was performed as part of the Australian Solar Thermal Research Institute (ASTRI), a project supported by the Australian Government, through the Australian Renewable Energy Agency (ARENA). We would like to thank Dr Bruce Cowie from the Australian Synchrotron for his assistance. The authors acknowledge the facilities, and the scientific and technical assistance, of Microscopy Australia (formerly known as AMMRF) and the Australian National Fabrication Facility (ANFF) at Flinders University. The authors acknowledge Flinders Microscopy and Microanalysis and their expertise. The authors thank A/Prof Vladimir Golovko (Canterbury University) for providing access to the $Au_9(PPh_3)_8(NO_3)_3$ clusters. The authors acknowledge Dr Benjamin Wade at Adelaide Microscopy (University of Adelaide).

Conflicts of Interest: The authors declare no conflict of interest.

References

1. Daghrir, R.; Drogui, P.; Robert, D. Modified TiO_2 For Environmental Photocatalytic Applications: A Review. *Ind. Eng. Chem. Res.* **2013**, *52*, 3581–3599. [CrossRef]
2. Linsebigler, A.L.; Lu, G.; Yates, J.T. Photocatalysis on TiO_2 Surfaces: Principles, Mechanisms, and Selected Results. *Chem. Rev.* **1995**, *95*, 735–758. [CrossRef]
3. Brinker, C.J.; Harrington, M.S. Sol-gel Derived Antireflective Coatings for Silicon. *Sol. Energy Mater.* **1981**, *5*, 159–172. [CrossRef]
4. Lottiaux, M.; Boulesteix, C.; Nihoul, G.; Varnier, F.; Flory, F.; Galindo, R.; Pelletier, E. Morphology and Structure of TiO_2 Thin Layers vs. Thickness and Substrate Temperature. *Thin Solid Film.* **1989**, *170*, 107–126. [CrossRef]
5. Yeung, K.S.; Lam, Y.W. A Simple Chemical Vapour Deposition Method for Depositing Thin TiO_2 Films. *Thin Solid Film.* **1983**, *109*, 169–178. [CrossRef]
6. Aarik, J.; Aidla, A.; Uustare, T.; Kukli, K.; Sammelselg, V.; Ritala, M.; Leskelä, M. Atomic Layer Deposition of TiO_2 Thin Films from TiI4 and H2O. *Appl. Surf. Sci.* **2002**, *193*, 277–286. [CrossRef]
7. Dakka, A.; Lafait, J.; Abd-Lefdil, M.; Sella, C. Optical Study of Titanium Dioxide Thin Films Prepared by RF Sputtering. *Moroc. J. Condens. Matter* **1999**, *2*, 153–156.
8. Tanemura, S.; Miao, L.; Wunderlich, W.; Tanemura, M.; Mori, Y.; Toh, S.; Kaneko, K. Fabrication and Characterization of Anatase/Rutile–TiO_2 Thin Films by Magnetron Sputtering: A Review. *Sci. Technol. Adv. Mater.* **2005**, *6*, 11–17. [CrossRef]
9. Miao, L.; Jin, P.; Kaneko, K.; Terai, A.; Nabatova-Gabain, N.; Tanemura, S. Preparation and characterization of polycrystalline anatase and rutile TiO_2 thin films by rf magnetron sputtering. *Appl. Surf. Sci.* **2003**, *212–213*, 255–263. [CrossRef]
10. Wang, T.M.; Zheng, S.K.; Hao, W.C.; Wang, C. Studies on photocatalytic activity and transmittance spectra of TiO_2 thin films prepared by r.f. magnetron sputtering method. *Surf. Coat. Technol.* **2002**, *155*, 141–145. [CrossRef]
11. Ye, Q.; Liu, P.Y.; Tang, Z.F.; Zhai, L. Hydrophilic properties of nano-TiO_2 thin films deposited by RF magnetron sputtering. *Vacuum* **2007**, *81*, 627–631. [CrossRef]
12. Huang, C.H.; Tsao, C.C.; Hsu, C.Y. Study on the photocatalytic activities of TiO_2 films prepared by reactive RF sputtering. *Ceram. Int.* **2011**, *37*, 2781–2788. [CrossRef]
13. Yoo, D.; Kim, I.; Kim, S.; Hahn, C.H.; Lee, C.; Cho, S. Effects of annealing temperature and method on structural and optical properties of TiO_2 films prepared by RF magnetron sputtering at room temperature. *Appl. Surf. Sci.* **2007**, *253*, 3888–3892. [CrossRef]
14. Hou, Y.-Q.; Zhuang, D.-M.; Zhang, G.; Zhao, M.; Wu, M.-S. Influence of annealing temperature on the properties of titanium oxide thin film. *Appl. Surf. Sci.* **2003**, *218*, 98–106. [CrossRef]
15. Grilli, M.L.; Yilmaz, M.; Aydogan, S.; Cirak, B.B. Room temperature deposition of XRD-amorphous TiO_2 thin films: Investigation of device performance as a function of temperature. *Ceram. Int.* **2018**, *44*, 11582–11590. [CrossRef]
16. Mathews, N.R.; Morales, E.R.; Cortés-Jacome, M.A.; Toledo Antonio, J.A. TiO_2 Thin Films—Influence of Annealing Temperature on Structural, Optical and Photocatalytic Properties. *Sol. Energy* **2009**, *83*, 1499–1508. [CrossRef]
17. Meng, F.; Xiao, L.; Sun, Z. Thermo-induced hydrophilicity of nano-TiO_2 thin films prepared by RF magnetron sputtering. *J. Alloy. Compd.* **2009**, *485*, 848–852. [CrossRef]
18. Taherniya, A.; Raoufi, D. Thickness dependence of structural, optical and morphological properties of sol-gel derived TiO_2 thin film. *Mater. Res. Express* **2018**, *6*, 016417. [CrossRef]
19. Çörekçi, Ş.; Kizilkaya, K.; Asar, T.; Öztürk, M.; Çakmak, M.; Ozcelik, S. Effects of Thermal Annealing and Film Thickness on the Structural and Morphological Properties of Titanium Dioxide Films. *Acta Phys. Pol. A* **2012**, *121*, 247–248. [CrossRef]
20. Pichugina, D.A.; Kuz'menko, N.E.; Shestakov, A.F. Ligand-protected gold clusters: The structure, synthesis and applications. *Russ. Chem. Rev.* **2015**, *84*, 1114–1144. [CrossRef]
21. Adnan, R.H.; Madridejos, J.M.L.; Alotabi, A.S.; Metha, G.F.; Andersson, G.G. A Review of State of the Art in Phosphine Ligated Gold Clusters and Application in Catalysis. *Adv. Sci.* **2022**, *2022*, 2105692. [CrossRef] [PubMed]
22. Daughtry, J.; Alotabi, A.S.; Howard-Fabretto, L.; Andersson, G.G. Composition and Properties of RF-Sputter Deposited Titanium Dioxide Thin Films. *Nanoscale Adv.* **2021**, *3*, 1077–1086. [CrossRef]

23. Yuan, X.; Ye, Y.; Lian, M.; Wei, Q. Structural Coloration of Polyester Fabrics Coated with Al/TiO$_2$ Composite Films and Their Anti-Ultraviolet Properties. *Materials* **2018**, *11*, 1011. [CrossRef] [PubMed]

24. Diamanti, M.V.; Del Curto, B.; Pedeferri, M. Interference Colors of Thin Oxide Layers on Titanium. *Color Res. Appl.* **2008**, *33*, 221–228. [CrossRef]

25. Wen, F.; Englert, U.; Gutrath, B.; Simon, U. Crystal Structure, Electrochemical and Optical Properties of [Au9(PPh3)8](NO3)3. *Eur. J. Inorg. Chem.* **2008**, *2008*, 106–111. [CrossRef]

26. Alotabi, A.S.; Gibson, C.T.; Metha, G.F.; Andersson, G.G. Investigation of the Diffusion of Cr2O3 into Different Phases of TiO$_2$ upon Annealing. *ACS Appl. Energy Mater.* **2021**, *4*, 322–330. [CrossRef]

27. Acres, R.G.; Ellis, A.V.; Alvino, J.; Lenahan, C.E.; Khodakov, D.A.; Metha, G.F.; Andersson, G.G. Molecular Structure of 3-Aminopropyltriethoxysilane Layers Formed on Silanol-Terminated Silicon Surfaces. *J. Phys. Chem. C* **2012**, *116*, 6289–6297. [CrossRef]

28. Alotabi, A.S.; Yin, Y.; Redaa, A.; Tesana, S.; Metha, G.F.; Andersson, G.G. Cr2O3 layer inhibits agglomeration of phosphine-protected Au9 clusters on TiO$_2$ films. *J. Chem. Phys.* **2021**, *155*, 164702. [CrossRef]

29. Ali, S.; Granbohm, H.; Lahtinen, J.; Hannula, S.-P. Titania Nanotubes Prepared by Rapid Breakdown Anodization for Photocatalytic Decolorization of Organic Dyes under UV and Natural Solar Light. *Nanoscale Res. Lett.* **2018**, *13*, 179. [CrossRef]

30. Chandra Sekhar, M.; Kondaiah, P.; Jagadeesh Chandra, S.V.; Mohan Rao, G.; Uthanna, S. Substrate temperature influenced physical properties of silicon MOS devices with TiO$_2$ gate dielectric. *Surf. Interface Anal.* **2012**, *44*, 1299–1304. [CrossRef]

31. Anderson, D.P.; Alvino, J.F.; Gentleman, A.; Al Qahtani, H.; Thomsen, L.; Polson, M.I.; Metha, G.F.; Golovko, V.B.; Andersson, G.G. Chemically-Synthesised, Atomically-Precise Gold Clusters Deposited and Activated on Titania. *Phys. Chem. Chem. Phys.* **2013**, *15*, 3917–3929. [CrossRef] [PubMed]

32. Anderson, D.P.; Adnan, R.H.; Alvino, J.F.; Shipper, O.; Donoeva, B.; Ruzicka, J.-Y.; Al Qahtani, H.; Harris, H.H.; Cowie, B.; Aitken, J.B. Chemically Synthesised Atomically Precise Gold Clusters Deposited and Activated on Titania. Part II. *Phys. Chem. Chem. Phys.* **2013**, *15*, 14806–14813. [CrossRef] [PubMed]

33. Al Qahtani, H.S.; Kimoto, K.; Bennett, T.; Alvino, J.F.; Andersson, G.G.; Metha, G.F.; Golovko, V.B.; Sasaki, T.; Nakayama, T. Atomically resolved structure of ligand-protected Au9 clusters on TiO$_2$ nanosheets using aberration-corrected STEM. *J. Chem. Phys.* **2016**, *144*, 114703. [CrossRef] [PubMed]

34. Al Qahtani, H.S.; Metha, G.F.; Walsh, R.B.; Golovko, V.B.; Andersson, G.G.; Nakayama, T. Aggregation Behavior of Ligand-Protected Au9 Clusters on Sputtered Atomic Layer Deposition TiO$_2$. *J. Phys. Chem. C* **2017**, *121*, 10781–10789. [CrossRef]

35. Howard-Fabretto, L.; Andersson, G.G. Metal Clusters on Semiconductor Surfaces and Application in Catalysis with a Focus on Au and Ru. *Adv. Mater.* **2020**, *32*, 1904122. [CrossRef] [PubMed]

36. Alotabi, A.S.; Osborn, D.J.; Ozaki, S.; Kataoka, Y.; Negishi, Y.; Tesana, S.; Metha, G.F.; Andersson, G.G. Suppression of phosphine-protected Au9 cluster agglomeration on SrTiO3 particles using a chromium hydroxide layer. *Mater. Adv.* **2022**, *3*, 3620–3630. [CrossRef]

37. Smoluchowski, M.V. Drei vortrage uber diffusion, brownsche bewegung und koagulation von kolloidteilchen. *Z. Fur Phys.* **1916**, *17*, 557–585.

38. Ostwald, W. Blocking of Ostwald ripening allowing long-term stabilization. *Phys. Chem.* **1901**, *37*, 385.

39. Wilcoxon, J.P.; Provencio, P. Etching and Aging Effects in Nanosize Au Clusters Investigated Using High-Resolution Size-Exclusion Chromatography. *J. Phys. Chem. B* **2003**, *107*, 12949. [CrossRef]

40. Moulder, J.F. Handbook of X-ray photoelectron spectroscopy: A reference book of standard data for use in x-ray photoelectron spectroscopy. In *Handbook of X-ray Photoelectron Spectroscopy*; Perkin-Elmer Corporation: Waltham, MA, USA, 1992.

41. Biesinger, M.C.; Brown, C.; Mycroft, J.R.; Davidson, R.D.; McIntyre, N.S. X-ray photoelectron spectroscopy studies of chromium compounds. *Surf. Interface Anal.* **2004**, *36*, 1550–1563. [CrossRef]

42. Kawawaki, T.; Kataoka, Y.; Hirata, M.; Akinaga, Y.; Takahata, R.; Wakamatsu, K.; Fujiki, Y.; Kataoka, M.; Kikkawa, S.; Alotabi, A.S.; et al. Creation of High-Performance Heterogeneous Photocatalysts by Controlling Ligand Desorption and Particle Size of Gold Nanocluster. *Angew. Chem. Int. Ed.* **2021**, *60*, 21340–21350. [CrossRef] [PubMed]

43. Madridejos, J.M.L.; Harada, T.; Falcinella, A.J.; Small, T.D.; Golovko, V.B.; Andersson, G.G.; Metha, G.F.; Kee, T.W. Optical Properties of the Atomically Precise C4 Core [Au9(PPh3)8]$^{3+}$ Cluster Probed by Transient Absorption Spectroscopy and Time-Dependent Density Functional Theory. *J. Phys. Chem. C* **2021**, *125*, 2033–2044. [CrossRef]

44. Krishnan, G.; Al Qahtani, H.S.; Li, J.; Yin, Y.; Eom, N.; Golovko, V.B.; Metha, G.F.; Andersson, G.G. Investigation of Ligand-Stabilized Gold Clusters on Defect-Rich Titania. *J. Phys. Chem. C* **2017**, *121*, 28007–28016. [CrossRef]

nanomaterials

MDPI

Article

Novel InGaSb/AlP Quantum Dots for Non-Volatile Memories

Demid S. Abramkin [1,2] and Victor V. Atuchin [3,4,5,6,*]

[1] Laboratory of Molecular Beam Epitaxy of III–V Semiconductor Compounds, Institute of Semiconductor Physics, SB RAS, 630090 Novosibirsk, Russia

[2] Department of Physics, Novosibirsk State University, 630090 Novosibirsk, Russia

[3] Laboratory of Optical Materials and Structures, Institute of Semiconductor Physics, SB RAS, 630090 Novosibirsk, Russia

[4] Research and Development Department, Kemerovo State University, 650000 Kemerovo, Russia

[5] Department of Industrial Machinery Design, Novosibirsk State Technical University, 630073 Novosibirsk, Russia

[6] R&D Center "Advanced Electronic Technologies", Tomsk State University, 634034 Tomsk, Russia

* Correspondence: atuchin@isp.nsc.ru

Abstract: Non-volatile memories based on the flash architecture with self-assembled III–V quantum dots (SAQDs) used as a floating gate are one of the prospective directions for universal memories. The central goal of this field is the search for a novel SAQD with hole localization energy (E_{loc}) sufficient for a long charge storage (10 years). In the present work, the hole states' energy spectrum in novel InGaSb/AlP SAQDs was analyzed theoretically with a focus on its possible application in non-volatile memories. Material intermixing and formation of strained SAQDs from a $Ga_xAl_{1-x}Sb_yP_{1-y}$, $In_xAl_{1-x}Sb_yP_{1-y}$ or an $In_xGa_{1-x}Sb_yP_{1-y}$ alloy were taken into account. Critical sizes of SAQDs, with respect to the introduction of misfit dislocation as a function of alloy composition, were estimated using the force-balancing model. A variation in SAQDs' composition together with dot sizes allowed us to find that the optimal configuration for the non-volatile memory application is GaSbP/AlP SAQDs with the 0.55–0.65 Sb fraction and a height of 4–4.5 nm, providing the E_{loc} value of 1.35–1.50 eV. Additionally, the hole energy spectra in unstrained InSb/AlP and GaSb/AlP SAQDs were calculated. E_{loc} values up to 1.65–1.70 eV were predicted, and that makes unstrained InGaSb/AlP SAQDs a prospective object for the non-volatile memory application.

Keywords: QD-Flash; self-assembled quantum dots; quantum dots memories; non-volatile memories; universal memories; hole localization; quaternary alloy; strain

Citation: Abramkin, D.S.; Atuchin, V.V. Novel InGaSb/AlP Quantum Dots for Non-Volatile Memories. *Nanomaterials* **2022**, *12*, 3794. https://doi.org/10.3390/nano12213794

Academic Editor: Orion Ciftja

Received: 22 September 2022
Accepted: 25 October 2022
Published: 27 October 2022

Publisher's Note: MDPI stays neutral with regard to jurisdictional claims in published maps and institutional affiliations.

Copyright: © 2022 by the authors. Licensee MDPI, Basel, Switzerland. This article is an open access article distributed under the terms and conditions of the Creative Commons Attribution (CC BY) license (https://creativecommons.org/licenses/by/4.0/).

1. Introduction

Semiconductor self-assembled quantum dots (SAQDs) are nanocrystals of a narrow bandgap material grown into a matrix or a shell of a wider bandgap material. In contrast to colloidal SAQDs [1,2], the epitaxial SAQDs formation occurs in Stranski–Krastanov growth mode as a result of surface relief reorganization [3]. One of the crucial advantages of this self organized growth mode is the formation of a nanoscale objects array without using nanoscale lithography. The flexibility of the SAQD energy spectrum, caused by quantum confinement effects and the alloy composition variation, makes them widely applicable in broad areas of modern electronics and optoelectronics [4–11]. The energy bands offset at an SAQD/matrix heterointerface provide a charge carrier localization into SAQDs, and that opens up the prospective of using SAQDs for charge storage in non-volatile memory cells. The most interesting object in this research field is the construction of a flash memory with a III–V SAQD array used as a floating gate. The basic principles of such memory cells were formulated in [12,13]. The basic idea consists in the hole localization in the SAQD potential that permits controlling the underlying channel conductivity by the field effect, as shown in the schematic diagram presented in Figure 1 [12].

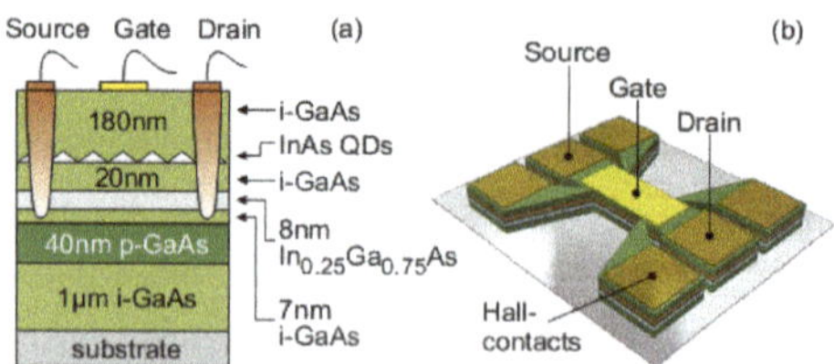

Figure 1. (**a**) Schematic cross section of the layered structure. (**b**) Sketch of the SAQD–Flash prototype. Hall-contacts were used for the transport measurements of 2DHG [12].

The fabrication of the first high-temperature SAQD flash memory prototype, based on InAs/AlGaAs SAQDs [13,14], reveals the main advantages of III–V SAQD flash cells in comparison with traditional Si/SiO$_2$-based ones. First of all, a significantly faster access time (about 20 ns) was mentioned [13], and that is comparable to a dynamic random-access memory (DRAM), in contrast to microsecond times common for a Si/SiO$_2$-based flash memory in the framework of planar technology. As is clearly seen in Figure 2 [12], the fast access is caused by (i) the direct hole capture into an SAQD at the writing mode with the rate limited by the hole energy relaxation and recharging of the structure capacitance ($f_{cutoff} \sim 1/(RC)$) only and (ii) the tunnel hole emission at the erase mode [13,15]. Miniaturization of the cell III–V of the SAQD-memory will minimize the limitation of the write/erase rate caused by recharging the capacitance of the structure and allow it to approach the limit determined by the hole capture time in the SAQD (less than 1 ps [16]). Second, in traditional Si/SiO$_2$-based flash memories, hot charge-carriers are used for the write/erase procedure, and it leads to a damaging of the structure and low endurance ($<10^6$ cycles). The direct hole capture into SAQDs, according to the SAQD flash concept, will solve this problem and allow a significant increase in endurance. Additionally, the advantage of using III–V materials is that III–V FETs are noticeably superior to Si FETs in their speed characteristics [17]. This allows us to hope for an increase in the speed of reading data from the memory cells due to the higher mobility of charge carriers in the transistor channel. In addition, the similarity of the structure of materials III–V and Si, as well as close values of technological norms of lithography and other post-growth processes for Si and III–V chips [18], suggest that: (i) the planar cell density of III–V SAQD-memory can reach that of planar Si flash and DRAM, and (ii) III–V flash technology can walk the path that Si flash technology has traveled and take full advantage of the 3D memory architecture. All these SAQD flash memory features are a perspective for the creation of universal memory, which combines the advantages of a fast DRAM memory and a long storage time of a non-volatile flash memory, and that allows for expecting a revolution in computer architecture. It is also necessary to note that some of the III–V materials are characterized by a high probability of optical interband transitions. This allows us to hope for the development of memory with optical access, which will accelerate the transition to computing systems based on the principles of photonics.

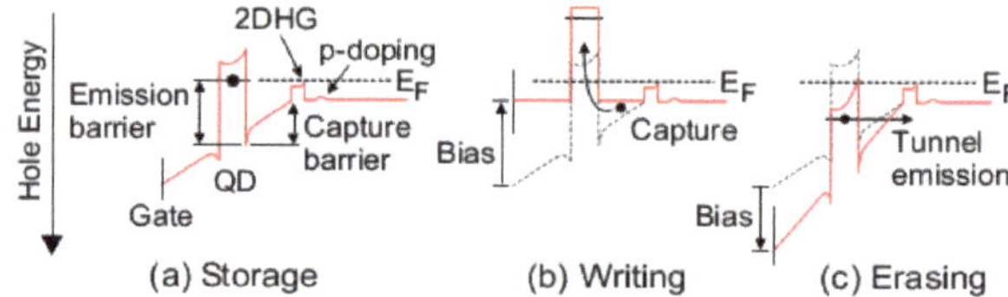

Figure 2. Schematic illustration of the (**a**) storage, (**b**) write and (**c**) erase operations in the SAQD–Flash concept [12].

Nevertheless, developing an SAQD-based non-volatile memory is still far from being finished. The main problem arising in this way is the insufficient hole localization energy in known SAQDs (E_{loc}) resulting in a low charge storage time. Actually, the first high-temperature prototype based on InAs/AlGaAs SAQDs [13,14] was characterized by a storage time of about few milliseconds, caused by E_{loc} lower than 0.8 eV. Thus, it became clear that the main weak point of this technology is the insufficient hole localization energy in presently available III–V SAQDs. Further investigations were focused on the SAQD formation in III–V heterosystems with a higher E_{loc} value. This trend has involved such well-known SAQDs system as GaSb/GaAs [19–21] and relatively novel heterosystems, for instance, GaSb/AlGaAs [22,23], GaSb/AlAs [24], InSb/AlAs [25], InGaAs/GaP [26,27], GaSb/GaP [28,29], InGaSb/GaP [30,31] and others, as discussed in [32]. The highest E_{loc} value was obtained for GaSbP/GaP SAQDs (about 1.18 eV [29]), and it results in a storage time of about four days. As is mentioned in Ref. [32], a storage time longer than 10 years may be reached at $E_{\mathrm{loc}} > 1.3$–1.5 eV.

Despite the impressively long charge storage in GaSbP/GaP SAQDs, this is not yet sufficient for a non-volatile memory cell. As it appears, a further increase in E_{loc} can be realized by increasing SAQD sizes and/or by the increase in the Sb content in ternary alloy GaSbP. However, it inevitably leads to a rise in strain level and a risk of plastic SAQD relaxation. On the other hand, it is possible to increase the hole localization by downshifting the matrix valence band top using AlP instead of GaP. Indeed, the valence band top in AlP lies ~0.5 eV lower than the GaP valence band top [33]. According to the simplest estimations, this allows us to expect an increase in the localization energy by the same 0.5 eV. According to calculations [12,13,32], an increase in the localization energy by 50 meV provides an increase in the charge storage time at room temperature by one order of magnitude. That is, *ceteris paribus*, replacing the matrix material from GaP to AlP can increase the storage time by 10 orders of magnitude, which corresponds to 10^8 years. However, despite such optimistic forecasts, the InGaSb/AlP heterosystem remains unexplored either theoretically or experimentally. Earlier, the Sb-based SAQDs in the AlP matrix were just briefly mentioned in [29,32,34], and no detailed SAQD energy spectrum calculations and, especially, attempts of epitaxial growth can be found in the literature. Embedding AlGaP barriers in the heterostructures with InGaSb/GaP SAQDs [31] and AlP barriers growing close to InGaAs/GaP SAQDs [27] were only discussed.

In this contribution, we report on the InGaSb/AlP SAQDs' energy spectrum calculations, taking into account the material intermixing and SAQD formation from the alloy with Al and P atoms. It is shown that strained GaSbP/AlP SAQDs are optimal for the non-volatile memory fabrication and allow for expecting the hole localization energy (up to 2.04 eV) along with the minimal elastic strain. Additionally, the unstrained InGaSb/AlP SAQD hole energy spectrum was investigated. The hole localization energy values are predicted up to $E_{\mathrm{loc}} = 1.65$–1.70 eV, and it precipitates the prospects of using unstrained InGaSb/AlP SAQDs for non-volatile memory applications.

2. Calculation Procedure

As was mentioned, a significant variation in the SAQD energy spectrum, provided by a strong influence of SAQD geometry and alloy composition on the energy level position, is the key feature inducing the technological interest in SAQD heterostructures. On the other hand, the energy spectrum's sensitivity to SAQD parameters makes the spectrum prediction a complicated task, requiring a wide range of parameters' variation. The present work is aimed at the theoretical investigation of the energy spectrum for novel InGaSb/AlP SAQDs depending on SAQD sizes and the alloy composition with the focus on hole localization energy.

First, we need to discuss the SAQD shape, which was used for modeling. Here, it is necessary to observe the available experimental data related to the III–V SAQD formation. InAs/GaAs is the most investigated III–V heterosystem. There is a huge quantity of experimental data indicating that, for the InAs/GaAs SAQDs, the nanostructures of different shapes, including disks, truncated cones, lenses, pyramids or truncated pyramids, can be produced by Stranski–Krastanov growth techniques [6,35–40]. Comparatively, SAQDs formed in the III-phosphide matrixes were rarely studied. The experimental data related to InGaAs/GaP [26], GaSb/GaP [41] and InGaSb/GaP [30] SAQDs allow us to use the truncated pyramid shape for modeling InGaSb/AlP SAQDs. Following [41], where the dot shapes were discussed in detail, we select a pyramid with a square base oriented along [011]-like directions and limited by lateral (111) and top (100) facets with the SAQD aspect ratio of 1:4. The SAQD shape used for modeling is shown in Figure 3.

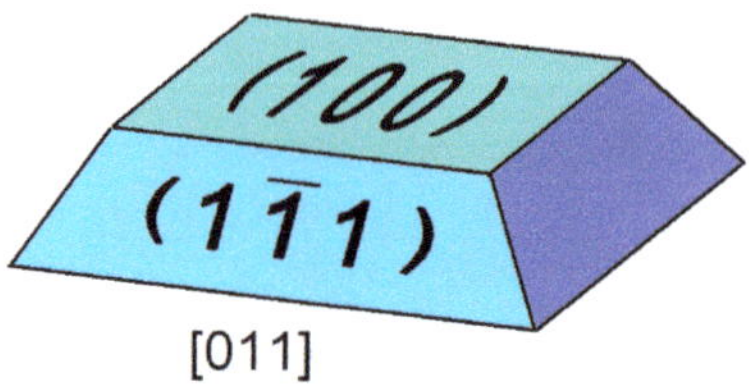

Figure 3. SAQD shape with the quadratic base along [011]-like directions and limited by (100) top/base planes and (111)-like side planes.

Besides the shape factor, the SAQD energy spectrum may be affected by elastic deformation. All III-phosphides and III-antimonides are of the same crystal structure (zinc blend) type, but the crystals are characterized by quite different lattice constants. In the most strained case of the InSb/AlP pair, the difference is as high as 15% [33]. Therefore, the nanoislands are inevitably formed with elastic strains. According to the model-solid theory [42], a strain leads to an energy band shift that, consequently, has its effect on the SAQD energy spectrum. Unfortunately, when the elastic energy of SAQD exceeds some critical level, plastic relaxation processes can occur and threading dislocations can be generated [43–46]. A distortion of translation symmetry in the dislocation core results in the appearance of deep centers close to the SAQDs, and it could lead to the tunnel hole escape from an SAQD and, consequently, to a dramatic shortening of hole storage time. Moreover, threading dislocations provide an electric bypass in the heterostructure that hinders the band-bending control by the gate bias. Thus, the calculation of critical SAQD sizes, accounting for the effects related to the introduction of dislocations, is an important task in our modeling.

Furthermore, material intermixing and SAQDs formation from the solid alloy result in the change in basic material parameters, such as lattice constant, bandgap and valence band offset [33], that also influence the strain value and energy band positions. The $In_xGa_{1-x}Sb$ deposition on AlP can, in general, result in the $In_xGa_{1-x}Sb_yP_{1-y}$/AlP SAQDs formation. The material intermixing can occur during the III-Sb deposition stage due to bulk interdiffusion [47–50], as well as during the growth of cap AlP layers due to the material segregation and exchange reaction between the SAQD and the AlP matrix [51–59]. In order to simplify calculations, we considered SAQDs consisting of quaternary alloys $Ga_xAl_{1-x}Sb_yP_{1-y}$, $In_xAl_{1-x}Sb_yP_{1-y}$ and $In_xGa_{1-x}Sb_yP_{1-y}$. As is shown below, this simplification does not decrease the interest in our results, but allows us to demonstrate the main regularity of the SAQD energy spectrum depending on the alloy composition. Thus, the calculation procedure can be subdivided into the following stages: (i) calculation of critical SAQD size depending on the SAQD alloy composition, (ii) taking into account the effect of material mixing on the SAQD material parameters, (iii) calculations of strain distribution in SAQDs, (iv) SAQD band alignment calculations and (v) charge carrier energy level calculations. Let us describe this algorithm in detail.

2.1. Critical Sizes Calculation

The introduction of the dislocation occurs when the critical SAQD elastic energy level is reached [43–46], and it is governed by SAQD sizes and SAQD/matrix lattice constant mismatch f dependent on the SAQD alloy composition. Thus, we need to calculate the critical sizes of SAQD as a function of the alloy composition. The force-balance approach [60,61] was used for the critical SAQD sizes calculation. This approach was developed by Fischer for the prediction of critical thickness when the dislocations are introduced in strained films, and, also, it was successfully used for the critical thickness prediction in the GeSi/Si layers [61]. For the III–V SAQD, the approach was first used in [60]. In the framework of this approach, the equilibrium situation is provided by the competition of two forces: (i) strain-assisted force that leads to the dislocation introduction and (ii) dislocation loop tension directed to the minimization of the total dislocation length. To simplify the calculations, in our case, the problem was simplified to the interaction between two 90° Lomer dislocations. Indeed, as is presented by the SAQD sketch given in Figure 4 [60], 60° dislocations in the SAQD can join and form 90° Lomer dislocations. These Lomer dislocations, lying at the bottom and top SAQD heterointerfaces, form the dislocation dipole (i.e., dislocations with equal, but opposite Burgers vectors).

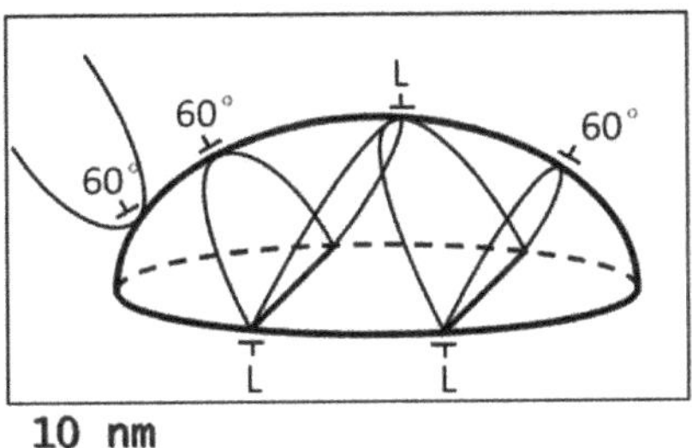

Figure 4. Location of dislocation loops around the SAQD; 60° dislocations tend to combine into Lomer dislocations (L). A dislocation loop (left) fails to wrap around the dot and threads towards the surface [60].

A distance between dislocations in a dipole corresponding to a dot height can be obtained from the equilibrium condition on the forces acting on dislocation loops, and it leads to the absence of excess shear stress τ_{exc}:

$$\tau_{exc} = \cos\lambda \ \cos\varphi \ \frac{2G(1+\nu)}{1-\nu}\left(f - \frac{b \ \cos\lambda}{2R_{h,p}}(1+\beta)\right) = 0 \tag{1}$$

with

$$\beta = \frac{1 - \frac{\nu}{4}}{4\pi\cos^2\lambda \ (1+\nu)} \ln\left(\frac{R_{h,p}}{b}\right) \tag{2}$$

$$R_{h,p} = \left(\frac{4}{h^2} + \frac{4}{p^2}\right)^{-\frac{1}{2}} \tag{3}$$

where $\lambda = 60°$ is the angle between the Burgers vector and the direction in the interface plane that is normal to the dislocation line, $\varphi = 35.3°$ is the angle between the slip plane and the strained interface normal, G is the shear modulus, $\nu = 0.31$ is the Poisson ratio (close for all binary antimonides and phosphides), f is the lattice mismatch between an SAQD alloy and AlP matrix, $b = a/\sqrt{2}$ is the Burgers vector magnitude (a is the SAQD lattice constant), h is the separation between two segments of the dislocation dipole and p is the lateral separation between two dislocations. Since the introduction of one dislocation was considered, the p value was increased to the infinity. Solving this transcendent equation by the graphical method (see sample plot in Figure S1 in the Supplementary Materials) and varying f and b according to the alloy composition by the linear Vegard law for quaternary alloys of generalized composition $A_xB_{1-x}C_yD_{1-y}$ ($a_{ABCD} = xy \cdot a_{AC} + (1-x)y \cdot a_{BC} + (1-x)(1-y) \cdot a_{BD} + x(1-y) \cdot a_{AD}$), we obtained the critical SAQD height (h_c) as a function of alloy composition. The obtained h_c values were used for consequent calculations.

2.2. Effect of Alloy Composition

The formation of a solid alloy from different atom species results in a disorder and material parameter fluctuation on a micro scale, comparable with few interatomic distances [62–65]. However, typical SAQD sizes of several nanometers [3–11] significantly exceed this distance, and this fact allows us to neglect the disorder and consider a solid alloy as a material with averaged constant parameters governed by the alloy composition. Material parameters for quaternary alloys of the $A_xB_{1-x}C_yD_{1-y}$ type were calculated in the quadratic approach using expression [33]:

$$P_{ABCD} = xy \cdot P_{AC} + (1-x)y \cdot P_{BC} + (1-x)(1-y) \cdot P_{BD} + x(1-y) \cdot P_{AD} + \\ +x(1-x)y \cdot C_{ABC} + (1-x)y(1-y) \cdot C_{BCD} + x(1-x)(1-y) \cdot C_{ABD} + xy(1-y) \cdot C_{ACD} \tag{4}$$

where x and y are the fractions of corresponding atoms, P_{ij} is the parameter value for binary compounds and C_{ijk} is the bowing parameters for the corresponding ternary alloys.

During the SAQD formation, material intermixing processes may result in the distribution of alloy composition over the SAQD bulk. The studies of composition variation in the well-known ternary alloy $In_xGa_{1-x}As/GaAs$ SAQDs yielded a reverse triangle distribution of the In content along the growth axis [35,36], where the In content rises to the SAQD top. However, quaternary and more complicated alloy SAQDs have not been investigated in such detail. In one of few contributions related to the consideration of multielement alloy SAQDs, the experimental data were reported for the atom distributions in InGaAsSb/GaP SAQDs grown on the GaAs sublayer [66]. The non-uniform element distributions along the growth direction were determined. The trends of element distribution were close to the triangle shapes, but the peak positions for In and Sb were shifted to the SAQD top in reference to the peak positions for Ga, P and As, and that was caused by drastically different segregation effects. However, the available data are not essential to see how the element profiles are changed as a function of average element content. Therefore, in order to simplify the calculation procedure, we used the SAQD model with a fixed alloy composition over the SAQD bulk.

2.3. Strain Distribution

Since the typical SAQD size drastically exceeds the interplanar distances in III–V materials, the SAQD material can be considered as a continuous matter. The strain was calculated in the framework of the model-solid approach [42]. A thin 2D layer coherently strained to the surrounding matrix and arranged along the (100) plane is schematically presented in Figure 5. Both layer and matrix materials are related to the zinc blend type. The lateral lattice constant of the thin layer is equal to the lattice constant of the unstrained matrix:

$$a_{||} = a_{mat} \tag{5}$$

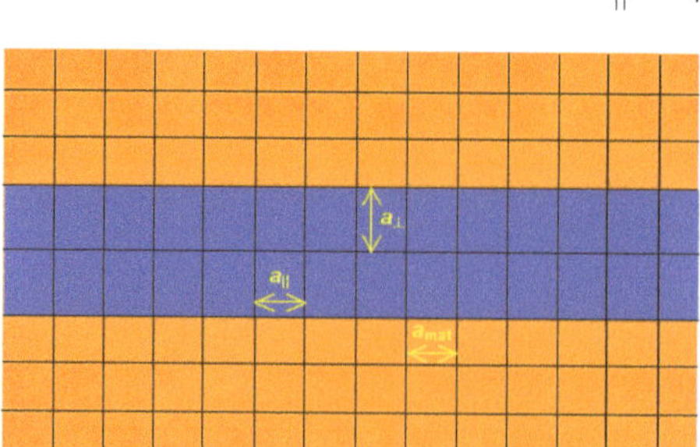

Figure 5. Thin layer of the material with a high lattice constant coherently strained to the matrix of a material with a low lattice constant.

Therefore, the lateral component of deformation tensor ε_{xx} can be written as:

$$\varepsilon_{xx} = \varepsilon_{yy} = \frac{a_{||} - a_0}{a_0} = \frac{a_{mat}}{a_0} - 1 = f \tag{6}$$

where a_0 is the unstrained lattice constant of thin layer material. According to the model-solid approach, the in-plane deformation induces the corresponding deformation along the growth axis. The lattice constant of the thin layer in the growth axis direction is:

$$a_\perp = a_0 \left(1 - 2\frac{C_{12}}{C_{11}}\varepsilon_{xx} \right) \tag{7}$$

where C_{12} and C_{11} are the elastic constants of a thin layer material. Thus, the component of the deformation tensor along the growth axis is:

$$\varepsilon_{zz} = \frac{a_\perp}{a_0} - 1 = -2\frac{C_{12}}{C_{11}}f \tag{8}$$

In this simple case, the deformation is almost localized in the thin strained layer, in contrast to the case of 3D SAQD, where the surrounding matrix layers are also strained [67–70]. The strain distribution over the SAQD volume and attached matrix material was calculated by the elastic energy minimization technique. This technique is based on the mesh point positions' variation until the total elastic energy of the system reaches a minimum. The calculations were performed using the Nextnano++ program package [71].

2.4. Band Alignment Calculation

The SAQD band alignment was calculated in two steps [42]. First, the band alignment for an unstrained SAQD was obtained using the valence band offsets (VBO) and bandgap energy values for the solid alloy [33]. Then, the band edge shift due to the elastic strain was taken into account using deformation potentials [42]. The strain itself can be divided into the hydrostatic strain $H = \varepsilon_{xx} + \varepsilon_{yy} + \varepsilon_{zz}$, that controls the change in the unit cell volume, and biaxial strain $I = \varepsilon_{zz} - \varepsilon_{xx}$, that accounts for the distortion of the unit cell shape. A hydrostatic strain leads to a change in charge density in a crystal, and that has the effect on bandgaps. The biaxial strain results in the reduction in the unit cell symmetry, and it affects the degenerated energy bands' splitting in X electron valleys and heavy-, light- and spin-orbital splitting hole sub-bands at the   point of the Brillouin zone for the zinc blend type crystals. Note that L electron valleys are not subject to splitting in the case of the (100) oriented interface. The band edge shifting due to hydrostatic strain H can be calculated by the following simple expression:

$$\Delta E_i^H = a_i\, H \tag{9}$$

where index i means the electron , X or L valley, or valence band and a_i, is the corresponding hydrostatic deformation potential. The X electron band edge splits into X_Z sub-bands, where the electron quasi-momentum is orthogonal to the interface, and X_{XY} sub-band, where the electron quasi-momentum is parallel to the interface plane. The corresponding energy shifts were determined by expressions:

$$\Delta E_Z = \frac{2}{3}b_X\, I \tag{10}$$

$$\Delta E_{XY} = -\frac{1}{3}b_X\, I \tag{11}$$

where b_X is the shear deformation potential for the X electron valley. Note that, for the typical case of a narrow bandgap material layer with high lattice constant embedded into a wide bandgap material matrix with low lattice constant, the value of biaxial strain I is positive. In combination with the positive values of b_X for all III–V materials (see *Section 2.6*

Materials Parameters and *Supplementary Materials*), this fact makes ΔE_Z positive and ΔE_{XY} negative. The energy splitting for hole subbands is described by relations:

$$\Delta E_{lh} = -\frac{1}{6}\Delta_0 + \frac{1}{4}\delta E_{001} + \frac{1}{2}\left(\Delta_0^2 + \Delta_0 \delta E_{001} + \frac{9}{4}(\delta E_{001})^2\right)^{\frac{1}{2}} \tag{12}$$

$$\Delta E_{hh} = \frac{1}{3}\Delta_0 - \frac{1}{2}\delta E_{001} \tag{13}$$

$$\Delta E_{SO} = -\frac{1}{6}\Delta_0 + \frac{1}{4}\delta E_{001} - \frac{1}{2}\left(\Delta_0^2 + \Delta_0 \delta E_{001} + \frac{9}{4}(\delta E_{001})^2\right)^{\frac{1}{2}} \tag{14}$$

where $\delta E_{001} = 2b_v I$, with valence band shear deformation potential b_v and Δ_0—spin-orbit splitting. Note that, in these expressions, energy shifts were calculated in relation to the averaged valence band energy, which lies $\Delta_0/3$ lower than the valence band top of an unstrained zinc blende crystal. For clarity, in order to illustrate the band alignment formation for a strained thin layer, we present all the above-described stages in Figure 6 by the example of GaSb/AlP. The band diagram for the unstrained heterojunction is presented in Figure 6a. The energy shifts due to the hydrostatic strain are accounted for in Figure 6b. As is clearly seen from the figure, the hydrostatic strain leads to the shifts in electron and hole band edges without splitting. The total effect of the hydrostatic and biaxial strains on the band alignment is shown in Figure 6c. The biaxial strain leads to energy band splitting, as it is governed by Expressions (10) and (11).

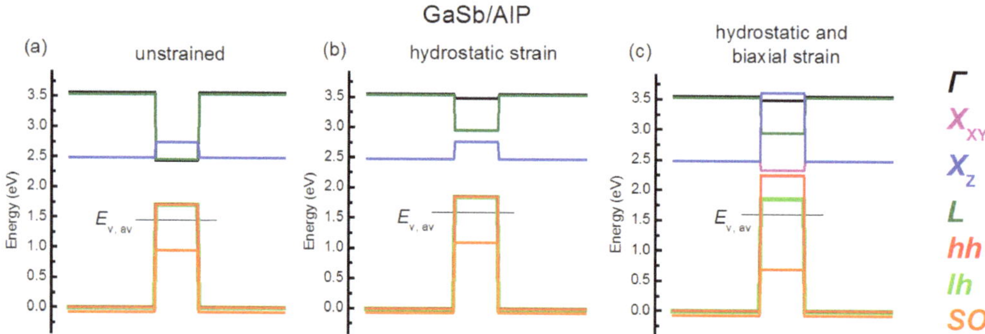

Figure 6. Band alignment for the thin GaSb layer coherently strained to the AlP matrix oriented along the (100) plane. The calculation is performed for 300 K. The band alignment for the unstrained GaSb/AlP layer (**a**), energy shifts due to the hydrostatic strain (**b**) and the both of the hydrostatic and the biaxial strains (**c**) are presented.

2.5. Energy Levels

The crucial feature of SAQDs, that distinguishes them from other low-dimensional semiconductor heterostructures such as quantum wells or quantum wires, is the 3D charge-carrier localization due to small SAQD sizes, which is comparable with the de Broglie wavelengths of electrons and holes. In order to calculate the charge carrier energy levels in an SAQD, we need to solve the 3D Schrödinger equation for a potential well formed by the SAQD band alignment. The simplest way is a calculation in the simple band approach, when charge carriers are considered as quasi-particles and where the effective mass and the interband interaction is not accounted. However, the interband interaction might result in a shift of energy levels, and it may be significant. The interaction between electrons and heavy-, light- and spin-splitting holes can be taken into account in the framework of the 8-band $k \times p$ approach [72]. We perform the test calculation of energy levels in the unstrained GaSb/AlP SAQD for different sizes using simple band and 8-band $k \times p$ approaches for comparison. The obtained results are shown in Figure 7. The calculations were performed using the Nextnano++ program package. This program package is commonly used for III–V SAQD energy spectrum calculations [25,28,73,74].

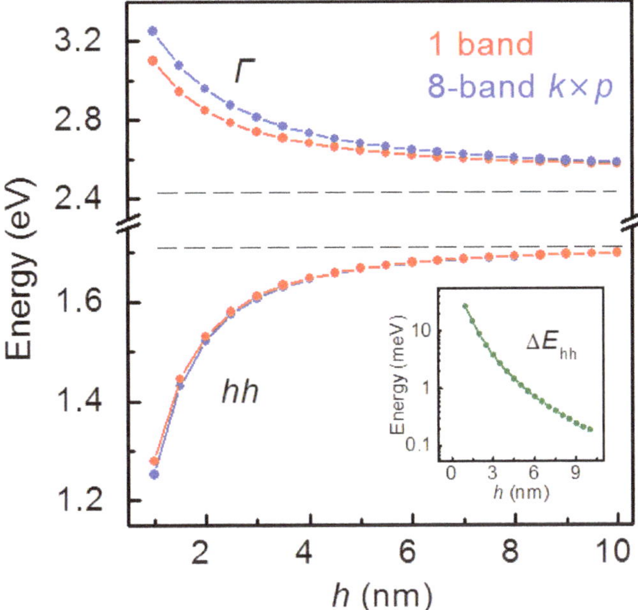

Figure 7. Energy levels for the ground state electrons and holes in unstrained GaSb/AlP SAQDs as a function of SAQD height. The calculation was performed in the simple band approach (red line-dots) and in the 8-band $k \times p$ approach (blue line-dots). The horizontal dashed lines show the band edge energies of the valley of the conduction band and the heavy hole sub-band for GaSb. The energy difference between the hole states, calculated in different approaches, is presented in the inset.

We need to note that the calculations were performed taking into account electrons' states and heavy-, light- and spin-splitting holes' states into SAQDs, without a consideration of the X electrons' states into the AlP matrix. Nevertheless, in the comparison of the energy of electrons with the X band edge in AlP shows, the ground electron states of unstrained GaSb/AlP SAQDs lie in the X valley of a conduction band of AlP. As is clearly seen from the curve behavior, the energy levels are shifted to the GaSb band edge in the SAQD, as shown in Figure 7 by dashed lines, with the increase in SAQD sizes. It is caused by reducing the quantum confinement effects. It is necessary to note that the difference in the electron energy level position, calculated in different approaches, exceeds 150 meV at the minimal SAQD height of 1 nm, and the value decreases with the dot size increase. It is topical to compare this energy difference with the uncertainty of material parameters reported for III–V compounds. The measurements of the GaSb direct bandgap at 300 K, performed by the photoreflectance and absorption spectroscopy techniques [75–77], yield the values in the range of 0.723–0.727 eV. Comparatively, VBO for the GaSb/AlP heterointerface is known with an accuracy of about 50 meV, as was discussed in [78]. Since this uncertainty is lower than the obtained energy difference, we conclude that, for a correct prediction of electron level position, the 8-band $k \times p$ approach is more useful. However, as can be seen in Figure 7, the energy difference between the results found by the simple band and 8-band $k \times p$ approaches for hole states is about 26 meV at the minimal SAQD height and even falls down lower than 1 meV with a dot size increase. This difference (i) does not exceed the VBO uncertainty and (ii) is not so significant in comparison with the hole localization energy, which is about 1.2–1.65 eV. This allows for the conclusion that, for the hole state energy calculation with the focus on E_{loc}, simple band and 8-band $k \times p$ approaches yield close results. Taking into account the easier and faster calculations with the use of the simple band approach, just this model was used for the following calculations.

2.6. Material Parameters

The material parameters, such as lattice constants, elastic constants, VBO, bandgap energies for , X and L valleys, spin-orbit splitting energy, effective charge-carrier masses, hydrostatic and shear deformation potentials and band parameters for 8-band $k \times p$ calculations for AlP, GaP, InP, AlSb, GaSb and InSb, and corresponding bowing parameters for ternary alloys, which were used for calculations, were taken from [33,79–81]. All used parameters are presented in Tables S1 and S2 in the Supplementary Materials.

3. Results

3.1. Critical SAQD Sizes

First of all, the critical SAQD sizes calculated for different quaternary alloy compositions are presented in Figure 8.

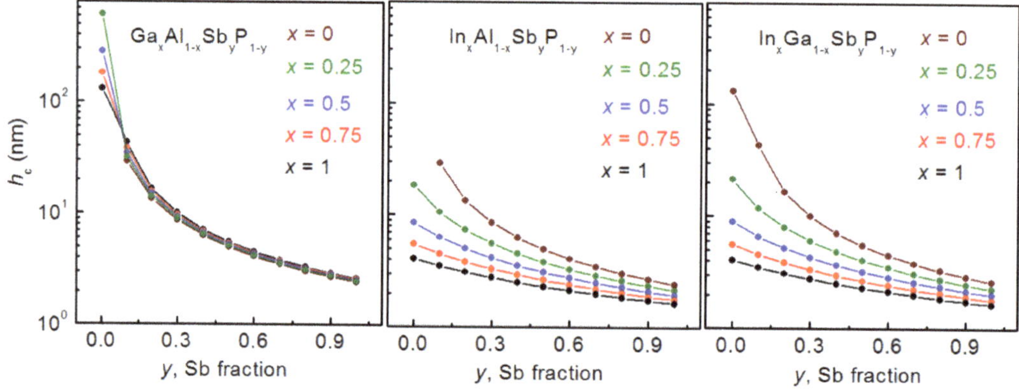

Figure 8. Dependences of h_c on the Sb fraction (y) for different contents of group III element (x) in Ga$_x$Al$_{1-x}$Sb$_y$P$_{1-y}$, In$_x$Al$_{1-x}$Sb$_y$P$_{1-y}$ and In$_x$Ga$_{1-x}$Sb$_y$P$_{1-y}$ SAQDs embedded in the AlP matrix.

As shown in the figure, increasing the Sb fraction in the SAQD composition leads to a drastic decrease in h_c, and this is caused by the increase in related lattice constants mismatch. At the same time, h_c is practically not changed with the variation in Ga fraction in Ga$_x$Al$_{1-x}$Sb$_y$P$_{1-y}$ SAQDs because of close lattice constants in the GaP and AlP and in the GaSb and AlSb pairs [33]. On the contrary, increasing the In content in In$_x$Al$_{1-x}$Sb$_y$P$_{1-y}$ and In$_x$Ga$_{1-x}$Sb$_y$P$_{1-y}$ SAQDs results in the appreciable h_c reduction governed by the trend of lattice constants mismatch. Note that the h_c value, for the most strained InSb, GaSb and AlSb SAQDs in the AlP matrix, is about 1.6, 2.6 and 2.6 nm, respectively. It is well-known that the localization energy in the SAQDs decreases with the SAQD size reduction due to the quantum confinement effect and that E_{loc} also increases with the increase in the fraction of narrow bandgap material in the alloy composition. Based on the strong dependences of h_c on the alloy composition (x,y), we expect a competition between the quantum confinement effect and the alloy composition effect in the determination of the hole energy level's position and, consequently, hole localization energy. This competition would manifest itelf in the nonmonotonic character of E_{loc} (x,y) dependencies. However, our expectations are not justified, and the E_{loc} (x,y) dependencies are monotonic in general, as will be demonstrated further.

3.2. Strain Distribution and Band Alignment

The obtained critical SAQD sizes were used for the strain calculations. The calculated strain distribution in GaSb$_{0.65}$P$_{0.35}$/AlP SAQD with the height of 4 nm is presented in Figure 9. As is shown below, this alloy composition provides the minimal elastic strain along with the E_{loc} value sufficient for application in non-volatile memories (~1.50 eV [12,13]). As is clearly seen in Figure 9, the strain distribution is strongly heterogeneous, and the deformation is not localized in the SAQD bulk. This provides a partial strain relaxation into the SAQD and reduces absolute peak values of deformation tensor components down to ε_{xx} = 6.78% and ε_{zz} = 4.49%, compared to 6.83%, as governed by the lattice constants mismatch. The results are in good agreement with the III–V SAQD strains discussed in the literature [67–70]. Note that the strain distribution is not significantly changed by the alloy composition variation.

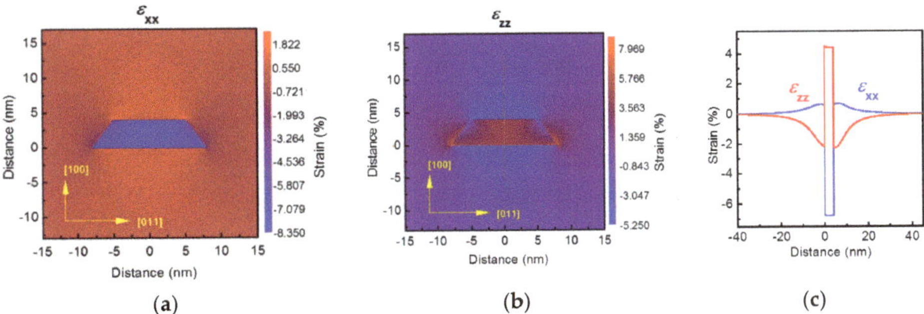

Figure 9. Strain distributions calculated for GaSb$_{0.65}$P$_{0.35}$/AlP SAQD with the height of 4 nm: (**a**) 2D map of ε_{xx} component, (**b**) 2D map of ε_{zz} component and (**c**) ε_{xx} and ε_{zz} trends along the central vertical SAQD axis marked by a thin yellow dashed line in panel (**b**).

3.3. Band Alignment and Localization Energy

The calculated band lineups of GaSb$_{0.65}$P$_{0.35}$/AlP SAQD are presented in Figure 10. The band edge trends along the central SAQD axis for the , X_Z, X_{XY} and L valleys of the conduction band and the heavy-, light- and spin-orbital splitting hole sub-bands are presented in the figure. The ground hole and electron states of the SAQD belong to the heavy hole band and the X_{XY} valley of conduction band, respectively, forming a band alignment of type-I. It is known that the heterostructures with the band alignment of type-I have a potential that localizes electrons and holes into the central narrow-bandgap material, while the heterostructures with the band alignment of type-II can localize only one kind of charge-carriers, electrons or holes in the central SAQD region and, for nonlocalized carriers, the potential works as a barrier [82,83]. The variation in SAQD alloy composition and size shows that the ground hole state of SAQDs lies in the subband of heavy holes, independently of the SAQD parameters. However, the composition and/or size variations can lead to a shift in the ground electron state to the X_Z valley of AlP conduction band, and it leads to the change in the type-I SAQD band alignment to type-II. It is necessary to note that the similar change in the heterostructure band alignment type was observed experimentalty for GaSb/GaP [28], GaAs/GaP [84] and InSb/AlAs [85] heterostrucrutes. Accordingly, the control of the type of heterostructure band alignment is important for the prediction of the absorption and recombination properties, which are governed by interband transitions. Since the present study is focused on the hole energy spectrum, a detailed determination of SAQD parameters controlling the band alignment type is beyond the scope of this work.

The localization energy E_{loc} is determined as an energy difference between the ground hole state and AlP valence band edge with the minimal energy in the SAQD vicinity, as shown in Figure 10. The dependence of E_{loc} on the Sb fraction (y) in Ga$_x$Al$_{1-x}$Sb$_y$P$_{1-y}$,

$In_xAl_{1-x}Sb_yP_{1-y}$ and $In_xGa_{1-x}Sb_yP_{1-y}$ SAQDs, at several fixed Ga or In fractions (x), is presented in Figure 11.

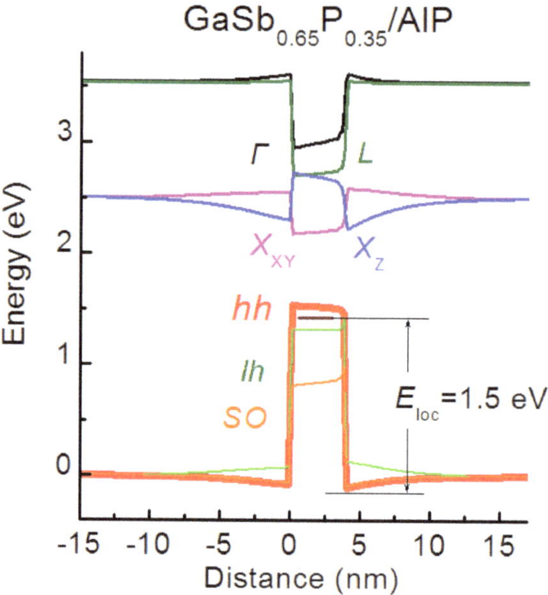

Figure 10. Calculated band alignment for the GaSb$_{0.65}$P$_{0.35}$/AlP SAQD.

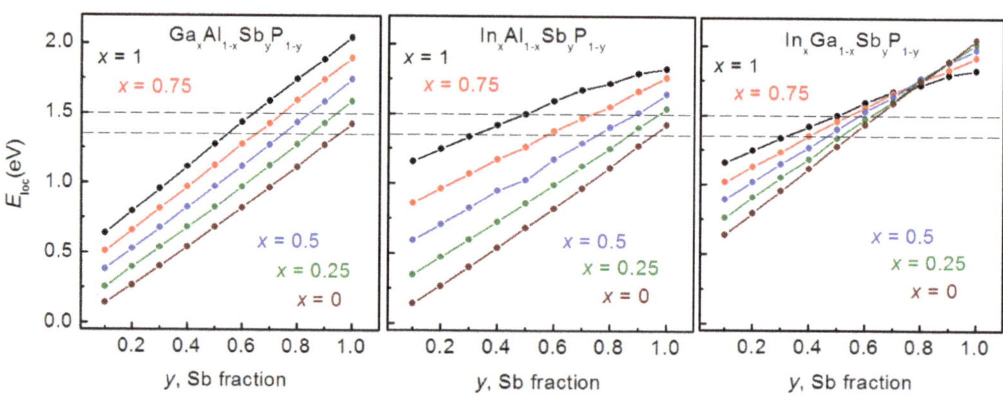

Figure 11. E_{loc} (y) dependences for different contents of group III elements, as calculated for $Ga_xAl_{1-x}Sb_yP_{1-y}$, $In_xAl_{1-x}Sb_yP_{1-y}$ and $In_xGa_{1-x}Sb_yP_{1-y}$ SAQDs embedded in the AlP matrix. Horizontal dashed lines indicate the energy range of E_{loc} = 1.35–1.5 eV required for 10 years of hole storage [12,13,32,34].

As is evident from the curve observation, with the Sb fraction increase, E_{loc} monotonically increases in all three systems, and a value as high as ~2.04 eV is observed for pure GaSb/AlP SAQDs. The increase in Ga and In contents in $Ga_xAl_{1-x}Sb_yP_{1-y}$ and $In_xAl_{1-x}Sb_yP_{1-y}$ SAQDs, respectively, also results in the E_{loc} increase. These results indicate that the variation in E_{loc} value is mainly determined by the change in alloy composition due to the changes in VBO. Indeed, as was shown by the SAQD critical size calculations, the increase in the fraction of narrow band gap material (InSb or GaSb) in the alloy is accompanied by a reduction in h_c. Nevertheless, the continuous E_{loc} growth with the increase in Sb or Ga/In fraction is observed in Figure 11, and it indicates a relatively weak role of the quantum confinement effect in the formation of the SAQD energy spectrum due to the high effective mass values for heavy holes. Thus, the nearly linear shape of the E_{loc} (y) dependences is governed by the alloy composition variation. This factor is dominant because the energy position of the SAQD valence band top is proportional to the alloy composition [33]. The sublinear behavior of the E_{loc} (y) functions observed for the InSb$_y$P$_{1-y}$/AlP SAQDs at $y > 0.6$ may be explained by a weak contribution of the quantum confinement effect for SAQDs with $h_c < 2.5$ nm. In the case of In$_x$Ga$_{1-x}$Sb$_y$P$_{1-y}$ SAQDs, the E_{loc} (y) dependence changes the character smoothly from GaSb$_y$P$_{1-y}$ to InSb$_y$P$_{1-y}$ with an x increase. Note that, for $y > 0.8$, E_{loc} is decreased, with a rise of the In content in an SAQD.

4. Discussion

Let us discuss the calculated results in the light of a possible application of the SAQDs under consideration in non-volatile memory cells. As was shown, SAQDs formed in the InGaSb/AlP heterosystem are characterized by a high hole localization energy up to 2.04 eV, and, accordingly, they are prospective objects for non-volatile memory cells. The variations in solid alloy compositions and sizes of SAQDs allow us to estimate the optimal SAQDs configuration.

First of all, we need to compare two extreme cases of GaSb/AlP and InSb/AlP SAQDs. As was obtained, GaSb/AlP SAQD, with the height of 2.6 nm and base lateral size of 10.4 nm, is characterized by $E_{loc} = 2.04$ eV, while the 1.6 nm high and 6.4 nm wide InSb/AlP SAQD has $E_{loc} = 1.82$ eV. The lower localization energy in the InSb/AlP SAQD is caused by a lower valence band discontinuity and stronger quantum confinement effect, as is clearly seen in Figure 12, where the corresponding band diagrams are given. The higher E_{loc} value, in combination with a significantly lower lattice mismatch (10.5% in GaSb/AlP vs. 15.6% in InSb/AlP [33]), indicates that GaSb/AlP SAQDs are more attractive for the creation of non-volatile memory cells. However, from the technological point of view, it is very difficult to prepare pure GaSb SAQDs embedded into the AlP matrix due to the unavoidable material intermixing during the SAQD formation [28] and the risk of plastic relaxation of the strain [86]. Moreover, according to the predictions of the hole storage time [12,13,32,34], the value of $E_{loc} \sim 2$ eV corresponds to the giant storage times of 10^8–10^{12} years at room temperature, depending on the hole capture cross-section. Such giant storage times (>10 years) are evidently redundant for non-volatile memories. Thus, without loss of memory functionality, an appropriate reduction in E_{loc} by the material intermixing is acceptable, with evident benefits of decreasing the SAQD strains. Thus, let us try to estimate the optimal SAQD configuration under the constraints of required storage time of 10 years and a minimal strain.

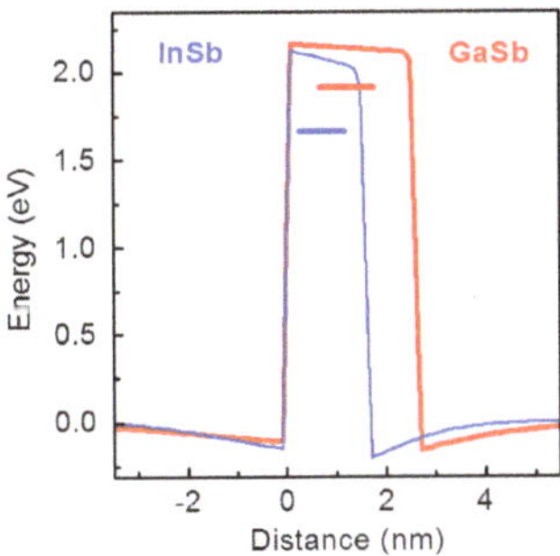

Figure 12. Valence band top diagrams and energy levels calculated for GaSb/AlP (red lines) and InSb/AlP (blue lines) SAQDs with critical sizes.

According to [12,13], the hole storage time can be estimated by relation:

$$t_s = \frac{e^{\frac{E_a}{kT}}}{\gamma T^2 \sigma_{\text{inf}}} \tag{15}$$

where T is the temperature, σ_{inf} is the capture cross-section at a high temperature and γ is the coefficient independent of temperature. Localization energies of 1.35–1.50 eV are high enough for the hole storage times of ~10 years, depending on the capture cross-section σ_{inf} which varied in the range of 10^{-12}–10^{-9} cm^2 in GaSb/GaP [29], InGaAs/GaP [27] and InGaSb/GaP [31] SAQDs, according to the available experimental results. Since the hole capture cross-section in SAQD is determined not only by the SAQD sizes but also by the hole–phonon interaction efficiency, Auger scattering and other factors [29], the prediction of the σ_{inf} value for the SAQD with specified alloy composition and size is a complicated task. Therefore, to simplify the task, the SAQDs were considered with an E_{loc} value lying in the range of 1.35–1.50 eV and with the critical dot size. The possible variation in σ_{inf} was not accounted for in the calculations. In Figure 11, the horizontal dashed lines designate the targeted E_{loc} range. Crossing these lines with E_{loc} (y) curves determined for different x values spotlights a possibility to estimate the alloy composition (x,y) related to the targeted E_{loc} level. In Figure 13, the composition parameters (x,y) for Ga$_x$Al$_{1-x}$Sb$_y$P$_{1-y}$, In$_x$Al$_{1-x}$Sb$_y$P$_{1-y}$ and In$_x$Ga$_{1-x}$Sb$_y$P$_{1-y}$ SAQDs obtained by this algorithm are shown with blue dots.

As is seen in Figure 13, in all alloys under consideration, y decreases on the x increase. The increase in x from 0 to 1 induces the y decrease from 1 to 0.55/0.65 in Ga$_x$Al$_{1-x}$Sb$_y$P$_{1-y}$ and from 1 to 0.3/0.5 in In$_x$Al$_{1-x}$Sb$_y$P$_{1-y}$, for E_{loc} fixed at 1.35 or 1.50 eV, respectively. As to the In$_x$Ga$_{1-x}$Sb$_y$P$_{1-y}$ solid solution, the x variation from 0 to 1 results in the y decrease from 0.5 to 0.3 or from 0.65 to 0.5, depending on the E_{loc} value fixed at 1.35 or 1.50 eV, respectively. The dependences of the lattice constants mismatch f in the SAQD alloys and AlP matrix on the x value are also presented in Figure 13, and they are marked with red dot-lines. The behavior of f with the x variation in different alloys is principally different. Indeed, in Ga$_x$Al$_{1-x}$Sb$_y$P$_{1-y}$/AlP, the absolute f value reduces monotonically from 10.5–10% to 6–7%, depending on the E_{loc} value, on the x variation from 0 to 1. The minimal absolute f values correspond to GaSb$_y$P$_{1-y}$/AlP SAQDs, with y lying in the range of 0.55–0.65. Note that these absolute f values of 6–7% are close to the lattice constants mismatch in the well-known InAs/GaAs SAQDs [33]. This fact is promising for the epitaxial growth of GaSb$_y$P$_{1-y}$/AlP SAQDs because the lattice mismatch is one of the important parameters controlling the possibility of SAQD formation. The absolute lattice mismatch levels in In$_x$Al$_{1-x}$Sb$_y$P$_{1-y}$ SAQD are as high as 10–12%, and they are weakly dependent on (x,y) values, as is clear in Figure 13. This effect is in a good agreement with the higher strain in the InSb/AlP system in reference to that in the GaSb/AlP system, as was discussed above. Moreover, in In$_x$Ga$_{1-x}$Sb$_y$P$_{1-y}$ SAQD, the absolute f value increases up to 9.5–11.5% at $x = 1$, as shown in the bottom panel in Figure 13. Thus, the calculations allow us to point to the GaSb$_y$P$_{1-y}$/AlP SAQD with $y = 0.55$–0.65, height of 4.0–4.5 nm and a base size of about 16–18 nm as the optimal configuration with minimal strain for providing the required $E_{\text{loc}} = 1.35$–1.50 eV.

It is necessary to discuss the perspectives of the epitaxial GaSbP/AlP SAQD growth. The novel GaSb/AlP heterosystem is most similar to the more investigated GaSb/GaP system. As was mentioned in [28,29,41], the GaSb deposition on GaP results in the GaSbP/GaP SAQD formation, with the Sb fraction lying in the range of 0.3–0.7, depending on the growth conditions. This information is optimistic for the successful growth of the GaSbP/AlP SAQDs with the required Sb content. Moreover, as shown above, the optimal configuration of GaSbP/AlP SAQDs with Sb content in the range of 0.55–0.65 is characterized by the f level being close to the lattice mismatch observed in the well-known InAs/GaAs heterosystem, where the perfectly strained SAQDs had already been obtained. However, some difficulties in the fabrication of the GaSbP/AlP SAQDs can be assumed due to the Al-Ga intermixing during the SAQD formation.

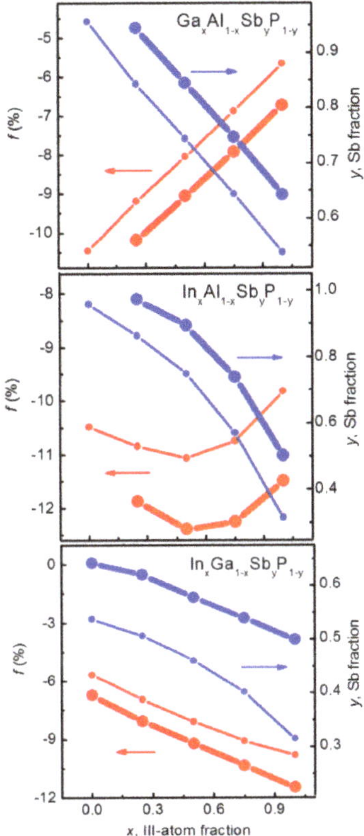

Figure 13. Sb fraction (y) as a function of group III element content (x) for the SAQD quaternary alloy (blue dots) calculated for the E_{loc} value equal to 1.35 (thin dot-lines) and 1.50 eV (thick dot-lines). Red dot-lines give the SAQD/matrix lattice constant mismatch f as a function of x with a corresponding y fraction.

Furthermore, the experimental data related to the formation of GaSb/GaP and GaAs/GaP SAQDs show a possibility of an exotic plastic strain relaxation mode. According to [84,86,87], at some growth conditions, a strain relaxation appeared due to the introduction of a system of 90° Lomer dislocations without 60° components. In this case, the dislocations lie at the interfaces, and they do not cross the SAQD bulk. Moreover, as was discussed in [88], the Lomer dislocation core does not contain dangling bonds, and that frees it from the deep level formation. The SAQD heterostructures, considered in [84,86,87], demonstrate a high luminescence intensity along with the radiative exciton recombination into SAQDs, and that confirms the absence of defect levels in and around SAQDs. Accordingly, it is expected that the SAQD system, where this strain relaxation mode is realized, may provide a long charge carrier storage time at appropriate E_{loc} values. It can be reasonably assumed that a similar strain relaxation mode can be realized in GaSb/AlP and InSb/AlP heterosystems. Thus, the calculations of E_{loc} for unstrained GaSb/AlP and InSb/AlP SAQDs are also interesting. Taking into account that this type of plastic relaxation leads to almost 100% relaxation of elastic deformations [84,86,87], the case of partial relaxation was not considered in this work.

The calculations were performed for a truncated pyramid SAQD consisting of pure GaSb and InSb in the AlP matrix. Since the calculation was implemented for the unstrained SAQDs, the strain effects were not accounted for. The energy level's position was calculated in the framework of the simple band approach. The band lineup diagram of unstrained

GaSb/AlP SAQD was already provided in Figure 6a. This one, for the InSb/AlP system, is presented in Figure 14a. In the diagrams, it is clearly seen that these lineups are very similar, especially in the valence band part, and this is explained by the close values of VBO for the unstrained GaSb/AlP and InSb/AlP heterointerfaces [33]. The hole localization energy in the unstrained GaSb/AlP and InSb/AlP SAQDs lies in the range of 1.30–1.75 eV (see Figure 14b), which makes these SAQDs interesting objects for non-volatile memories, along with the strained InGaSb/AlP SAQDs.

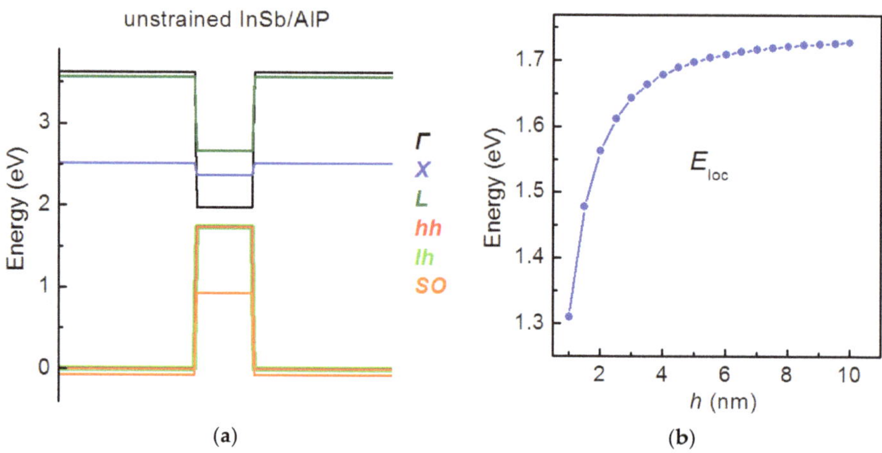

Figure 14. (**a**) Band alignment for the unstrained InSb/AlP SAQD. (**b**) The dependence of E_{loc} on the SAQD height in the unstrained InSb/AlP SAQD.

Additionally, it should be noted that the considered InGaSb/AlP SAQDs might be attributed to the novel type of semiconductor low-dimensional heterostructures with the band alignment of type-I and indirect bandgap in some alloy compositions. These SAQDs are of interest as objects for investigations of localized exciton spin dynamics [89,90] due to the extremely long radiative lifetime of indirect excitons [91] lying in the microsecond range. As was demonstrated in [89], the fast bulk spin relaxation mechanisms are suppressed for trions in SAQDs, and this resulted in long, up to 30 µs, spin relaxation times. The coexistence of >10 years of hole storage time and microsecond exciton spin relaxation time in one structure opens the way for the creation of novel prospective hybrid memory devices based on a positive trion into SAQDs, combining a long charge storage in the floating gate with fast data processing using a trion spin.

5. Conclusions

The energy spectrum of novel InGaSb/AlP SAQDs was investigated theoretically with the focus on possible non-volatile memory applications. The detailed calculations of the energy spectra of strained $Ga_xAl_{1-x}Sb_yP_{1-y}$, $In_xAl_{1-x}Sb_yP_{1-y}$ and $In_xGa_{1-x}Sb_yP_{1-y}$ SAQDs formed in the AlP matrix were performed for different alloy compositions and SAQD sizes. The variations were performed, keeping SAQD sizes as critical, with respect to the introduction of dislocations for different alloy compositions. It was theoretically elucidated that, among all SAQDs under consideration, the GaSbP/AlP SAQDs with the Sb fraction in the range of 0.55–0.65 have a minimal SAQD/matrix lattice constant mismatch and provided E_{loc} of 1.35–1.50 eV. These E_{loc} values are sufficient for non-volatile memory applications. This makes the GaSb/AlP heterosystem the most appropriate one for non-volatile memory cell fabrication. Additionally, the energy spectra of unstrained InSb/AlP and GaSb/AlP SAQDs were calculated. In these systems, the values of E_{loc} up to 1.65–1.70 eV were predicted, which also makes these SAQDs suitable for non-volatile memory applications.

Supplementary Materials: The following supporting information can be downloaded at: https://www.mdpi.com/article/10.3390/nano12213794/s1, Table S1: Materials parameters for AlP, GaP, InP, AlSb, GaSb and InSb at 300 K, which were used for the calculations. a_0—lattice constant, C_{11}, C_{12}, C_{44}—elastic constants, $E_g^{,X,L}$—bandgaps for , X and L valleys, $a^{,X,L}_v$—hydrostatic deformation potentials for the conduction band edge in , X and L points of the Brillouin zone and valence band, $b_{X,v}$—shear deformation potential for the conduction band at X point of the Brillouin zone and valence band, Δ_0—energy of spin-orbital splitting in the valence band, VBO—valence band offsets, m—electron effective mass in the point of the Brillouin zone, m_{Xt} and m_{Xl}—transversal and longitudinal electron effective mass in the X point of the Brillouin zone, m_{Lt} and m_{Ll}—transversal and longitudinal electron effective mass in the L point of the Brillouin zone, m_{hh}, m_{lh}, m_{SO}—effective masses for the heavy, light and spin-orbital splitting holes, F—Kane's parameter, E_P—Kane's matrix element, $\gamma_{1,2,3}$—Luttinger parameters for the valence band. Table S2: Bowing parameters for GaAlP, InGaP, AlInP, GaAlSb, InGaSb, AlInSb, AlSbP, GaSbP and InSbP [33]. Figure S1: Solving of the Equation (16) by the graphic method for the case of In0.5Ga0.5Sb0.5P0.5/AlP SAQD. Linear part of the equation depicted by the blue line, the logarithmic one by the red line. The black arrow points to the obtained hc value.

Author Contributions: Conceptualization, D.S.A.; methodology, D.S.A.; investigation, D.S.A.; writing—original draft preparation, D.S.A., V.V.A.; writing—review and editing, D.S.A., V.V.A.; visualization, D.S.A.; project administration, D.S.A.; funding acquisition, D.S.A. All authors have read and agreed to the published version of the manuscript.

Funding: This work was supported by the Russian Science Foundation, grant 22-22-20031 https://rscf.ru/project/22-22-20031/ (21 March 2022), and by the Novosibirsk Regional Government, grant No. r-14 (6 April 2022).

Institutional Review Board Statement: Not applicable.

Informed Consent Statement: Not applicable.

Data Availability Statement: Data are available from the authors on request.

Conflicts of Interest: The authors declare that they have no known competing financial interests or personal relations that could have appeared to influence the work reported in this paper.

References

1. Sanmartin-Matalobos, J.; Bermejo-Barrera, P.; Aboal-Somoza, M.; Fondo, M.; Garcia-Deibe, A.M.; Corredoira-Vazquez, J.; Alves-Iglesias, Y. Semiconductor Quantum Dots as Target Analytes: Properties, Surface Chemistry and Detection. *Nanomaterials* **2022**, *12*, 2501. [CrossRef]

2. Svit, K.A.; Zarubanov, A.A.; Duda, T.A.; Trubina, S.V.; Zvereva, V.V.; Fedosenko, E.V.; Zhuravlev, K.S. Crystal Structure and Predominant Defects in CdS Quantum Dots Fabricated by the Langmuir–Blodgett Method. *Langmuir* **2021**, *37*, 5651. [CrossRef]

3. Wang, Z.M. *Self Assembled Quantum Dots*; Springer: New York, NY, USA, 2008.

4. Wu, J.; Peng, J. Self-assembly of InAs quantum dots on GaAs(001) by molecular beam epitaxy. *Front. Phys.* **2015**, *10*, 7–58. [CrossRef]

5. Yakimov, A.I.; Kirienko, V.V.; Utkin, D.E.; Dvurechenskii, A.V. Light-Trapping-EnhancedPhotodetection in Ge/Si Quantum Dot Photodiodes Containing Microhole Arrays with Different Hole Depths Depths. *Nanomaterials* **2022**, *12*, 2993. [CrossRef]

6. Xu, B.; Wang, G.; Du, Y.; Miao, Y.; Li, B.; Zhao, X.; Lin, H.; Yu, J.; Su, J.; Dong, Y.; et al. Radamson Monolithic Integration of O-Band InAs Quantum Dot Lasers with Engineered GaAs Virtual Substrate Based on Silicon. *Nanomaterials* **2022**, *12*, 2704. [CrossRef]

7. Trevisi, G.; Seravalli, L.; Friger, P. Photoluminescence monitoring of oxide formation and surface state passivation on InAs quantum dots exposed to water vapor. *Nano Res.* **2016**, *9*, 3018. [CrossRef]

8. Tang, T.; Yu, T.; Yang, G.; Sun, J.; Zhan, W.; Xu, B.; Zhao, C.; Wang, Z. Investigation into the InAs/GaAs quantum dot material epitaxially grown on silicon for O band lasers. *J. Semicond.* **2022**, *43*, 012301. [CrossRef]

9. Yu, P.; Wu, J.; Gao, L.; Liu, H.; Wang, Z. InGaAs and GaAs quantum dot solar cells grown by droplet epitaxy. *Sol. Energy Mater. Sol. Cells* **2017**, *161*, 377. [CrossRef]

10. Huang, X.; Su, R.; Yang, J.; Rao, M.; Liu, J.; Yu, Y.; Yu, S. Wafer-Scale Epitaxial Low Density InAs/GaAs Quantum Dot for Single Photon Emitter in Three-Inch Substrate. *Nanomaterials* **2021**, *11*, 930. [CrossRef]

11. Von Helversen, M.; Haisler, A.V.; Daurtsev, M.P.; Dmitriev, D.V.; Toropov, A.I.; Rodt, S.; Haisler, V.A.; Derebezov, I.A.; Reitzenstein, S. Triggered Single-Photon Emission of Resonantly Excited Quantum Dots Grown on (111)B GaAs Substrate. *Phys. Status Solidi RRL* **2022**, *16*, 2200133. [CrossRef]

12. Marent, A.; Nowozin, T.; Gelze, J.; Luckert, F.; Bimberg, D. Hole-based memory operation in an InAs/GaAs quantum dot heterostructure. *Appl. Phys. Lett.* **2009**, *95*, 242114. [CrossRef]

13. Marent, A.; Nowozin, T.; Geller, M.; Bimberg, D. The QD-Flash: A quantum dot-based memory device, Semicond. *Sci. Technol.* **2011**, *26*, 014026. [CrossRef]

14. Nowozin, T.; Narodovitch, M.; Bonato, L.; Bimberg, D.; Ajour, M.N. Khaled Daqrouq and Abdullah Balamash, Room-Temperature Hysteresis in a Hole-Based Quantum Dots Memory Structure. *J. Nanotech.* **2013**, *2013*, 797964. [CrossRef]

15. Arikan, I.F.; Cottet, N.; Nowozin, T.; Bimberg, D. Transparency Engineering in Quantum Dot-Based Memories. *Phys. Stat. Solidi A* **2018**, *215*, 1800018. [CrossRef]

16. Geller, M.; Marent, A.; Stock, E.; Bimberg, D.; Zubkov, V.I.; Shulgunova, I.S.; Solomonov, A.V. Hole capture into self-organized InGaAs quantum dots. *Appl. Phys. Lett.* **2006**, *89*, 232105. [CrossRef]

17. Chau, R.; Datta, S.; Majumdar, A. Opportunities and Challenges of III-V Nanoelectronics for Future High-Speed, Low-Power Logic Applications. In Proceedings of the Technical Digest-IEEE Compound Semiconductor Integrated Circuit Symposium, Palm Springs, CA, USA, 30 October 2005–2 November 2005; p. 4. [CrossRef]

18. Riel, H.; Wernersson, L.-E.; Hong, M.; del Alamo, J.A. III–V compound semiconductor transistors—From planar to nanowire structures. *MRS Bull.* **2014**, *39*, 668. [CrossRef]

19. Nowozin, T.; Marent, A.; Bonato, L.; Schliwa, A.; Bimberg, D.; Smakman, E.P.; Garleff, J.K.; Koenraad, P.M.; Young, R.J.; Hayne, M. Linking structural and electronic properties of high-purity self-assembled GaSb/GaAs quantum dots. *Phys. Rev. B* **2012**, *86*, 035305. [CrossRef]

20. Hayne, M.; Young, R.J.; Smakman, E.P.; Nowozin, T.; Hodgson, P.; Garleff, J.K.; Rambabu, P.; Koenraad, P.M.; Marent, A.; Bonato, L.; et al. The structural, electronic and optical properties of GaSb/GaAs nanostructures for charge-based memory. *J. Phys. D Appl. Phys.* **2013**, *46*, 264001. [CrossRef]

21. Baik, M.; Kyhm, J.-h.; Kang, H.-K.; Jeong, K.-S.; Kim, J.S.; Cho, M.-H.; Song, J.D. Optical characteristics of type-II hexagonal-shaped GaSb quantum dots on GaAs synthesized using nanowire self-growth mechanism from Ga metal droplet. *Sci. Rep.* **2021**, *11*, 7699. [CrossRef]

22. Nowozin, T.; Bonato, L.; Hogner, A.; Wiengarten, A.; Bimberg, D.; Lin, W.-H.; Lin, S.-Y.; Reyner, C.J.; Liang, B.L.; Huffaker, D.L. 800 meV localization energy in GaSb/GaAs/Al$_{0.3}$Ga$_{0.7}$As quantum dots. *Appl. Phys. Lett.* **2013**, *102*, 052115. [CrossRef]

23. Shoji, Y.; Tamaki, R.; Okada, Y. Temperature Dependence of Carrier Extraction Processes in GaSb/AlGaAs Quantum Nanostructure Intermediate-Band Solar Cells. *Nanomaterials* **2021**, *11*, 344. [CrossRef] [PubMed]

24. Shamirzaev, T.S.; Abramkin, D.S.; Gutakovskii, A.K.; Putyato, M.A. Novel Self-Assembled Quantum Dots in the GaSb/AlAs Heterosystem. *JETP Lett.* **2012**, *95*, 534–536. [CrossRef]

25. Abramkin, D.S.; Rumynin, K.M.; Bakarov, A.K.; Kolotovkina, D.A.; Gutakovskii, A.K.; Shamirzaev, T.S. Quantum Dots Formed in InSb/AlAs and AlSb/AlAs Heterostructure. *JETP Lett.* **2016**, *103*, 692–698. [CrossRef]

26. Stracke, G.; Glacki, A.; Nowozin, T.; Bonato, L.; Rodt, S.; Prohl, C.; Lenz, A.; Eisele, H.; Schliwa, A.; Strittmatter, A.; et al. Growth of In$_{0.25}$Ga$_{0.75}$As quantum dots on GaP utilizing a GaAs interlayer. *Appl. Phys. Lett.* **2012**, *101*, 223110. [CrossRef]

27. Bonato, L.; Sala, E.M.; Stracke, G.; Nowozin, T.; Strittmatter, A.; Ajour, M.N.; Daqrouq, K.; Bimberg, D. 230 s room-temperature storage time and 1.14eV hole localization energy in In$_{0.5}$Ga$_{0.5}$As quantum dots on a GaAs interlayer in GaP with an AlP barrier. *Appl. Phys. Lett.* **2015**, *106*, 042102. [CrossRef]

28. Abramkin, D.S.; Shamirzaev, V.T.; Putyato, M.A.; Gutakovskii, A.K.; Shamirzaev, T.S. Coexistence of type-I and type-II band alignment in Ga(Sb, P)/GaP heterostructures with pseudomorphic self-assembled quantum dots. *JETP Lett.* **2014**, *99*, 76–81. [CrossRef]

29. Bonato, L.; Arikan, I.F.; Desplanque, L.; Coinon, C.; Wallart, X.; Wang, Y.; Ruterana, P.; Bimberg, D. Hole localization energy of 1.18eV in GaSb quantum dots embedded in GaP. *Phys. Status Solidi B* **2016**, *253*, 1869. [CrossRef]

30. Sala, E.M.; Stracke, G.; Selve, S.; Niermann, T.; Lehmann, M.; Schlichting, S.; Nippert, F.; Callsen, G.; Strittmatter, A.; Bimberg, D. Growth and structure of In$_{0.5}$Ga$_{0.5}$Sb quantum dots on GaP(001). *Appl. Phys. Lett.* **2016**, *109*, 102102. [CrossRef]

31. Sala, E.M.; Arikan, I.F.; Bonato, L.; Bertram, F.; Veit, P.; Christen, J.; Strittmatter, A.; Bimberg, D. MOVPE-Growth of In-GaSb/AlP/GaP(001) Quantum Dots for Nanoscale Memory Applications. *Phys. Status Solidi B* **2018**, *225*, 1800182. [CrossRef]

32. Bimberg, D.; Mikolajick, T.; Wallart, X. Novel Quantum Dot Based Memories with Many Days of Storage Time. In Proceedings of the NVMTS 2019-Non-Volatile Memory Technology Symposium, Durham, NC, USA, 28–30 October 2019; p. 8986178. [CrossRef]

33. Vurgaftman, I.; Meyer, J.R.; Ram-Mohan, L.R. Band parameters for III–V compound semiconductors and their alloys. *J. Appl. Phys.* **2001**, *89*, 5815. [CrossRef]

34. Nowozin, T.; Bimberg, D.; Ajour, M.N.; Awedh, M. Materials for Future Quantum Dot-Based Memories. *J. Nanomater.* **2013**, *2013*, 215613. [CrossRef]

35. Liu, N.; Tersoff, J.; Baklenov, O.; Holmes, A.L., Jr.; Shih, C.K. Nonuniform Composition Profile in In$_{0.5}$Ga$_{0.5}$As Alloy Quantum Dots. *Phys. Rev. Lett.* **2000**, *84*, 334. [CrossRef] [PubMed]

36. Bruls, D.M.; Vugs, J.W.A.M.; Koenraad, P.M.; Salemink, H.W.M.; Hopkinson, J.H.W.M.; Skolnick, M.S.; Long, F.; Gill, S.P.A. Determination of the shape and indium distribution of low-growth-rate InAs quantum dots by cross-sectional scanning tunneling microscopy. *Appl. Phys. Lett.* **2002**, *81*, 1708. [CrossRef]

37. Abouzaid, O.; Mehdi, H.; Martin, M.; Moeyaert, J.; Salem, B.; David, S.; Souifi, A.; Chauvin, N.; Hartmann, J.-M.; Ilahi, B.; et al. O-Band Emitting InAs Quantum Dots Grown by MOCVD on a 300mm Ge-Buffered Si(001) Substrate. *Nanomaterials* **2020**, *10*, 2450. [CrossRef]

38. McCaffrey, J.P.; Robertson, M.D.; Poole, P.J.; Riel, B.J.; Fafard, S. Interpretation and modeling of buried InAs quantum dots on GaAs and InP substrates. *J. Appl. Phys.* **2001**, *90*, 1784. [CrossRef]

39. Walther, T.; Cullis, A.G.; Norris, D.J.; Hopkinson, M. Nature of the Stranski-Krastanow Transition during Epitaxy of InGaAs on GaAs. *Phys. Rev. Lett.* **2001**, *86*, 2381. [CrossRef]

40. Eisele, H.; Lenz, A.; Heitz, R.; Timm, R.; Dähne, M.; Temko, Y.; Suzuki, T.; Jacob, K. Change of InAs/GaAs quantum dot shape and composition during capping. *J. Appl. Phys.* **2008**, *104*, 124301. [CrossRef]

41. Desplanque, L.; Coinon, C.; Troadec, D.; Ruterana, P.; Patriarche, G.; Bonato, L.; Bimberg, D.; Wallart, X. Morphology and valence band offset of GaSb quantum dots grown on GaP (001) and their evolution upon capping. *Nanotechnology* **2017**, *28*, 225601. [CrossRef]

42. Van de Walle, C.G. Band lineups and deformation potentials the model-solid theory. *Phys. Rev. B* **1989**, *39*, 1871. [CrossRef]

43. LeGoues, F.K.; Tersoff, J.; Reuter, M.C.; Hammar, M.; Tromp, R. Relaxation mechanism of Ge islands/Si(001) at low temperature. *Appl. Phys. Lett.* **1995**, *67*, 2317. [CrossRef]

44. Knelangen, M.; Consonni, V.; Trampert, A.; Riechert, H. *In situ* analysis of strain relaxation during catalyst-free nucleation and growth of GaN nanowires. *Nanotechnology* **2010**, *21*, 245705. [CrossRef] [PubMed]

45. Li, W.; Wang, L.; Chai, R.; Wen, L.; Wang, Z.; Guo, W.; Wang, H.; Yang, S. Anisotropic Strain Relaxation in Semipolar (11-22) InGaN/GaN Superlattice Relaxed Templates. *Nanomaterials* **2022**, *12*, 3007. [CrossRef] [PubMed]

46. Du, Y.; Xu, B.; Wang, G.; Miao, Y.; Li, B.; Kong, Z.; Dong, Y.; Wang, W.; Radamson, H.H. Review of Highly Mismatched III-V Heteroepitaxy Growth on (001) Silicon. *Nanomaterials* **2022**, *12*, 741. [CrossRef] [PubMed]

47. Djie, H.S.; Wang, D.-N.; Ooi, B.S.; Hwang, J.C.M.; Fang, X.-M.; Wu, Y.; Fastenau, J.M.; Liu, W.K. Intermixing of InGaAs quantum dots grown by cycled, monolayer deposition. *J. Appl. Phys.* **2006**, *100*, 033527. [CrossRef]

48. Jiang, C.; Sakaki, H. Sb/As intermixing in self-assembled GaSb/GaAs type II quantum dot systems and control of their photoluminescence spectra. *Phys. E* **2005**, *26*, 180–184. [CrossRef]

49. Lever, P.; Fu, L.; Jagadish, C.; Gal, M.; Tan, H.H. Interdiffusion in Semiconductor Quantum Dot Structure. *Mat. Res. Soc. Symp. Proc.* **2003**, *744*, 65. [CrossRef]

50. Walther, T. Measurement of Diffusion and Segregation in Semiconductor Quantum Dots and Quantum Wells by Transmission Electron Microscopy: A Guide. *Nanomaterials* **2019**, *9*, 872. [CrossRef]

51. Muraki, K.; Fukatsu, S.; Shiraki, Y.; Ito, R. Surface segregation of In atoms during molecular beam epitaxy and its influence on the energy levels in InGaAs/GaAs quantum wells. *Appl. Phys. Lett.* **1992**, *61*, 557. [CrossRef]

52. Fukatsu, S.; Fujita, K.; Yaguchi, H.; Shiraki, Y.; Ito, R. Self-limitation in the surface segregation of Ge atoms during Si molecular beam epitaxial growth. *Appl. Phys. Lett.* **1991**, *59*, 2103. [CrossRef]

53. Hong, K.-B.; Kuo, M.-K. Effects of Segregation on the Strain Fields and Electronic Structures of InAs Quantum Dots. In Proceedings of the ASME 2009 International Mechanical Engineering Congress and Exposition, Lake Buena Vista, FL, USA, 13–19 November 2009; Micro and Nano Systems. Volume 12, pp. 57–62. [CrossRef]

54. Haxha, V.; Drouzas, I.; Ulloa, J.M.; Bozkurt, M.; Koenraad, P.M.; Mowbray, D.J.; Liu, H.Y.; Steer, M.J.; Hopkinson, M.; Migliorato, M.A. Role of segregation in InAs/GaAs quantum dot structures capped with a GaAsSb strain-reduction layer. *Phys. Rev. B* **2009**, *80*, 165334. [CrossRef]

55. Mura, E.E.; Gocalinska, A.; Juska, G.; Moroni, S.T.; Pescaglini, A.; Pelucchi, E. Tuning InP self-assembled quantum structures to telecom wavelength: A versatile original InP(As) nanostructure "workshop". *Appl. Phys. Lett.* **2017**, *110*, 113101. [CrossRef]

56. Brasil, M.J.S.P.; Nahory, R.E.; Tamargo, M.C.; Schwarz, S.A. Roughness at the interface of thin InP/InAs quantum wells. *Appl. Phys. Lett.* **1993**, *63*, 2688. [CrossRef]

57. Taskinen, M.; Sopanen, M.; Lipsanen, H.; Tulkki, J.; Tuomi, T.; Ahopelto, J. Self-organized InAs islands on (100) InP by metalorganic vapor-phase epitaxy. *Surf. Sci.* **1997**, *376*, 60. [CrossRef]

58. Wang, B.; Zhao, F.; Peng, Y.; Jin, Z.; Li, Y.; Liu, S. Self-organized InAs quantum dots formation by As/P exchange reaction on (001) InP substrate. *Appl. Phys. Lett.* **1998**, *72*, 2433. [CrossRef]

59. Steindl, P.; Sala, E.M.; Alen, B.; Marron, D.F.; Bimberg, D.; Klenovsky, P. Optical response of (InGa)(AsSb)/GaAs quantum dots embedded in a GaP matrix. *Phys. Rev. B.* **2019**, *100*, 195407. [CrossRef]

60. Xie, H.; Prioli, R.; Fischer, A.M.; Ponce, F.A.; Kawabata, R.M.S.; Pinto, L.D.; Jakomin, R.; Pires, M.P.; Souza, P.L. Improved optical properties of InAs quantum dots for intermediate band solar cells by suppression of misfit strain relaxation. *J. Appl. Phys.* **2016**, *120*, 034301. [CrossRef]

61. Fischer, A.; Kuhne, H.; Richter, H. New Approach in Equilibrium Theory for Strained Layer Relaxation. *Phys. Rev. Lett.* **1994**, *73*, 2712. [CrossRef]

62. Roble, A.A.; Patra, S.K.; Massabuau, F.; Frentrup, M.; Leontiadou, M.A.; Dawson, P.; Kappers, M.J.; Oliver, R.A.; Graham, D.M.; Schulz, S. Impact of alloy fluctuations and Coulomb effects on the electronic and optical properties of c-plane GaN/AlGaN quantum wells. *Sci. Rep.* **2019**, *9*, 18862. [CrossRef]

63. Luo, S.; Wang, Y.; Liang, B.; Wang, C.; Wang, S.; Fu, G.; Mazur, Y.I.; Ware, M.E.; Salamo, G.J. Photoluminescence study of exciton localization in InGaAs bulk and InGaAs/InAlAs wide quantum well on InP (001) substrate. *J. Lumin.* **2022**, *246*, 118827. [CrossRef]

64. Singh, J.; Bajaj, K.K. Role of interface roughness and alloy disorder in photoluminescence in quantum-well structures. *J. Appl. Phys.* **1985**, *57*, 5433. [CrossRef]

65. Mooney, P.M.; Theis, T.N.; Calleja, E. Effect of local alloy disorder on the emission kinetics of deep donors (DX centers) in $Al_xGa_{1-x}As$. *J. Electron. Mater.* **1991**, *20*, 23. [CrossRef]

66. Gajjela, R.S.R.; Hendriks, A.L.; Douglas, J.O.; Sala, E.M.; Steindl, P.; Klenovský, P.; Bagot, P.A.J.; Moody, M.P.; Bimberg, D.; Koenraad, P.M. Structural and compositional analysis of (InGa)(AsSb)/GaAs/GaP Stranski–Krastanov quantum dots. *Light Sci. Appl.* **2021**, *10*, 125. [CrossRef]

67. Liu, X.F.; Luo, Z.J.; Zhou, X.; Wei, J.M.; Wang, Y.; Guo, X.; Lang, Q.Z.; Ding, Z. Theoretical study of stress and strain distribution in coupled pyramidal InAs quantum dots embedded in GaAs by finite element method. *Eur. Phys. J. B* **2019**, *92*, 138. [CrossRef]

68. Kosarev, A.N.; Chaldyshev, V.V.; Cherkashin, N. Experimentally-Verified Modeling of InGaAs Quantum Dots. *Nanomaterials* **2022**, *12*, 1967. [CrossRef]

69. Zhang, L.; Song, Y.; Chen, Q.; Zhu, Z.; Wang, S. InPBi Quantum Dots for Super-Luminescence Diodes. *Nanomaterials* **2018**, *8*, 705. [CrossRef]

70. Grundmann, M.; Stier, O.; Bimberg, D. InAs/GaAs pyramidal quantum dots: Strain distribution, optical phonons, and electronic structure. *Phys. Rev. B* **1995**, *52*, 11969. [CrossRef]

71. Birner, S.; Zibold, T.; Andlauer, T.; Kubis, T.; Sabathil, M.; Trellakis, A.; Vogl, P. nextnano: General Purpose 3-D Simulations. *IEEE Trans. Electron. Devices* **2007**, *54*, 2137–2142. [CrossRef]

72. Stier, O.; Grundmann, M.; Bimberg, D. Electronic and optical properties of strained quantum dots modeled by 8-band k–p theory. *Phys. Rev. B* **1999**, *59*, 5688. [CrossRef]

73. Aleksandrov, I.A.; Gutakovskii, A.K.; Zhuravlev, K.S. Influence of shape of GaN/AlN quantum dots on luminescence decay law. *Phys. Status Solidi A* **2012**, *209*, 653–656. [CrossRef]

74. Llorens, J.M.; Taboada, A.G.; Ripalda, J.M.; Alonso-Alvarez, D.; Alen, B.; Martın-Sanchez, J.; Garcia1, J.M.; Gonzalez, Y.; Sanchez, A.M.; Beltran, A.M.; et al. Theoretical modelling of quaternary GaInAsSb/GaAs self-assembled quantum dots. *J. Phys. Conf. Ser.* **2010**, *245*, 012081. [CrossRef]

75. Joullie, A.; Eddin, A.Z.; Girault, B. Temperature dependence of the L^c_6-$^c_6$ energy gap in gallium antimonide. *Phys. Rev. B* **1981**, *23*, 928. [CrossRef]

76. Ghezzi, C.; Magnanini, R.; Parisini, A.; Rotelli, B.; Tarricone, L.; Bosacchi, A.; Franchi, S. Optical absorption near the fundamental absorption edge in Gasb. *Phys. Rev. B* **1995**, *52*, 1463. [CrossRef] [PubMed]

77. Bellani, V.; di Lemia, S.; Geddo, M.; Guizzetti, G.; Bosacchi, A.; Franchi, S.; Magnanini, R. Thermoreflectance study of the direct energy gap of GaSb. *Solid State Commun.* **1997**, *104*, 81. [CrossRef]

78. Wei, S.-H.; Zunger, A. Calculated natural band offsets of all II–VI and III–V semiconductors: Chemical trends and the role of cation *d* orbitals. *Appl. Phys. Lett.* **1998**, *72*, 2011. [CrossRef]

79. Muñoz, M.C.; Armelles, G. X-point deformation potentials of III-V semiconductors in a tight-binding approach. *Phys. Rev. B* **1993**, *48*, 2839. [CrossRef] [PubMed]

80. Wei, S.-H.; Zunger, A. Predicted band-gap pressure coefficients of all diamond and zinc-blende semiconductors: Chemical trends. *Phys. Rev. B* **1999**, *60*, 5404–5411. [CrossRef]

81. Madelung, O. (Ed.) *Semiconductors-Basic Data*, 2nd ed.; Springer: New York, NY, USA, 1996.

82. Abramkin, D.S.; Gutakovskii, A.K.; Shamirzaev, T.S. Heterostructures with diffused interfaces: Luminescent technique for ascertainment of band alignment type. *J. Appl. Phys.* **2018**, *123*, 115701. [CrossRef]

83. Wang, J.; Xu, F.; Zhang, X.; An, W.; Li, X.; Song, J.; Ge, W.; Tian, G.; Lu, J.; Wang, X.; et al. Evidence of Type-II Band Alignment in III-nitride Semiconductors: Experimental and theoretical investigation for $In_{0.17}Al_{0.83}N$/GaN heterostructures. *Sci. Rep.* **2014**, *4*, 6521. [CrossRef]

84. Abramkin, D.S.; Putyato, M.A.; Budennyy, S.A.; Gutakovskii, A.K.; Semyagin, B.R.; Preobrazhenskii, V.V.; Kolomys, O.F.; Strelchuk, V.V.; Shamirzaev, T.S. Atomic structure and energy spectrum of Ga(As,P)/GaP heterostructures. *J. Appl. Phys.* **2012**, *112*, 083713. [CrossRef]

85. Abramkin, D.S.; Bakarov, A.K.; Gutakovskii, A.K.; Shamirzaev, T.S. Spinodal Decomposition in InSb/AlAs Heterostructures. *Semiconductors* **2018**, *52*, 1392. [CrossRef]

86. Abramkin, D.S.; Putyato, M.A.; Gutakovskii, A.K.; Semyagin, B.R.; Preobrazhenskii, V.V.; Shamirzaev, T.S. New System of Self Assembled GaSb/GaP Quantum Dots. *Semiconductors* **2012**, *46*, 1534. [CrossRef]

87. Shamirzaev, T.S.; Abramkin, D.S.; Gutakovskii, A.K.; Putyato, M.A. High quality relaxed GaAs quantum dots in GaP matrix. *Appl. Phys. Lett.* **2010**, *97*, 023108. [CrossRef]

88. Stirman, J.N.; Crozier, P.A.; Smith, D.J.; Phillipp, F.; Brill, G.; Sivananthan, S. Atomic-scale imaging of asymmetric Lomer dislocation cores at the Ge/Si (001) heterointerface. *Appl. Phys. Lett.* **2004**, *84*, 2530. [CrossRef]

89. Dunker, D.; Shamirzaev, T.S.; Debus, J.; Yakovlev, D.R.; Zhuravlev, K.S.; Bayer, M. Spin relaxation of negatively charged excitons in (In,Al)As/AlAs quantum dots with indirect band gap and type-I band alignment. *Appl. Phys. Lett.* **2012**, *101*, 142108. [CrossRef]

90. Shamirzaev, T.S.; Rautert, J.; Yakovlev, D.R.; Debus, J.; Gornov, A.Y.; Glazov, M.M.; Ivchenko, E.L.; Bayer, M. Spin dynamics and magnetic field induced polarization of excitons in ultrathin GaAs/AlAs quantum wells with indirect band gap and type-II band alignment. *Phys. Rev. B* **2017**, *96*, 035302. [CrossRef]

91. Shamirzaev, T.S.; Debus, J.; Abramkin, D.S.; Dunker, D.; Yakovlev, D.R.; Dmitriev, D.V.; Gutakovskii, A.K.; Braginsky, L.S.; Zhuravlev, K.S.; Bayer, M. Exciton recombination dynamics in an ensemble of (In,Al)As/AlAs quantum dots with indirect band-gap and type-I band alignment. *Phys. Rev. B* **2011**, *84*, 155318. [CrossRef]

 *nanomaterials*

Article

Causes and Consequences of Ordering and Dynamic Phases of Confined Vortex Rows in Superconducting Nanostripes

Benjamin McNaughton [1,2], Nicola Pinto [1,3], Andrea Perali [4] and Milorad V. Milošević [2,*]

1. School of Science and Technology, Physics Division, University of Camerino, 62032 Camerino, Italy
2. Department of Physics, University of Antwerp, Groenenborgerlaan 171, B-2020 Antwerp, Belgium
3. Advanced Materials Metrology and Life Science Division, INRiM (Istituto Nazionale di Ricerca Metrologica), Strade delle Cacce 91, 10135 Turin, Italy
4. School of Pharmacy, Physics Unit, University of Camerino, 62032 Camerino, Italy
* Correspondence: milorad.milosevic@uantwerpen.be

Abstract: Understanding the behaviour of vortices under nanoscale confinement in superconducting circuits is important for the development of superconducting electronics and quantum technologies. Using numerical simulations based on the Ginzburg–Landau theory for non-homogeneous superconductivity in the presence of magnetic fields, we detail how lateral confinement organises vortices in a long superconducting nanostripe, presenting a phase diagram of vortex configurations as a function of the stripe width and magnetic field. We discuss why the average vortex density is reduced and reveal that confinement influences vortex dynamics in the dissipative regime under sourced electrical current, mapping out transitions between asynchronous and synchronous vortex rows crossing the nanostripe as the current is varied. Synchronous crossings are of particular interest, since they cause single-mode modulations in the voltage drop along the stripe in a high (typically GHz to THz) frequency range.

Keywords: superconducting; nanostripes; vortex; confinement; critical current; flux

Citation: McNaughton, B.; Pinto, N.; Perali, A.; Milošević, M.V. Causes and Consequences of Ordering and Dynamic Phases of Confined Vortex Rows in Superconducting Nanostripes. *Nanomaterials* **2022**, *12*, 4043. https://doi.org/10.3390/nano12224043

Academic Editor: Orion Ciftja

Received: 19 October 2022
Accepted: 15 November 2022
Published: 17 November 2022

Publisher's Note: MDPI stays neutral with regard to jurisdictional claims in published maps and institutional affiliations.

Copyright: © 2022 by the authors. Licensee MDPI, Basel, Switzerland. This article is an open access article distributed under the terms and conditions of the Creative Commons Attribution (CC BY) license (https://creativecommons.org/licenses/by/4.0/).

1. Introduction

Superconducting nanostripes (SNs) are a fundamental component in superconducting electronics, and they are crucial for various applications in the field of quantum technology. Superconducting nanostripe single-photon detectors (SNSPDs), for instance, are used for quantum communication and applications in astronomy and spectroscopy [1–4]. Other superconducting electronics include prototypical logic devices [5–7], flux qubits used in quantum computers [8–10], diodes [11–13], and electromagnetic resonators [14–16]. Narrow SNs experience an enhancement of critical parameters [17–20] due to confinement forces acting on the superconducting condensate [21–26]. Such confinement in narrow SNs can cause large magnetoresistance oscillations [27–29], where the time-averaged voltage/resistance, as a function of the applied magnetic field, exhibits pronounced peaks at alternating transitions between static and dynamic vortex phases. At higher applied fields, with multiple rows of vortices or high currents, a continuous motion of vortices causes a monotonic background on which the resistance oscillations due to entries of additional vortices are superimposed [27,30]. Commensurate effects between the SN width w and the number of vortex rows n are observed in the critical current as a function of the out-of-plane magnetic field H (for fixed w) or w (for fixed H) [31,32]. Optimised operation of some of the suggested superconducting electronics may be achieved on a specific geometry of vortices. For example, a single row of vortices was found to be preferable in [7], producing a giant non-local electrical resistance from vortices moving very far (several microns) from the local current drive. This effect appears to be important for a feasible long-range information transfer by vortices that are unaltered by the passing current. Moving vortices, however, exceed the bare transfer of information in importance. For example, vortices coherently crossing SNs can produce electromagnetic radiation [15,16], where higher radiation power is emitted when multiple vortices exit the SN simultaneously. In narrow SNs, rows of vortices can cross the SN asynchronously and synchronously [27,33], depending on competing forces (confinement, vortex–vortex interaction, Lorentzian forces), but the criteria for synchronous crossings are not yet well understood. In this respect, a study on the behaviour of vortices in SNs with small widths is important in order to reveal the favoured geometry of vortices for the static case (no sourced current) and the relation to the dynamic case (with sourced current). Understanding how a vortex lattice is affected by the interaction with the edge confining force and other dynamic forces is important when considering SNs for the applications mentioned above. Studying the dynamic dissipative states under strong confinement in the 1D–2D crossover regime can reveal how vortices cross a SN under different conditions (w, H, current intensity). Moreover, information on the possible vortex velocity under confinement [34,35] can be important for both fast information transfer and the frequency of radiation emitted by moving vortices.

In this work, we investigate how confinement in SNs affects the vortex configurations by using Ginzburg–Landau simulations [36]. We present a vortex row phase diagram as a function of H for a given w. An investigation of the dependence on the magnetic field of the average number of vortices reveals strong confinement effects. With increasing width, reconfiguration from the vortex rows to the vortex lattice takes place, offering a criterion for defining quasi-1D-to-2D dimensional crossover, where the SN effectively becomes a nanofilm in terms of its superconducting properties. Additionally, a commensurate behaviour of the critical current, $J_{c1}(H)$, was found when varying H by using a time-dependent GL approach in order to simulate the effects of the sourced current. We show that the values of the local minima in $J_{c1}(H)$ (defined as the onset of vortex motion and corresponding dissipation) are directly related to the row transitions shown in our vortex row phase diagram.

Further simulations of the current–voltage (I-V) characteristics in SNs evidenced transitions among different resistive regimes (Meissner, flux–flow, flux–flow instability, phase slips, normal state). I-V curves showing similar features to those of our simulations for SNs have been experimentally measured only for wider structures [37,38]. We find that for a SN with an average vortex density $\lesssim 1/80\xi^2$ (with ξ being the coherence length) in a flux–flow regime, vortices cross the SN in a periodic/continuous fashion, causing modulations in the voltage drop that are experimentally detectable. Such a periodic flow may produce electromagnetic radiation [15] and features characteristic power spectra [39], which we report by performing fast Fourier transformation of the calculated voltage drop as a function of time during vortex motion. The recorded average vortex velocity (up to 10s of km/s) was used to discuss the washboard frequencies [15,16] in the flux–flow regime for thin SNs of niobium [38].

Provided that the vortex density is sufficiently high, as the sourced current density is increased, we show transitions of vortex row crossings from quasi-synchronous to synchronous. Synchronised crossings are desirable for small-band electromagnetic emitters operating in the GHz or THz range. For typical ultra-thin niobium SNs [18], the range of modulation frequencies in the microwave regime was between 10 and 800 GHz. Asynchronous regimes are disruptive for coherent emissions, but host a number of local dynamic vortical transitions and transformations that are of fundamental importance for advanced devices and are unattainable otherwise. The article is organised as follows. We first introduce the theoretical framework and methods used for the numerical simulations. We then present the results and discussions of all of the above-listed phenomena by using both stationary and time-dependent Ginzburg–Landau approaches. The main conclusions of our work are already emphasised in the section that provides the results before being additionally commented on in the conclusions of the article.

2. Materials and Methods

The numerical simulations performed in this work were all conducted on SNs such as that exemplified in Figure 1. The SNs have dimensions with a length L, width w, and thickness $d \ll \xi, \lambda$, where ξ is the coherence length and λ is the magnetic field penetration depth of the superconducting state. For sufficiently large values of H, vortices form in the sample with a normal core of radius ξ and penetration of the magnetic field up to a characteristic length of λ. In the samples of our interest, which are very thin, the effective penetration depth $\Lambda = \lambda^2/d$ by far exceeds the dimensions of the SN, such that the magnetic response of the superconductor is negligibly small compared to the applied magnetic field. Simulations of such SNs were performed using the stationary (SGL) and time-dependent Ginzburg–Landau (TDGL) formalism. In the SGL approach, we self-consistently solve the coupled equations

$$(-i\nabla - \mathbf{A})^2\Psi = \Psi\left(1 - |\Psi|^2\right), \tag{1}$$

$$\vec{j} = -\kappa^2\nabla^2\mathbf{A} = \frac{1}{2i}(\Psi^*\nabla\Psi - \Psi\nabla\Psi^*) - |\Psi|^2\mathbf{A} \tag{2}$$

where Ψ is the superconducting order parameter, $\mathbf{A}$ is the vector potential, and $\kappa = \Lambda/\xi$ is the effective Ginzburg–Landau parameter. We work with dimensionless units, where the length is given in units of the temperature-dependent coherence length $\xi(T) = \xi$, the vector potential $\mathbf{A}$ is given in units of $c\hbar/2e\xi$, the magnetic field $\vec{H}$ is given in units of the bulk upper critical field $H_{c2} = c\hbar/2e\xi^2$, the current is given in units of the GL current $j_{GL} = c\Phi_0/(8\pi^2\lambda^2\xi)$, and the order parameter Ψ is normalised to its value in the absence of an applied field or sourced current (Ψ_0). We impose the Neumann boundary condition at the superconductor–insulator boundary at the lateral edges of the SN:

$$\vec{n} \cdot (-i\nabla - \mathbf{A})\Psi|_{boundary} = 0. \tag{3}$$

Along the length of the SN (x-axis), we enforce periodic boundary conditions for $\mathbf{A}$ and Ψ (for the unit cell length 32ξ, which is sufficient to capture the physics of interest in this work), of the form [40]:

$$\mathbf{A}(x_0 + L_x) = \mathbf{A}(x) + \nabla\chi_f(x) \tag{4}$$

$$\Psi(x_0 + L_x) = \Psi(x)\exp\left[i\frac{2e}{\hbar c}\chi_f(x)\right], \tag{5}$$

where $\nabla\chi_f$ respects the gauge used for the magnetic field. Equations (1) and (2) are solved numerically on a discretised Cartesian grid according to [36] by iteratively using the finite-difference method and the link–variable approach [41] until convergence within a prespecified error is achieved. Then, the supercurrent is calculated from the value of the order parameter and the vector potential (nearly entirely provided by the external magnetic field). With this method, we obtain the vortex row configuration–transition diagram as a function of H and w of the SN.

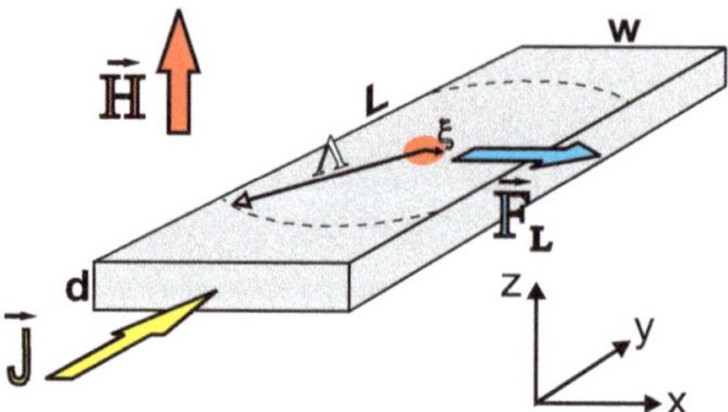

Figure 1. Schematic illustration of a superconducting nanostripe with width w, length L, and thickness d along the x, y, and z directions, respectively, in a homogeneous out-of-plane applied magnetic field, **H**. The SN contains an example of a single vortex, with a normal core of radius $\sim \xi$, and a distribution of magnetic field around it characterised by Λ. When a current density **J** is sourced, the vortex will experience the Lorentz force, $\mathbf{F}_L$.

The generalised time-dependent Ginzburg–Landau formalism [42,43] should instead be employed to properly study the dynamical properties of the superconducting condensate (with order parameter $\Psi(\mathbf{r}, t)$) in the presence of an external magnetic field **H** (with vector potential **A**) and sourced current density **J**, given by

$$\tau_{GL}N(0)\frac{u}{\sqrt{1-(\Gamma|\Psi|)^2}}\left[\frac{\delta\Psi}{\delta t} + i\frac{e^*}{\hbar}\varphi\Psi + \left(\frac{\Gamma}{\sqrt{2}}\right)^2\frac{\delta|\Psi|^2}{\delta t}\Psi\right]$$
$$= -\left(a + b|\Psi|^2\right)\Psi + \frac{\hbar^2}{2m^*}(\nabla - ie^*\mathbf{A})^*\Psi, \tag{6}$$

$$\nabla^2\varphi = \nabla[Im\{\Psi^*(\nabla - iA)\Psi\}], \tag{7}$$

where $a = \frac{\alpha}{2m^*\gamma}$, $b = \frac{\beta}{4m^{*2}\gamma^2}$, and $\Gamma = \frac{2\tau_i}{\hbar\sqrt{2m^*\gamma}}$. The Ginzburg–Landau order parameter's relaxation time is τ_{GL}; $N(0)$ is the density of states at the Fermi level; the parameter $u = 5.79$ in conventional superconductors; e^* is the effective charge; φ is the electrostatic potential; τ_i is the electron–phonon inelastic scattering time; α, β, γ are material parameters. Equation (6) is solved by being coupled with the equation for the electrostatic potential (Equation (7)) by using Neumann boundary conditions at all sample edges, except for the leads into which sourced current is injected, where $\Psi = 0$ and $\nabla\varphi = \pm J$. This theory is derived for dirty gapless superconductors, where Cooper-pair breaking occurs due to strong inelastic electron–phonon scattering, and the physical quantities Ψ and A must relax over a time scale much longer than τ_i. The distance over which an electric field can penetrate into the superconductor and the length over which relaxation processes occur are given by the characteristic inelastic diffusion length $L_i = \sqrt{D\tau_i}$, where D is the diffusion parameter, which is proportional to the electronic mean free path. In cases where $L_i << \xi$, our simulations require very fine grid spacing (reflected in a consequently smaller time step in the implicit Crank–Nicolson method used) to yield physically correct results. In general, superconducting materials at T close to the superconducting-to-normal transition temperature, T_c, satisfy the conditions for the slow temporal and spatial variations that are ideally required for the applicability of the GL formalism. In the TDGL formalism, distances are given in units of $\xi(T) = \xi$; time is given in units of $\tau_{GL} = \frac{\pi\hbar}{8k_BT_c(1-T/T_c)u}$; temperature is given in units of T_c; the order parameter Ψ is given in units of $\Delta(0) = 4k_BT_cu^{1/2}(1-T/T_c)^{1/2}/\pi$; φ is given in units of $\varphi_{GL} = \hbar/e^*\tau_{GL}$; the vector potential A is scaled to $A_0 = H_{c2}\xi$; the current density is scaled to $J_0 = \sigma_n\varphi_0/\xi$. The simulations are performed irrespective of the temperature T/T_c, and all physical quantities are scaled and normalised by reference quantities at a given temperature. Note that even though the GL approach is formally valid close to T_c, experiments have shown the possibility of extending the GL predictions to a finite T range below T_c (see, e.g., [18] and the references therein). Moreover, in the simulations, the heat generated by the Joule effect is lost on a time scale shorter than the inelastic scattering time, assuming that the heat transfer coefficient is large enough to allow a fast dissipation. The approach adopted is equivalent to the inclusion of a solution of the thermal balance equation [34]. Our findings are valid at any temperature, provided that the coherence length is known at a given temperature.

3. Results

In what follows, by using SGL and TDGL simulations, we study how confinement forces in narrow SNs affect the stationary vortex configurations and their dynamics under H and a sourced DC current density, J.

3.1. Equilibrium Vortex Configurations

We started by producing the vortex row phase diagram using the SGL approach, which showed the conditions for the formation of a number of vortex rows, n, as a function of H and w of the SN. Each dashed curve in the diagram shown in Figure 2, which plots the width of the SN versus H, represents the appearance of the nth vortex row ($n = 1 \div 5$) in the ground state of the system as the magnetic field is increased. Examples of the corresponding vortex configurations for a SN of $w = 12\zeta$ for different intensities of H are shown in Figure 3, and they correspond to the pinpointed dots (labelled a–h) in Figure 2.

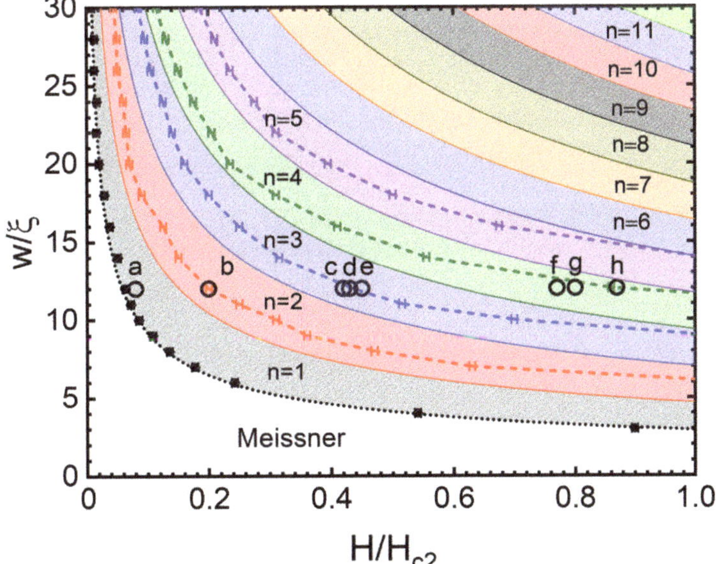

Figure 2. Equilibrium vortex row phase diagram that plots the SN's width (in units of ζ) as a function of the intensity of the applied magnetic field (in units of H_{c2}) for different numbers of formed vortex rows (n). The simulations were performed using the SGL approach with periodic boundary conditions along the length, with a unit cell length of $L = 32\zeta$. The dashed lines denote the threshold for the formation of an additional vortex row (shown here up to $n = 5$). The coloured regions represent the approximated regions for $n > 1$, which are delimited by solid lines given by the expression $H_{row}/H_{c2} = \frac{\pi n^2 \zeta^2}{\sqrt{3}w^2}$. The circles, which are labelled a–h, relate to the vortex configurations shown in Figure 3. The black dotted line corresponds to the analytical expression $H_0/H_{c2} = K\frac{\pi^2 \zeta^2}{2w^2}$ [44], with $K = 1.7$ [45].

To identify the threshold H for the transition to the vortex row configuration with a higher n, the ground states were first obtained for each SN at different values of H; then, the spatial distribution of the superconducting order parameter $|\Psi|^2$ was plotted (similarly to Figure 3) and carefully analysed, with a focus on the geometrical interpretation of the vortex configuration. In the SGL approach adopted in our simulations, the SN was considered periodic along its length, with a unit cell of $L = 32\zeta$. Several checks that were carried out by extending the unit cell length until 80ζ confirmed all of the following results.

Early theoretical works [44,46] showed that the magnetic field at which the surface barrier is suppressed and a single vortex can be stable in a SN is $H_0/H_{c2} = \pi^2\zeta^2/2w^2$. The subsequent experimental observations of vortex penetration fields by Stan et al. [45] showed a very good agreement with that expression, up to a multiplying constant K. Our numerical data (black dots in Figure 2) reconfirm that finding, as the vortex penetration fields were found to nearly ideally match the same functional dependence on w, with a multiplying constant of $K = 1.7$.

The approximate criteria for further reconfiguration of the vortex states and the appearance of additional vortex rows can be obtained in the following way. We consider an Abrikosov triangular lattice with the lattice parameter $a = 1.075\sqrt{\phi_0/H}$. The vortices are arranged in a body-centred hexagonal lattice, so the Wigner–Seitz unit cell is hexagonal with a unit area per flux quantum of $A = \frac{\sqrt{3}}{2}a^2$. For a narrow SN, to accommodate n rows of vortices, the spacing, w_v, among the vortex rows must obey the inequality $w_v \le w/n$. Using the previous expression for the Abrikosov vortex density, we substitute $A = \frac{\sqrt{3}}{2}w_v^2 = \frac{\sqrt{3}w^2}{2n^2}$ to obtain the zeroth-order approximation for the threshold magnetic field required for the formation of new rows, yielding $H_{row}/H_{c2} = \frac{\pi n^2\zeta^2}{\sqrt{3}w^2}$. These approximate threshold H values are shown in Figure 2 by the solid lines that delimit different coloured regions, indicating transitions among states with different numbers of vortex rows. In general, the behaviour of threshold H found using the SGL simulations agrees well with the prediction of the formula. The values are, however, mostly higher than the approximate ones, which is attributed to the role played by the edge barriers for vortex entry and exit (varying depending on w and H). In addition, the rearrangement of the vortex lattice with every vortex penetration is not taken into account in the latter basic analytical formula. Note that such effects of the vortex–vortex interactions and interactions with the edge Meissner currents (causing the confinement force) dominate the formation of the vortex configurations in narrow SNs and present the main point of interest in this work.

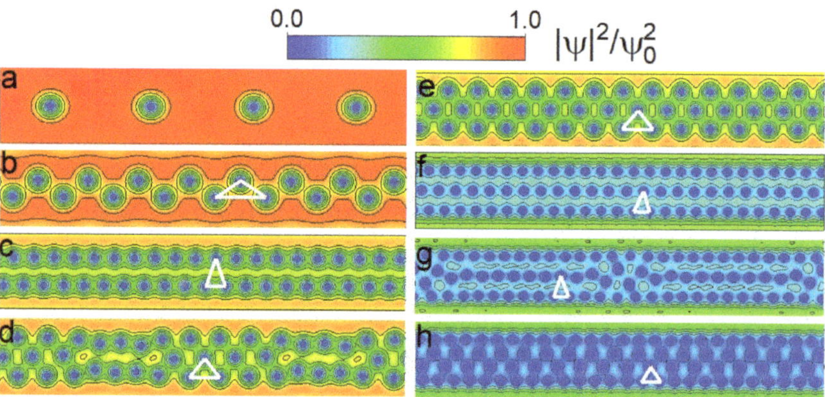

Figure 3. Calculated vortex configurations plotted as the Cooper-pair density for the ground state of a SN of width $w = 12\zeta$ in a periodic cell of $L = 64\zeta$ at different applied H/H_{c2} values: (**a**) 0.08; (**b**) 0.20; (**c**) 0.42; (**d**) 0.43; (**e**) 0.45; (**f**) 0.77; (**g**) 0.80; (**h**) 0.87 (cf. Figure 2). Panels (**b**,**d**,**g**) depict the vortex states at the nucleation of a second, third and fourth row, respectively. Panels (**c**,**e**,**h**) show the most lattice-like packing conditions for two, three and four vortex rows, respectively. The white lines connecting the cores of three neighbouring vortices illustrate the deformation of the Abrikosov lattice [47] in the SN. The colour bar denotes the values of the Cooper-pair density shown in the panels. Each depicted configuration is indicated in Figure 2 with an open dot and is labelled accordingly.

For a SN of $w = 12\xi$, we show different vortex row configurations in Figure 3 as they are formed in the ground state at different H values (marked by open dots in Figure 2). After the formation and growth of the population of the first vortex row (Figure 3a), H is increased, and the vortices are rearranged into a closely packed "zig-zag" state (Figure 3b). This close packing is emphasised by a white triangle that progressively deviates from the equilateral shape expected in the Abrikosov vortex lattice with the increase in H. Obviously, in this state, the Meissner currents will exert a strong repulsive and confining force on the vortices from the SN edges (i.e., a strong Bean–Livingston edge barrier [25]), resulting in a vortex spacing that is far smaller than the above rough analytical estimates (leading to the solid lines in Figure 2).

Starting from the one-row configuration (Figure 3a), further raising H and n strengthens the relevance of the vortex–vortex interaction forces for the resulting vortex configuration, which will increase the separation between the two rows (Figure 3c). At this point, we observe that additional vortices in the SN cannot uniformly balance the aforementioned competing force in the entire SN, leading to a local rearrangement of the vortex lattice to three rows (Figure 3d). Only by further increasing the field and having enough vortices in the SN can the full three-row state be formed (Figure 3e; notice the nearly ideal triangular lattice that is formed). For the considered width of the SN, the state with three vortex rows persists toward a much larger field due to quantum confinement, such that the vortices very strongly overlap in a closely packed structure (Figure 3f). Nevertheless, in the vicinity of the bulk upper critical field, a fourth row forms, first locally (Figure 3g) and eventually in the entire SN (Figure 3h), before the superconductivity is destroyed. No further rows of vortices can form with the higher field, and the existing vortex rows increasingly overlap until the normal state is established.

We reiterate that the transitions among rows of vortices and the final arrangement of vortices in the lattice are strongly affected by the competition of the two forces, which are both dependent on H. As the magnetic field is increased, the edge Meissner current also increases up to the penetration of new vortices, while every new vortex changes the landscape of the vortex–vortex interactions in the SN. As exemplified in Figure 4 for $w = 20\xi$, this nontrivial balance of competing forces can lead to a re-entrant behaviour in terms of the number of vortex rows formed. In such cases, the zig-zag instability of the vortex row can be "cured" back into a single row by increasing the Meissner currents, as the lateral confinement forces grow with the increase in the magnetic field. As H is increased further, the additional penetrating vortices tip the scale in favour of vortex interactions, and a definite reconfiguration into a state with two rows forms. This re-entrant behaviour was observed for nearly all SN widths considered in the range of $w = 20 \div 60\xi$ and only for the transition $n = 1 \rightarrow 2$. In such cases, we took the first onset of the zig-zag instability to mark the $n = 1 \rightarrow 2$ transition in Figure 2. Moreover, this range of widths in which such strong edge effects are detected marks the crossover from the quasi-1D to a 2D film-like behaviour.

As the magnetic field is increased, vortices penetrate the SNs of different widths, vortex rows are formed, and a gradual evolution from a quasi-1D row pattern into a 2D vortex lattice is expected. To evaluate this crossover, we calculated the average area occupied by a single vortex as a function of H in all of the states found and compared this with the expected behaviour of the Abrikosov vortex lattice area. The strong confinement in the narrowest SN [25,31] dominates the vortex–vortex interaction, leading to the compression of the vortices into fewer vortex rows and, consequently, a larger average area per vortex. This can be seen in Figure 5 for $w \leq 8\xi$. As the width of the SN is made larger, the confining force from the edge current (at a given H) becomes less dominant with respect to the vortex–vortex interaction, resulting in a progressively closer agreement with the expected behaviour of a triangular vortex lattice [47]. This tendency is clearly visible upon the formation of the third vortex row (cf. Figure 5).

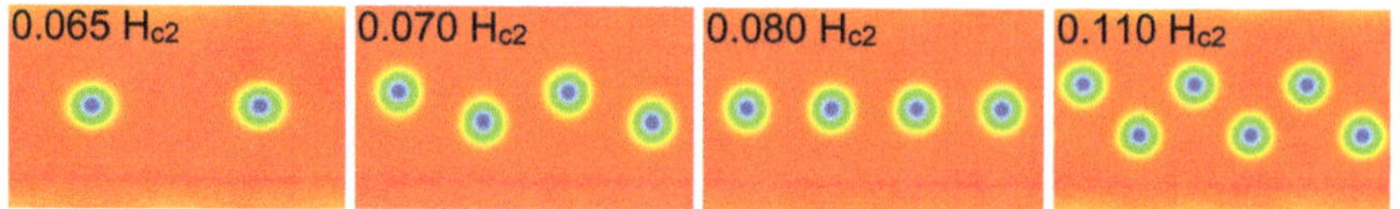

Figure 4. Example of a re-entrant transition between the one- and two-row vortex states in a SN of width $w = 20\xi$ caused by the competition between the confinement imposed by Meissner currents and the vortex density while both are changed with increasing H (values are indicated inside the panels).

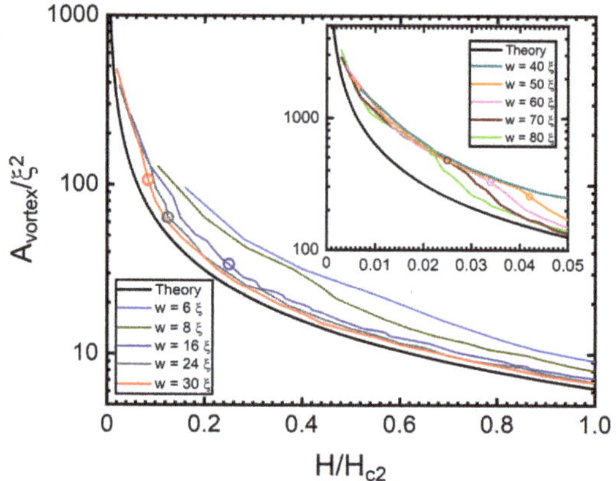

Figure 5. Area of the Wigner–Seitz unit cell containing a single vortex as a function of the applied magnetic field for a SN of width $w = 6 \div 30\xi$ and $w = 40 \div 80\xi$ in the inset, for small values of H. The analytical expression for the Abrikosov vortex lattice (AVL) area, $\frac{A}{\xi^2} = 2\pi \frac{H_{c2}}{H}$, is plotted as a black line. The open dots in each curve indicate the intensity of H for the formation of the third vortex row, above which the curves progressively approach the AVL expression when w is increased.

3.2. Vortex Dynamics under Sourced Current

All of the above results were obtained in the stationary case, where no current was sourced to the SN. A sourced current may change the stationary states or induce vortex dynamics that are specific to the nanoconfined regime. In what follows, we examine these non-equilibrium effects through TDGL simulations of time-dependent processes.

When a small transport current flows along a SN under an applied magnetic field $H \geq H_{c1}$, the present vortices experience a push across the SN, due to Lorentz-type force ($\propto \mathbf{J} \times \mathbf{H}$). As the current density is increased, vortices will continue to shift across the SN, finally leaving the stripe for a sufficiently large Lorentz force. This defines the first critical current density (J_{c1}) for which vortices are able to overcome the edge barrier [25,31] and start to cross the SN continuously, nucleating on one side, moving across the SN, and exiting at the opposing edge. The critical current depends on the magnetic field H for a given width w of the SN. The first critical current density as a function of H, $J_{c1}(H)$ is shown in Figure 6 for a SN of $w = 12\xi$. A commensurate effect is observed between H and n, where the minima in the curve correspond to a transition to a state with an additional vortex row. Previous works have reported similar behaviours using different theoretical approaches [48,49], including a comprehensive study that used the TDGL approach [32] and revealed the relations between n, w, and the applied magnetic field [31]. Additionally, a similar effect related to this commensurability could be observed in the dependence on the magnetic field of the resistance, and this effect was reported in [7]. We included the study of commensurate effects in our work to emphasise the relation between the local minima and the transition to a new vortex row. The increase in J_{c1} from the local minima as the applied field H is further increased is caused by the competition between vortex–vortex interactions and the confinement from the SN's edge. After a local minimum, when a new row is formed, the vortex–vortex interactions are strong and the confining edge currents that produce an entry/exit barrier are reduced and are more easily overcome with lower sourced currents. As H is increased, the Meissner currents induced at the edge increase [50], reinforcing the edge barrier. The vortex row phase diagram in Figure 2 can be used to predict the transition field value, where local minima occur in the $J_{c1}(H)$ curves, which is an experimentally verifiable feature. Note, however, that the threshold fields for the formation of new rows in the presence of a sourced current are somewhat different from those presented in Figure 2, since the Lorentz push exerted by the current, effectively increases the confinement experienced by vortices prior to the onset of their motion.

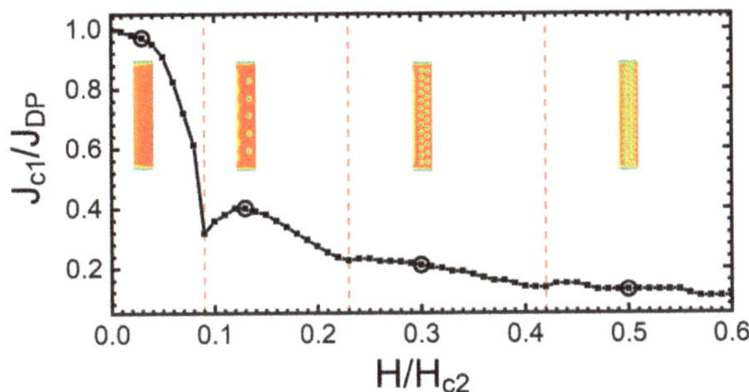

Figure 6. First critical current density normalised to $J_{DP} = 0.385J_{GL}$ as a function of the applied magnetic field normalised to H_{c2} for a SN of width $w = 12\xi$, which was obtained by using the TDGL approach. Vertical red lines mark the transition to one-, two- and three-vortex-row states with magnetic fields of $H/H_{c2} = 0.09, 0.23$ and 0.42, respectively. The insets illustrate the vortex row configurations with the selected magnetic fields (marked by open dots) for a sourced current density just below the critical one.

The TDGL approach also allowed us to simulate the voltage–current density (V-J) characteristics of SNs, which are presented in Figure 7 for stripes with $w = 6, 9, 12, 18\xi$, under an applied magnetic field $H = 0.25H_{c2}$. The analysis of the V-J characteristics reveals a number of features related to different resistive regimes in each curve. At low values of J, stationary vortices are shifted to a new position across the SN due to the Lorentz force produced by the sourced current, so the resulting voltage drop and resistance remain zero. An example of this can be seen in Figure 7 (for $w = 12\xi$) in the states labelled 1 and 2. When J_{c1} is reached, the vortices cross the SN, and their perpetual motion leads to a finite resistivity value. Snapshots of this flux–flow regime can be seen in the states labelled 3 and 4 in Figure 7. By further increasing J and being in the presence of vortex–vortex interaction forces, a SN in the dissipative state exhibits flux–flow instability, where vortex cores interact during dynamics and the ordered lattice structure is lost during motion (the state labelled 5 in Figure 7). At even higher values of J, the vortices align during motion in a slip-streamed geometry (the vortices tailgate, i.e., subsequent vortices crossing the SN move in the wake of the previous vortex [33,35]) before a Langer–Ambegaokar phase slip [51] occurs across the SN. The normal area covered by the phase slip grows laterally with further increases in J, and additional steps in the V-J curve appear, with every slip-stream being merged with the growing phase slip, as seen in the states labelled 6–10 in Figure 7. When J reaches roughly $0.65J_{DP}$, the SN transitions to a fully normal state with a linear ohmic behaviour. Similar V-I curves have been observed both numerically [52] and experimentally for Nb-C microstrips that were fabricated using focused-ion-beam-induced deposition [38]. In real materials, the presence of disorder and defects changes the behaviour described in this work. For example, edge defects are a favourable point for vortex entry, as current crowding occurs in the local defect region, which leads to favoured positions for vortex penetration. In addition, disorder on length scales larger than ξ may create bulk pinning regions [31].

On the other hand, small disordered regions (smaller than ξ) lead to increased inelastic scattering times, resulting in finite values of Γ. Hence, a viscous condensate will be formed, changing the dynamic behaviour of vortices and introducing additional resistive states, such as "vortex channels". The appearance of such resistive states can be advantageous for EM emitters, provided that the vortex rows are synchronised, which we discuss next in the case of negligible disorder and defects.

Next, we discuss how the observed vortex crossings modulate the voltage drop across the SN and how synchronous and asynchronous crossings affect the spectrum of frequencies as a consequence of those modulations. In Figures 8–11, we show the voltage as a function of time, $V(t)$, for different sourced currents J and their corresponding spectra of frequencies (obtained by the Fourier transform of $V(t)$) for SNs of $w = 6\xi$ and $w = 12\xi$ under an applied field $H = 0.25H_{c2}$. In each case, we first use the TDGL approach to find the ground states for each SN with the given magnetic field; then, we sweep J from 0 up to $\simeq J_{DP}$ in sufficiently small steps (typically $\simeq 0.025J_{DP}$). At each current step, the simulation was left to run for a sufficiently long time such that a dynamic equilibrium was reached (typically up to $t = 5 \times 10^3\,\tau_{GL}$) before recording the data. The thus-obtained $V(t)$ and the spatial distribution of the superconducting order parameter at each time step were used to produce Figures 8 and 10.

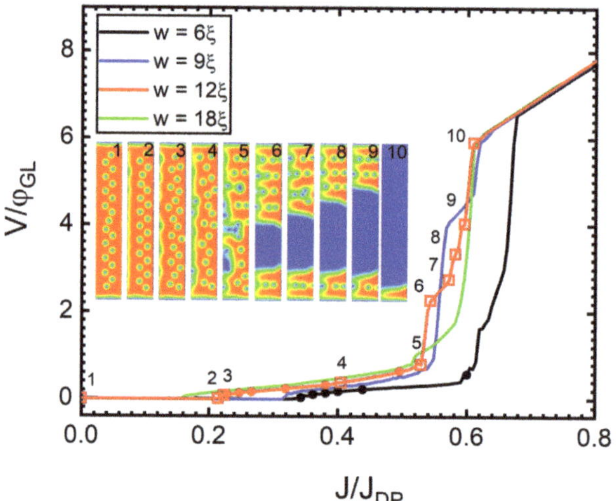

Figure 7. Normalised voltage drop as a function of the normalised current density for SNs of widths $w = 6, 9, 12$, and 18ξ with a magnetic field $H = 0.25H_{c2}$. The black and red dots (for $w = 6$ and 12ξ, respectively) mark the values of the current density at which the analysis of the modulation frequency spectra are presented in Figures 9 and 11. Inset: Snapshots of the Cooper-pair density for a SN of $w = 12\xi$, numbered 1–10 from left to right, are indicated by open red squares.

In the dissipative state, $V(t)$ increases as vortices move across the SN, with the maxima corresponding to the exit of a vortex and the minima corresponding to an entry of a vortex [34,53], leading to modulations of $V(t)$ for both SNs considered (Figures 8 and 10). Considering the SN of $w = 6\xi$, which is sourced with the lowest current that causes the vortex crossing ($J = 0.348J_{DP}$ in this case), $V(t)$ shows evidence of asynchronous vortex dynamics, with several distinct features having a periodicity of 486 τ_{GL}. Even though the vortices do not cross in synchronised rows, there is a quasi-synchronised behaviour that manifests in the repetition of vortex crossings in a given dynamic configuration. As J is increased from $0.348J_{DP}$ to $0.406J_{DP}$ (panels A–D in Figure 8), the modulations in the voltage evolve, and the number of modulations caused by quasi-synchronous crossings is reduced. Finally, beyond $J = 0.444J_{DP}$ (panel E), there is only one mode that repeats periodically, i.e., the vortex dynamics become fully synchronous, and they accelerate with the further increase in the current (panel F). The relative spectra of the frequencies for $V(t)$ are shown in Figure 9 in panels labelled correspondingly to the panels of Figure 8. The repetitive modes of vortex crossings within the particular dynamic configuration lead to peaks at specific frequencies. As the current density is increased, the spectra show an evolution to a single peak, corresponding to the frequency of $0.03\tau_{GL}^{-1}$.

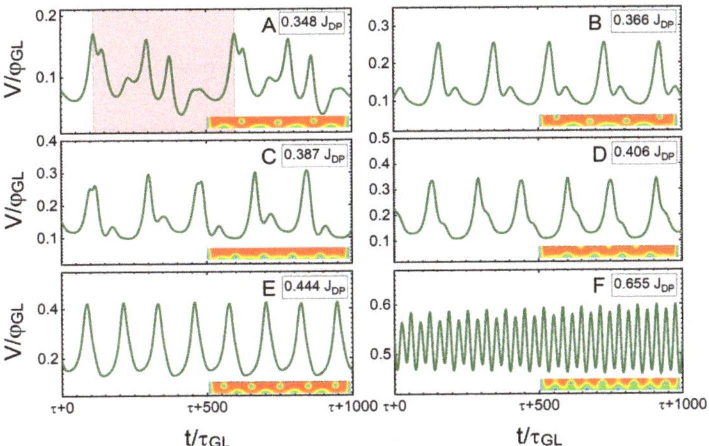

Figure 8. Normalised voltage drop as a function of time (normalised to τ_{GL}) for a SN of $w = 6\xi$ under a magnetic field of $H = 0.25H_{c2}$ sourced with different current densities of: (**A**) $0.348J_{DP}$, (**B**) $0.366J_{DP}$, (**C**) $0.387J_{DP}$, (**D**) $0.406J_{DP}$, (**E**) $0.444J_{DP}$, and (**F**) $0.655J_{DP}$. Each panel contains an illustrative snapshot of the spatial distribution of the Cooper-pair density during the dynamics. The pale red area in (**A**) shows the periodicity of the spectrum.

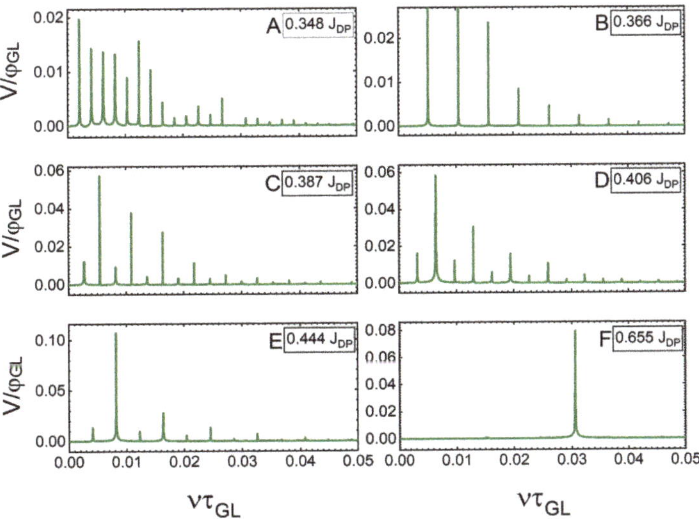

Figure 9. Spectra of the modulation frequencies ν (normalised to τ_{GL}^{-1}) of the temporal voltage signals shown in Figure 8. (**A**) $0.348J_{DP}$, (**B**) $0.366J_{DP}$, (**C**) $0.387J_{DP}$, (**D**) $0.406J_{DP}$, (**E**) $0.444J_{DP}$, and (**F**) $0.655J_{DP}$.

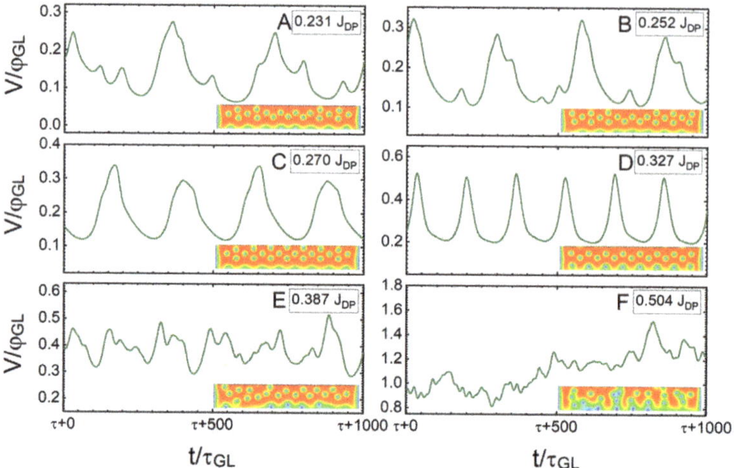

Figure 10. Normalised voltage drop as a function of time (normalised to τ_{GL}) for a SN of $w = 12\xi$ under a magnetic field of $H = 0.25H_{c2}$, sourced with different current densities: (**A**) $0.231J_{DP}$, (**B**) $0.252J_{DP}$, (**C**) $0.270J_{DP}$, (**D**) $0.327J_{DP}$, (**E**) $0.387J_{DP}$ and (**F**) $0.504J_{DP}$. Each panel contains an illustrative snapshot of the spatial distribution of the Cooper-pair density during the dynamics.

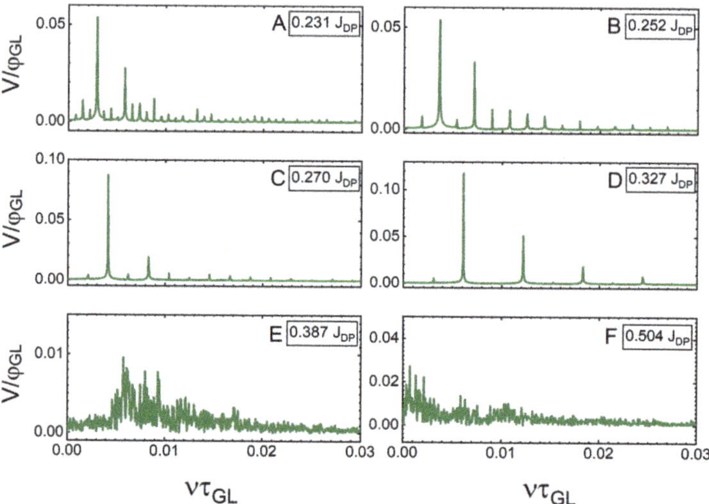

Figure 11. Spectra of the modulation frequencies ν (normalised to τ_{GL}^{-1}) of the temporal voltage signals shown in Figure 10. (**A**) $0.231J_{DP}$, (**B**) $0.252J_{DP}$, (**C**) $0.270J_{DP}$, (**D**) $0.327J_{DP}$, (**E**) $0.387J_{DP}$ and (**F**) $0.504J_{DP}$.

A similar analysis for a SN with $w = 12\xi$ is shown in Figure 10 because a wider nanostripe allows the formation of multiple vortex rows in the ground state. Panel A of Figure 10, shows $V(t)$ at $J = 0.231J_{DP}$, when the vortices start dissipatively crossing the SN in a quasi-synchronous fashion. As J is increased to $0.327J_{DP}$, the vortex crossings become increasingly synchronised (panels B–D). However, at $J = 0.387J_{DP}$, the flux–flow instability sets in (panel E) and causes an increasingly chaotic behaviour as J is increased to $J = 0.504J_{DP}$ (panel F). In this regime, the apparent chaotic behaviour is caused by the competition between the standard vortex–vortex repulsion and the effective attractive core–core interaction due to preferential tailgating at large vortex velocities, which interchange their dominance over each vortex during the collective dynamics. For $J > 0.52 J_{DP}$, a phase slip occurs, which will grow as sourced current density is raised (shown in Figure 7 for the states labelled 6–10), and the remaining vortices cross the stripe in tailgated rows. This case of tailgated vortices causes periodic modulations $V(\tau)$; however, at such high values of J, this regime is unstable and, therefore, not considered in the following discussion. So, we only consider the region of strict flow–flow during the discussion of synchronised vortex crossings.

The spectra of the frequency modulations (Figure 11) show an analogous behaviour to that of the narrower SN discussed previously. At low values of J, when the crossings are quasi-synchronous, we see many mode contributions (i.e., few dominant peaks accompanied by many additional smaller peaks). As J is increased and synchronicity improves, the smaller contributions disappear, and the frequency component with the largest contribution is strengthened. However, at the onset of the flux–flow instability ($J = 0.387 J_{DP}$), we observe a broad contribution centred around the frequency $\nu = 0.06\tau_{GL}^{-1}$ (corresponding to the median frequency of crossing of the vortex lattice as a whole, with many individual asynchronous crossings being superimposed). At $J = 0.504 J_{DP}$ (panel F), the spectrum loses all order, corresponding to the chaotic behaviour of vortex crossings.

Vortices that continuously cross the SN will cause oscillations in the electric and magnetic fields, leading to detectable emission of electromagnetic radiation [15,16,54]. The crossing of a single vortex releases a very small amount of energy, whereas multiple vortices moving coherently will emit a significant (and more easily detectable) amount of energy [54]. In a coherently moving lattice of vortices, periodic vortex crossing in the SN will cause the emission of radiation at a frequency of $\omega = 2\pi v/a$ (washboard frequency) and at harmonics of $\omega = 2\pi m v/a$ ($m = 2,3\ldots$), where v is the vortex speed and a is the lattice spacing (i.e., the distance between two parallel adjacent rows in our case) along the direction of motion [15]. The highest frequency emitted cannot exceed $\Delta/\hbar$, where Δ is the superconducting gap of the SN. The theoretically predicted existence of radiation has been experimentally confirmed [16].

The results of our simulations of the vortex velocity as a function of the sourced current density are shown in Figure 12a. They evidence a linear dependence at lower values of J for both of the narrow SNs considered above (which is similar to the behaviour seen in [52,55]). However, when J is increased to intermediate values, we find a deviation from the linear dependence, which is due to the increasingly facilitated vortex tailgating. We use these values of velocities and the frequency spectra to further discuss the potential for coherent radiation of vortices crossing the SN. In detail, considering the SN of width $w = 6\xi$, at $J = 0.348 J_{DP}$, the average velocity is $v \approx 0.03\xi\tau_{GL}^{-1}$. The corresponding spectrum of modulations (Figure 9—panel A) shows a number of contributions, with the first five occurring at $\nu_0 = 0.0021$, $\nu_1 = 0.0041$, $\nu_2 = 0.0062$, $\nu_3 = 0.0083$, and $\nu_4 = 0.0104\ \tau_{GL}^{-1}$, which are harmonics of the fundamental mode (ν_0). The period of the cycle of repeating vortex crossings in this case is T = 486 τ_{GL}, while the wavelength is 14.6ξ (obtained from $\lambda = vt$ by using the value of the vortex velocity in Figure 12). As the vortices move in a quasi-synchronous manner, the washboard frequency [15] is not applicable.

As the sourced current is increased, the vortices cross the SN in a more synchronised manner. In this case, we can apply the relation for the washboard frequency to the values of the average vortex velocity v and the frequency of the first harmonic ν_0 and obtain the value for the (virtual) lattice spacing a. By increasing J from $0.366 J_{DP}$ to $0.655 J_{DP}$, the values of this lattice spacing decrease from 6.2ξ to 2.8ξ, where the values are extrapolated from the washboard frequency relation. The combination of increasing Lorentz force and edge confining forces causes the reduction of a and an increase in the vortex density during the dynamics. However, in the wider SN ($w = 12\xi$), we do not observe the same behaviour. In this case, the value of a remains constant ($\simeq 3.6\xi$, obtained using the washboard frequency relation) as the current is increased in the dissipative state, until it transitions to asynchronous crossings for high values of J. At this point, the resultant force on the vortices no longer leads to a preferred geometrical dynamic configuration. A combination of the Lorentz force, edge Meissner current confining forces, vortex–vortex interactions, and vortex nucleation rate results in the asynchronous behaviour. This suggests that in the wider SN, the vortex–vortex repulsion within the lattice is more deterministic for the resulting lattice spacing a than the interactions with confining edges during the vortex dynamics.

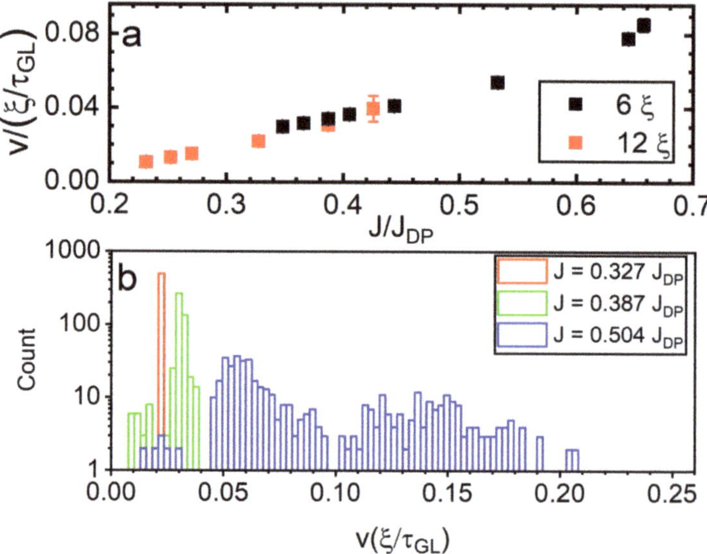

Figure 12. Panel (**a**)—Normalised average vortex velocity in units of ξ/τ_{GL} versus the normalised current density sourced to SNs of widths 6ξ and 12ξ under an applied magnetic field of $H = 0.25H_{c2}$. Panel (**b**)—Histogram of vortex velocities for different values of J relating to different vortex crossing regimes.

The synchronisation of vortex crossing in a fixed lattice with large sourced currents was observed and discussed in [33], albeit without identifying an exact regime as such.

To better understand the regime in which synchronous lattice crossings can occur, we performed a number of additional simulations. Figure 13 shows four other examples of the modulations in $V(t)$, corresponding to a SN with $w = 12\xi$ and $H = 0.2H_{c2}$ (panels a.i–iv), which realises a single row of vortices in the static case (Figure 2). At $J = 0.28J_{DP}$, the SN becomes dissipative with asynchronous vortex crossing behaviour (a.i), until the point at $J \simeq 0.36J_{DP}$, where quasi-synchronous crossings begin (a.ii). The latter continues (a.iii) until $J = 0.48J_{DP}$, where the flux–flow instability sets in (a.iv), achieving the normal state at $J_{c2} = 0.66J_{DP}$. Panels b.i–iv of Figure 13 show similar behaviours for a SN of width $w = 24\xi$ in an applied magnetic field $H = 0.12H_{c2}$ (realising two vortex rows), which does not transition to synchronised crossings as the sourced current is increased. However, after increasing the magnetic field applied to either SN, synchronous crossings will take place (panels c.i–iv and d.i–iv). For SNs of $w = 12$ and 24ξ at $H = 0.50$ and $0.15H_{c2}$, respectively, the vortex configuration comprises three parallel rows. For $w = 12\xi$ ($w = 24\xi$), synchronous crossings start at $J \simeq 0.20J_{DP}$ ($J = 0.28J_{DP}$) and continue until the onset of flux–flow instability at $J = 0.28J_{DP}$ ($J = 0.42J_{DP}$).

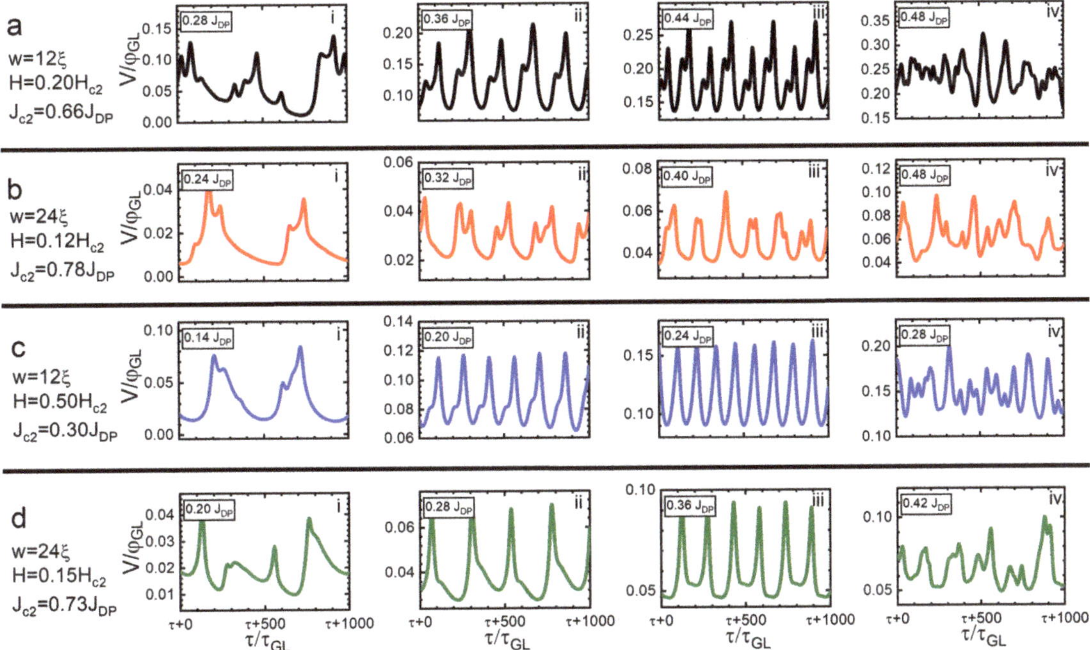

Figure 13. Temporal evolution of the normalised voltage drop with increasing (indicated) values of the sourced current density for two SNs of widths 12ξ and 24ξ. All plots exhibit the voltage modulations caused by vortex crossing. Panels (**a**). i–iv: $w = 12\xi$, $H = 0.20H_{c2}$ (single row of vortices). Panels (**b**). i–iv: $w = 24\xi$, $H = 0.12H_{c2}$ (two vortex rows). Panels (**c**). i–iv: $w = 12\xi$, $H = 0.50H_{c2}$ (three rows). Panels (**d**). i–iv: $w = 24\xi$, $H = 0.15H_{c2}$ (three rows). The first panel in each row corresponds to the onset of the dissipative state; the second and third belong to the synchronous/quasi-synchronous regime; the fourth is the onset of the flux–flow instability regime.

One can conclude that the frequency of the radiation stemming from coherent vortex crossings can be tuned with H and/or with J. The applied magnetic field H changes the vortex density (affecting the number of rows) and, hence, a, while the transport current directly changes the vortex velocity. Both factors influence the behaviour of vortex crossings, which, in turn, affect the electromagnetic radiation emitted at frequencies $\nu = v/a$ [33]. For an insight into the values expected in the experiment, we consider the parameters measured for Nb thin films by Pinto et al. [18]. For example, in a Nb film of thickness $d = 20$ nm, $\xi(0) \simeq 8.0$ nm, and $T_c = 8$ K, given a Ginzburg–Landau time of $\tau_{GL} \simeq 65$ fs, our results show an average velocity of vortices crossing the Nb SN of thickness 20 nm and width 50–100 nm to be in the range 1–10 km/s, with the first harmonic frequencies being in the range of 1–50 GHz. These values are similar to those reported by Dobrovolskiy et al. [16,38] and to those of Embon et al. [35]. A thin and narrow superconductor with a high value of T_c and a small value of τ_{GL} ($\simeq 1 - 10$ fs), in which faster vortex crossings could be realised, could be used as a terahertz radiation source. Such sources are highly sought for a variety of applications [56], including clinical applications [57] and terahertz time-domain spectroscopy [58].

4. Conclusions

In summary, our comprehensive study details how lateral quantum confinement in superconducting nanostripes (SNs) leads to experimentally detectable physical effects in the spatial configurations and dynamical phases of vortices and vortex rows. In the stationary case, we presented an equilibrium vortex row phase diagram in the parametric space delimited by the width of the nanostripe and the applied magnetic field. The diagram showed how the narrowest SNs support a lower number of vortex rows up to relatively large values of the applied magnetic field, thus exhibiting a lower average vortex density (for given values of H) that deviates more from the theoretical value attributed to the vortex density in an Abrikosov lattice. Although the phase diagram predicts the number of rows in an SN under a given magnetic field without referring to specific superconducting materials, our work provides an indication of the range of microscopic parameters that can be found in several families of superconductors, which deserves to be exploited in order to design novel and advanced quantum devices. At intermediate SN widths, i.e., $w = 20 \div 60\xi$, we pointed out that an interplay of vortex interactions, Meissner currents, and the related edge barrier can lead to a re-entrant behaviour of the vortex row states $(1 \rightarrow 2 \rightarrow 1)$ as the field is increased, which may lead to large magnetoresistance oscillations, as discussed in [30]. In relation to transport and dynamics, we reported that the critical current (for the very onset of dissipation) as a function of the magnetic field will exhibit oscillations commensurate with the formation of every additional vortex row. With sourced current beyond the critical value, the vortices crossing the SN modulate the voltage drop across the SN, which is also linked to the emission of electromagnetic radiation. More vortices crossing in synchronicity causes more radiation power to be emitted for that crossing frequency. We showed that depending on its width and the applied magnetic field a SN, under a sourced current, enters the dissipating state, with the vortices crossing relatively slowly and quasi-synchronously at first. As the current intensity is increased, the crossings become increasingly synchronous, until flux–flow instability occurs and synchronicity is gradually lost. Not every example showed the evolution toward synchronised vortex crossings; however, a regime for the occurrence of the crossing of vortex rows in a fixed lattice was observed for states with an average vortex area of $A \lesssim 80\xi^2$. This is also related to effective confinement, as a large vortex density is required so that the edge confining forces can adequately act on the vortex rows and lock them into a dynamically synchronous lattice, with typical crossing frequencies in GHz-to-THz range, which is beneficial for electromagnetic radiation in the corresponding frequency bandwidth.

Author Contributions: The project was conceived and supervised by A.P. and M.V.M. and executed by B.M., who performed all of the calculations and data analysis and wrote the manuscript. N.P. critically screened the results and discussion with respect to typical experimental observables. All authors contributed to the final scientific statement of the article and agreed to the published version of the manuscript.

Funding: This research was supported by the Research Foundation—Flanders (FWO).

Acknowledgments: The authors acknowledge the input from N. De Leo and M. Fretto of INRiM, Turin, Italy.

Conflicts of Interest: The authors declare no conflict of interest.

References

1. Hadfield, R.H. Single-photon detectors for optical quantum information applications. *Nat. Photonics* **2009**, *3*, 696–705. [CrossRef]
2. Natarajan, C.M.; Tanner, M.G.; Hadfield, R.H. Superconducting nanowire single-photon detectors: Physics and applications. *Supercond. Sci. Technol.* **2012**, *25*, 063001. [CrossRef]
3. Gol'Tsman, G.; Okunev, O.; Chulkova, G.; Lipatov, A.; Semenov, A.; Smirnov, K.; Voronov, B.; Dzardanov, A.; Williams, C.; Sobolewski, R. Picosecond superconducting single-photon optical detector. *Appl. Phys. Lett.* **2001**, *79*, 705–707. [CrossRef]
4. Dauler, E.A.; Robinson, B.S.; Kerman, A.J.; Yang, J.K.; Rosfjord, K.M.; Anant, V.; Voronov, B.; Gol'tsman, G.; Berggren, K.K. Multi-element superconducting nanowire single-photon detector. *IEEE Trans. Appl. Supercond.* **2007**, *17*, 279–284. [CrossRef]

5. Vlasko-Vlasov, V.; Colauto, F.; Buzdin, A.I.; Rosenmann, D.; Benseman, T.; Kwok, W.K. Magnetic gates and guides for superconducting vortices. *Phys. Rev. B* **2017**, *95*, 144504. [CrossRef]

6. Vlasko-Vlasov, V.; Colauto, F.; Buzdin, A.I.; Rosenmann, D.; Benseman, T.; Kwok, W.K. Manipulating Abrikosov vortices with soft magnetic stripes. *Phys. Rev. B* **2017**, *95*, 174514. [CrossRef]

7. Córdoba, R.; Orús, P.; Jelić, Ž.L.; Sesé, J.; Ibarra, M.R.; Guillamón, I.; Vieira, S.; Palacios, J.J.; Suderow, H.; Milosević, M.V.; et al. Long-range vortex transfer in superconducting nanowires. *Sci. Rep.* **2019**, *9*, 12386. [CrossRef] [PubMed]

8. Brooke, J.; Bitko, D.; Rosenbaum; Aeppli, G. Quantum annealing of a disordered magnet. *Science* **1999**, *284*, 779–781. [CrossRef] [PubMed]

9. Johnson, M.; Bunyk, P.; Maibaum, F.; Tolkacheva, E.; Berkley, A.; Chapple, E.; Harris, R.; Johansson, J.; Lanting, T.; Perminov, I.; et al. A scalable control system for a superconducting adiabatic quantum optimization processor. *Supercond. Sci. Technol.* **2010**, *23*, 065004. [CrossRef]

10. Strauch, F.W.; Johnson, P.R.; Dragt, A.J.; Lobb, C.; Anderson, J.; Wellstood, F. Quantum logic gates for coupled superconducting phase qubits. *Phys. Rev. Lett.* **2003**, *91*, 167005. [CrossRef]

11. Reichhardt, C.; Reichhardt, C.O. Jamming and diode effects for vortices in nanostructured superconductors. *Phys. C Supercond.* **2010**, *470*, 722–725. [CrossRef]

12. Wambaugh, J.; Reichhardt, C.; Olson, C.; Marchesoni, F.; Nori, F. Superconducting fluxon pumps and lenses. *Phys. Rev. Lett.* **1999**, *83*, 5106. [CrossRef]

13. Daido, A.; Ikeda, Y.; Yanase, Y. Intrinsic Superconducting Diode Effect. *Phys. Rev. Lett.* **2022**, *128*, 037001. [CrossRef] [PubMed]

14. Samkharadze, N.; Bruno, A.; Scarlino, P.; Zheng, G.; DiVincenzo, D.; DiCarlo, L.; Vandersypen, L. High-kinetic-inductance superconducting nanowire resonators for circuit QED in a magnetic field. *Phys. Rev. Appl.* **2016**, *5*, 044004. [CrossRef]

15. Bulaevskii, L.; Chudnovsky, E. Electromagnetic radiation from vortex flow in type-II superconductors. *Phys. Rev. Lett.* **2006**, *97*, 197002. [CrossRef] [PubMed]

16. Dobrovolskiy, O.; Bevz, V.; Mikhailov, M.Y.; Yuzephovich, O.; Shklovskij, V.; Vovk, R.; Tsindlekht, M.; Sachser, R.; Huth, M. Microwave emission from superconducting vortices in Mo/Si superlattices. *Nat. Commun.* **2018**, *9*, 4927. [CrossRef]

17. Perali, A.; Bianconi, A.; Lanzara, A.; Saini, N.L. The gap amplification at a shape resonance in a superlattice of quantum stripes: A mechanism for high Tc. *Solid State Commun.* **1996**, *100*, 181–186. [CrossRef]

18. Pinto, N.; Rezvani, S.J.; Perali, A.; Flammia, L.; Milošević, M.V.; Fretto, M.; Cassiago, C.; De Leo, N. Dimensional crossover and incipient quantum size effects in superconducting niobium nanofilms. *Sci. Rep.* **2018**, *8*, 4710. [CrossRef] [PubMed]

19. Guidini, A.; Flammia, L.; Milošević, M.V.; Perali, A. BCS-BEC crossover in quantum confined superconductors. *J. Supercond. Nov. Magn.* **2016**, *29*, 711–715. [CrossRef]

20. Saraiva, T.; Cavalcanti, P.; Vagov, A.; Vasenko, A.; Perali, A.; Dell'Anna, L.; Shanenko, A. Multiband material with a quasi-1D band as a robust high-temperature superconductor. *Phys. Rev. Lett.* **2020**, *125*, 217003. [CrossRef] [PubMed]

21. Moshchalkov, V.; Gielen, L.; Strunk, C.; Jonckheere, R.; Qiu, X.; Haesendonck, C.V.; Bruynseraede, Y. Effect of sample topology on the critical fields of mesoscopic superconductors. *Nature* **1995**, *373*, 319–322. [CrossRef]

22. Datta, S. *Electronic Transport in Mesoscopic Systems*; Cambridge University Press: Cambridge, UK, 1997.

23. Cren, T.; Fokin, D.; Debontridder, F.; Dubost, V.; Roditchev, D. Ultimate vortex confinement studied by scanning tunneling spectroscopy. *Phys. Rev. Lett.* **2009**, *102*, 127005. [CrossRef] [PubMed]

24. Marrocco, N.; Pepe, G.; Capretti, A.; Parlato, L.; Pagliarulo, V.; Peluso, G.; Barone, A.; Cristiano, R.; Ejrnaes, M.; Casaburi, A.; et al. Strong critical current density enhancement in NiCu/NbN superconducting nanostripes for optical detection. *Appl. Phys. Lett.* **2010**, *97*, 092504. [CrossRef]

25. Bean, C.; Livingston, J. Surface barrier in type-II superconductors. *Phys. Rev. Lett.* **1964**, *12*, 14. [CrossRef]

26. Flammia, L.; Zhang, L.F.; Covaci, L.; Perali, A.; Milošević, M. Superconducting nanoribbon with a constriction: A quantum-confined Josephson junction. *Phys. Rev. B* **2018**, *97*, 134514. [CrossRef]

27. Berdiyorov, G.; Chao, X.; Peeters, F.; Wang, H.; Moshchalkov, V.; Zhu, B. Magnetoresistance oscillations in superconducting strips: A Ginzburg-Landau study. *Phys. Rev. B* **2012**, *86*, 224504. [CrossRef]

28. Anderson, P.; Dayem, A. Radio-frequency effects in superconducting thin film bridges. *Phys. Rev. Lett.* **1964**, *13*, 195. [CrossRef]

29. Córdoba, R.; Baturina, T.; Sesé, J.; Yu Mironov, A.; De Teresa, J.; Ibarra, M.; Nasimov, D.; Gutakovskii, A.; Latyshev, A.; Guillamón, I.; et al. Magnetic field-induced dissipation-free state in superconducting nanostructures. *Nat. Commun.* **2013**, *4*, 1437. [CrossRef]

30. Berdiyorov, G.; Milošević, M.; Latimer, M.; Xiao, Z.; Kwok, W.; Peeters, F. Large magnetoresistance oscillations in mesoscopic superconductors due to current-excited moving vortices. *Phys. Rev. Lett.* **2012**, *109*, 057004. [CrossRef]

31. Kimmel, G.J.; Glatz, A.; Vinokur, V.M.; Sadovskyy, I.A. Edge effect pinning in mesoscopic superconducting strips with non-uniform distribution of defects. *Sci. Rep.* **2019**, *9*, 211. [CrossRef]

32. Vodolazov, D. Vortex-induced negative magnetoresistance and peak effect in narrow superconducting films. *Phys. Rev. B* **2013**, *88*, 014525. [CrossRef]

33. Vodolazov, D.Y.; Peeters, F. Rearrangement of the vortex lattice due to instabilities of vortex flow. *Phys. Rev. B* **2007**, *76*, 014521. [CrossRef]

34. Jelić, Ž.; Milošević, M.; Silhanek, A. Velocimetry of superconducting vortices based on stroboscopic resonances. *Sci. Rep.* **2016**, *6*, 35687. [CrossRef] [PubMed]

35. Embon, L.; Anahory, Y.; Jelić, Ž.L.; Lachman, E.O.; Myasoedov, Y.; Huber, M.E.; Mikitik, G.P.; Silhanek, A.V.; Milošević, M.V.; Gurevich, A.; et al. Imaging of super-fast dynamics and flow instabilities of superconducting vortices. *Nat. Commun.* **2017**, *8*, 85. [CrossRef] [PubMed]
36. Milošević, M.; Geurts, R. The Ginzburg–Landau theory in application. *Phys. C Supercond.* **2010**, *470*, 791–795. [CrossRef]
37. Carapella, G.; Sabatino, P.; Barone, C.; Pagano, S.; Gombos, M. Current driven transition from Abrikosov-Josephson to Josephson-like vortex in mesoscopic lateral S/S'/S superconducting weak links. *Sci. Rep.* **2016**, *6*, 35694. [CrossRef]
38. Dobrovolskiy, O.; Vodolazov, D.Y.; Porrati, F.; Sachser, R.; Bevz, V.; Mikhailov, M.Y.; Chumak, A.; Huth, M. Ultra-fast vortex motion in a direct-write Nb-C superconductor. *Nat. Commun.* **2020**, *11*, 3291. [CrossRef]
39. Hebboul, S.; Johnson, D.; Rokhlin, M. Radio-frequency oscillations in two-dimensional superconducting In- InO x: A possible evidence for vortex density waves. *Phys. Rev. Lett.* **1999**, *82*, 831. [CrossRef]
40. Doria, M.M.; Gubernatis, J.; Rainer, D. Virial theorem for Ginzburg-Landau theories with potential applications to numerical studies of type-II superconductors. *Phys. Rev. B* **1989**, *39*, 9573. [CrossRef]
41. Kato, R.; Enomoto, Y.; Maekawa, S. Effects of the surface boundary on the magnetization process in type-II superconductors. *Phys. Rev. B* **1993**, *47*, 8016. [CrossRef] [PubMed]
42. Ivlev, B.; Kopnin, N. Electric currents and resistive states in thin superconductors. *Adv. Phys.* **1984**, *33*, 47–114. [CrossRef]
43. Kramer, L.; Watts-Tobin, R. Theory of dissipative current-carrying states in superconducting filaments. *Phys. Rev. Lett.* **1978**, *40*, 1041. [CrossRef]
44. Clem, J. Paper K36. 06. *Bull. Am. Phys. Soc.* **1998**, *43*, 411.
45. Stan, G.; Field, S.B.; Martinis, J.M. Critical field for complete vortex expulsion from narrow superconducting strips. *Phys. Rev. Lett.* **2004**, *92*, 097003. [CrossRef]
46. Maksimova, G. Mixed state and critical current in narrow semiconducting films. *Phys. Solid State* **1998**, *40*, 1607–1610. [CrossRef]
47. Abrikosov, A.A. Nobel Lecture: Type-II superconductors and the vortex lattice. *Rev. Mod. Phys.* **2004**, *76*, 975. [CrossRef]
48. Mawatari, Y.; Yamafuji, K. Critical current density in thin films due to the surface barrier. *Phys. C Supercond.* **1994**, *228*, 336–350. [CrossRef]
49. Carneiro, G. Equilibrium vortex-line configurations and critical currents in thin films under a parallel field. *Phys. Rev. B* **1998**, *57*, 6077. [CrossRef]
50. Vodolazov, D.Y.; Maksimov, I.; Brandt, E. Vortex entry conditions in type-II superconductors.: Effect of surface defects. *Phys. C Supercond.* **2003**, *384*, 211–226. [CrossRef]
51. Langer, J.S.; Ambegaokar, V. Intrinsic resistive transition in narrow superconducting channels. *Phys. Rev.* **1967**, *164*, 498. [CrossRef]
52. Cadorim, L.R.; de Oliveira Junior, A.; Sardella, E. Ultra-fast kinematic vortices in mesoscopic superconductors: The effect of the self-field. *Sci. Rep.* **2020**, *10*, 18662. [CrossRef]
53. Jelić, Ž.; Milošević, M.; Van de Vondel, J.; Silhanek, A. Stroboscopic phenomena in superconductors with dynamic pinning landscape. *Sci. Rep.* **2015**, *5*, 14604. [CrossRef]
54. Dolgov, O.; Schopohl, N. Transition radiation of moving Abrikosov vortices. *Phys. Rev. B* **2000**, *61*, 12389. [CrossRef]
55. Berdiyorov, G.; Milošević, M.; Peeters, F. Kinematic vortex-antivortex lines in strongly driven superconducting stripes. *Phys. Rev. B* **2009**, *79*, 184506. [CrossRef]
56. Hafez, H.; Chai, X.; Ibrahim, A.; Mondal, S.; Férachou, D.; Ropagnol, X.; Ozaki, T. Intense terahertz radiation and their applications. *J. Opt.* **2016**, *18*, 093004. [CrossRef]
57. Son, J.H.; Oh, S.J.; Cheon, H. Potential clinical applications of terahertz radiation. *J. Appl. Phys.* **2019**, *125*, 190901. [CrossRef]
58. Gowen, A.A.; O'Sullivan, C.; O'Donnell, C. Terahertz time domain spectroscopy and imaging: Emerging techniques for food process monitoring and quality control. *Trends Food Sci. Technol.* **2012**, *25*, 40–46. [CrossRef]

 *nanomaterials*

Article

Complex Phase-Fluctuation Effects Correlated with Granularity in Superconducting NbN Nanofilms

Meenakshi Sharma [1,†] , Manju Singh [2], Rajib K. Rakshit [2], Surinder P. Singh [3], Matteo Fretto [4], Natascia De Leo [4], Andrea Perali [5] and Nicola Pinto [1,4,*,†]

1 School of Science and Technology, University of Camerino, 62032 Camerino, Italy
2 CNR-Institute for the Study of Nanostructured Materials, 40129 Bologna, Italy
3 National Physical Laboratory, CSIR, New Delhi 110012, India
4 Advanced Materials Metrology and Life Science Division, INRiM, 10135 Torino, Italy
5 School of Pharmacy, Physics Unit, University of Camerino, 62032 Camerino, Italy
* Correspondence: nicola.pinto@unicam.it
† These authors contributed equally to this work.

Abstract: Superconducting nanofilms are tunable systems that can host a 3D–2D dimensional crossover leading to the Berezinskii–Kosterlitz–Thouless (BKT) superconducting transition approaching the 2D regime. Reducing the dimensionality further, from 2D to quasi-1D superconducting nanostructures with disorder, can generate quantum and thermal phase slips (PS) of the order parameter. Both BKT and PS are complex phase-fluctuation phenomena of difficult experiments. We characterized superconducting NbN nanofilms thinner than 15 nm, on different substrates, by temperature-dependent resistivity and current–voltage (I-V) characteristics. Our measurements evidence clear features related to the emergence of BKT transition and PS events. The contemporary observation in the same system of BKT transition and PS events, and their tunable evolution in temperature and thickness was explained as due to the nano-conducting paths forming in a granular NbN system. In one of the investigated samples, we were able to trace and characterize the continuous evolution in temperature from quantum to thermal PS. Our analysis established that the detected complex phase phenomena are strongly related to the interplay between the typical size of the nano-conductive paths and the superconducting coherence length.

Keywords: NbN; ultrathin films; BKT transition; phase slips; granular superconductivity

Citation: Sharma, M.; Singh, M.; Rakshit, R.K.; Singh, S.P.; Fretto, M.; De Leo, N.; Perali, A.; Pinto, N. Complex Phase-Fluctuation Effects Correlated with Granularity in Superconducting NbN Nanofilms. *Nanomaterials* **2022**, *12*, 4109. https://doi.org/10.3390/nano12234109

Academic Editor: Orion Ciftja

Received: 11 November 2022
Accepted: 19 November 2022
Published: 22 November 2022

Publisher's Note: MDPI stays neutral with regard to jurisdictional claims in published maps and institutional affiliations.

Copyright: © 2022 by the authors. Licensee MDPI, Basel, Switzerland. This article is an open access article distributed under the terms and conditions of the Creative Commons Attribution (CC BY) license (https://creativecommons.org/licenses/by/4.0/).

1. Introduction

Effects related to thermal and quantum fluctuations in low-dimensional superconductors, such as phase slips [1–5], quantum criticality [6], superconductor-insulator transition [7] and quantum-phase transitions [8,9], have been studied for several decades. These quantum and many-body effects are controlled by several film properties, such as spatial dimensions, electronic disorder and structural inhomogeneities [10,11].

In recent years, the scientific interest has been focused on quasi 2D and 1D systems, where the presence of resistive states close to the superconducting transition temperature T_c has been found to produce detectable effects in the transport properties [10,11].

In such systems, the emerging resistive states are a fundamental phenomenon, which involves the understanding of advanced concepts as topological excitations, phase disorder and interplay between different length scales of the superconducting state. These have important applications in the field of quantum technologies, ultrasonic detectors of radiation, single photon detector and nanocalorimeters [10,12].

Efforts have been made to study low-dimensional systems theoretically, and from the experimental side, the attention has been focused on thin films in which a crossover from 3D to 2D Berezinskii–Kosterlitz–Thouless (BKT) transition [13–16] occurs, lowering the thickness to few nanometers, as is the case, for instance, for NbN films [15,17–20].

A superconducting-to-normal-state transition in a 2D-XY model was introduced to explain the formation of thermally excited vortex-anti-vortex pairs (VAP) in ultrathin films, even in the absence of an applied magnetic field [13], resulting in the BKT topological phase transition [13,21]. The nature of a BKT transition is completely different from the standard second-order phase transition given by the Landau paradigm. It is driven by the binding of topological excitations without any symmetry breaking associated with the onset of the order parameter [22].

A BKT-like transition is expected to occur even in dirty superconductors, at a reduced film thickness d, under some physical constraints. A required condition is that the Pearl length $\Lambda = 2\lambda^2/d$ (λ being the London penetration depth), exceeds the sample size with negligible screening effects due to charged supercurrents [23]. Furthermore, $\lambda \gg w$ or $d \ll \xi_{GL}$ (ξ_{GL} being the Gintzburg–Landau coherence length) must be fulfilled in order to detect BKT effects, though some experiments have reported the BKT transition is also outside the theoretically established limits [24].

To observe this remarkable phenomenon in real systems, two approaches have been explored. In the first, the BKT theory predicts a universal jump in the film's superfluid stiffness, n_s, at the characteristic temperature $T_{BKT} < T_{MF}$, T_{MF} being the mean-field superconducting transition temperature. This jump is related to the VAP binding through a logarithmic interaction potential between free vortices. Sourced current can break bound pairs, producing free vortices, inducing nonlinear effects in I-V curves [24]. For the second approach, transition is observed in the correlation length, which diverges exponentially at T_{BKT}, in contrast to the power-law dependence expected within the Ginzburg–Landau (GL) theory [25,26].

When further reducing film size, fluctuations can result in the formation of multiple resistive states in I-V curves, at both low and high T ranges. These intermediate states form due to the change in phase of the order parameter by 2π, and they will result in a discontinuous voltage jump, forming phase slips. These PS are characteristic of a quasi-1D system, which form by a river of fast moving vortices (kinematic vortices) driven by the topological excitations, annihilating in the middle of the sample [13,27]. However, several superconductors have been explored experimentally to study such topological effects, including NbSe$_2$ [27], NbN [13], Nb$_2$N [11], Nb [28] and many more.

In particular, NbN is a known and well-studied material, belonging to the family of strongly coupled type-II superconductors, and it is potentially interesting for several technological applications, due to its relatively high value of bulk T_c ($\simeq$16 K). Its small coherence length, ξ ($\approx$4 nm), requires fabrication of extremely thin films (few nanometers of thickness) with fine control of their properties to achieve the 2D superconducting regime [29].

In this work, a detailed experimental study was carried out about the electrical properties of superconducting, NbN ultra-thin films, aimed at investigating the crossover regime from a quasi-2D BKT phenomenon to a quasi-1D PS mechanism. Outcomes evidenced the presence of large fluctuation effects mostly close to T_c, and an exponential decrease at lower temperatures. In one case, the freezing of thermal fluctuations at the lowest temperatures gave rise to a quantum phase slip (QPS) phenomenon, whereas in the other case two distinct resistive transitions, below T_c, were detected for one of the thinnest NbN films, suggesting the unexpected coexistence of a BKT transition and phase-slip phenomena at the dimensional crossover from 2D to 1D.

Therefore, findings on superconducting to resistive-state transition features in investigated thin films of NbN have been questioning if their origin is due to thermal fluctuations (e.g., thermal PS), quantum fluctuations (e.g., QPS) [30] or a proximity effect mechanism among coupled nano-sized superconducting grains [11]. Our study suggests that the thickness threshold where quantum phase fluctuation effects can start to appear is not yet clearly defined, since BKT transition or QPS phenomena have been detected even in NbN films that are nominally 10 nm thick. Hence, specific conditions to detect BKT transitions and PS events in quasi-2D systems, especially those being granular in nature, deserve to be further studied both theoretically and experimentally.

2. Materials And Methods

2.1. Deposition

NbN films with nominal thicknesses of 5, 10 and 15 nm were deposited on several substrates, such as MgO, Al_2O_3 and SiO_2 (see Table 1), by using DC magnetron sputtering. The optimized deposition rate was $\simeq 0.4$ nm/s, at a substrate temperature of 600 °C and at 200 W of discharge power. The N_2/Ar ratio was fixed at 1:7 during the deposition process.

Table 1. NbN film's properties. Starting from the left, columns are: film acronym (SC: Al_2O_3 c-cut; SR: Al_2O_3 r-cut; MO: MgO; SO: SiO_2; the number following the two letters refers to the film thickness in nm units); resistivity value at 15 K; superconducting transition temperature; superconducting transition width; superconducting critical-current density at 0 K.

Sample	ρ (Ω cm)	T_c (K)	ΔT_c (K)	J_{c0} (MA/cm^2)
MO5a *	8.0×10^{-5}	10.072	0.08	0.40 ± 0.014
MO5b **	5.8×10^{-4}	11.02	0.79	2.24 ± 0.013
MO10	1.2×10^{-4}	13.29	0.27	9.98 ± 0.15
MO15	2.4×10^{-4}	13.83	0.23	11.40 ± 0.016
SC5	8.0×10^{-5}	10.64	0.43	0.90 ± 0.010
SC10	2.4×10^{-4}	13.50	0.40	10 ± 0.028
SC15	1.7×10^{-4}	12.73	0.24	5.29 ± 0.08
SR5 **	1.1×10^{-4}	11.76	0.68	0.63 ± 0.013
SR10	1.7×10^{-4}	12.43	0.30	8.3 ± 0.18
SR15	2.3×10^{-4}	12.58	0.38	6.14 ± 0.13
SO5	9.3×10^{-7}	9.40	0.46	0.89 ± 0.024

* Both MO5a and MO5b belonged to the same deposition run, but the fabrication process of their Hall bar was carried out by using slightly different parameters. ** The Hall bar width of this film was 10 µm.

2.2. Fabrication of the Hall Bar

For the electrical characterization, a suitable 8-contact Hall bar geometry was designed (see inset in Figure 1), by using LibreCAD software. The corresponding optical mask was realized by using direct laser lithography (at a wavelength of 375 nm) exploiting a µPG101 laser writer from Heidelberg Instruments. The Hall bar length was fixed at 1000 µm, and bar widths of 10 and 50 µm were chosen. The Hall bars were patterned by optical lithography with a KarlSuss mask aligner (mod. MJB3) by using a reversible photoresist (AZ 5214E Photoresists MicroChemicals GmbH), spun at 4000 rpm, resulting in a nominal thickness of 1.2 µm. Later, an etching step was exploited to define the final geometry of the NbN Hall bars. In particular, the etching process was carried out by using deep reactive ion etching (PlasmaPro 100 Cobra Inductively Coupled Plasma Etching System, Oxford Instrument). The fabrication parameters (etching time, ICP power, substrate temperature, etc.) were optimized while taking into account the film thicknesses and typology of substrates. The best etching selectivity between the optical resist and the NbN thin film was obtained with a mixture of CF4 and Ar with fluxes of 90 and 10 sccm, respectively, at an operating pressure of 50 mTorr. ICP power and RIE RF power were 500 and 30 W, respectively. All the samples were cleaned with a light-oxygen plasma before the etching, to eliminate any organic or lithographic residuals. Further, the samples were glued to a copper holder with a high-thermal-conductivity glue (GE Varnish) in order to have better temperature control and stability in the cryostat. Aluminum wires (see inset of Figure 1) were bridge bonded to the pads of the Hall bar and to the pads of a small printed circuit board, placed close to the sample. A scanning electron microscope (SEM) was used for the visual inspection to check the intermediate steps of the fabrication process.

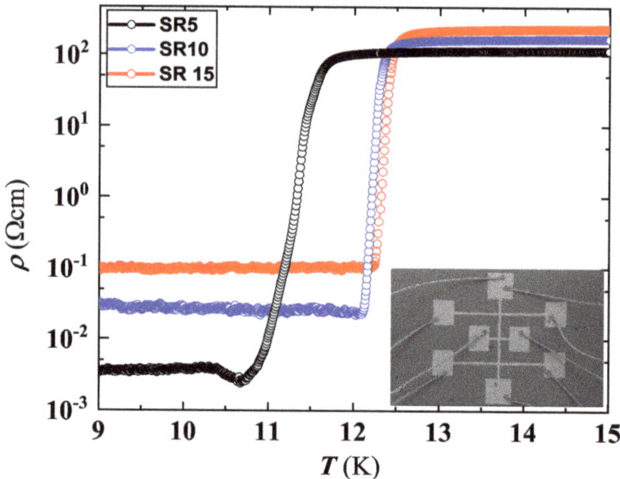

Figure 1. NbN resistivity behavior around T_c, for the set deposited on the Al_2O_3 r−cut substrate (see Table 1). Inset: Scanning electron microscopy of a typical Hall bar shaped film with Al wires bonded to sample pads. Current carrying contacts were located on the top and bottom, along the vertical line; and a couple of lateral contacts from the same side were used to detect the voltage drop.

2.3. Electrical Characterization

Resistivity, $\rho(T)$, and current–voltage (I-V) characteristics were measured as a function of temperature in a liquid-free He cryostat (Advanced Research System mod. DE210) equipped with two Si diode thermometers (Lakeshore mod. DT-670)—one out of two calibrated [31,32]. A temperature controller Lakeshore mod. 332 was used to read the temperature of the uncalibrated thermometer, thermally anchored to the second stage of the cryostat. The film temperature was measured by using one channel of a double source-meter (Keysight mod. B2912A). The other channel of the instrument was earmarked for electrical characterization of NbN film properties (ρ and I-V) in the 4-contact geometry, sourcing the current (0.1 ÷ 100 μA, typically 1 μA for ρ) and detecting the voltage drop. Either dc or a pulsed mode technique was used.

I-V characteristics was carried out both in dc mode [33] and in pulse mode. For measurements executed by the pulse-mode technique [34], a sweep of current pulses of increasing intensities, ranging from tens of μA to few mA, each of duration 1.1 ms, was used. Due to the high thermal inertia of the cryostat, data were collected without any thermal stabilization in the whole range of temperatures ($\approx 5 \div 300$ K) upon sample cooling. The maximum T change, detected at the lowest T during data acquisition (e.g., ρ), was around 15 mK. For each data point of the resistivity curve, typically 30 values were averaged, and a suitably selection of the working parameters of the source-meter allowed us to capture several values during the transition from the superconducting to the normal state. For I-V characteristics, typically 200 points were collected in a few seconds for each curve, and the maximum T variation during each I-V curve acquisition was $\lesssim$50 mK.

3. Results

3.1. Superconducting State Properties

NbN film properties were investigated by resistivity and current–voltage characteristics as a function of the temperature.

At the superconducting (SC) transition, the investigated films showed a resistivity jump spanning from ≈2 to ≈5 orders of magnitude; T_c depends on the film thickness and substrate type, and its value rapidly decreases at $d \leq 10$ nm (see Table 1).

The $\rho(T)$ curves around T_c for the set deposited on the Al$_2$O$_3$ r-cut substrate is reproduced in Figure 1. The SC transition lowers with the decrease of the film thickness, becoming substantial when passing from $d = 10$ to $d = 5$ nm. The normal state resistivity does not show, in general, a clear correlation with the value of d, appearing even reversed with respect to the d value, for the set deposited on the r-cut sapphire (Figure 1). However, normal-state resistivity variation is confined to within a half order of magnitude for $5 \leq d \leq 15$ nm. The behavior of $\rho(T)$ for SR5 is analyzed in detail in the next section. Generally, a higher T_c together with a narrow SC transition width, ΔT_c, were measured on MgO substrates above $d = 10$ nm (Figure 2 and Table 1), whereas a consistent worsening of these parameters occurred on SiO$_2$ (this behavior was detected also in films of higher thicknesses not reported in the present work). Intermediate values of both T_c and ΔT_c were measured on Al$_2$O$_3$ substrates; T_c on the c-cut type was slightly higher at $d > 5$ nm (Figure 2). Compared to NbN films deposited on Al$_2$O$_3$ r-cut, those on the c-cut type showed a wider range of T_c variation with d and a tendency toward the narrowing of ΔT_c (Table 1). An increase in d denotes an improvement in the film's quality, in agreement with findings reported by other groups [35,36].

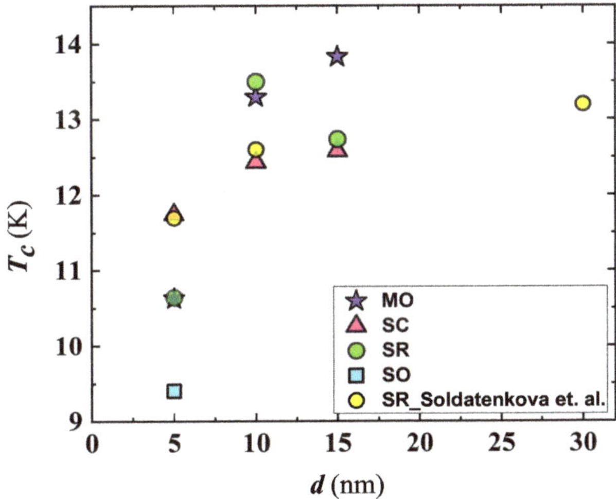

Figure 2. Thickness dependence of the superconducting transition temperature of NbN films deposited on MgO (MO), SiO$_2$ (SO) and Al$_2$O$_3$ substrates, both c-cut (SC) and r-cut (SR) types. For comparison, T_c data of films deposited on Al$_2$O$_3$ r-cut substrates by Soldatenkova et al. [37] were plotted.

While the general trend of T_c as a function of d was reported in the literature by several groups [38,39], T_c spreading depends also on the crystal structure [40], deposition technique, the partial pressure of nitrogen used during the film fabrication process, etc., making a direct comparison of results measured by different groups difficult [14,38,39,41]. Anyway, taking into account NbN films deposited on Al$_2$O$_3$ r-cut substrates, our T_c values, while appearing scattered at $d \leq 10$ nm, follow a trend similar to that found by Soldatenkova et al. on the same substrate type (see Figure 2) [37]. Scattering of T_c in NbN films reflects inhomogeneity issues that are characteristic of this superconducting system [16].

The NbN film's properties were studied further through current–voltage characteristics. Temperature dependent I-V curves exhibit hysteresis and a well-defined transition from the superconducting to the normal state for $d > 10$ nm (Figure 3); and at 5 and 10 nm of thickness, several NbN films evidenced the presence of small steps along the SC transition branch of the I-V curve, which is better detailed in the next section. For these films, we assumed as critical current I_c, the value at which the first step occurs. The value of the critical current, I_c, progressively reduces with the rise in T, and smoothing of the transition occurs approaching T_c. The temperature dependence of the superconducting critical-current density, $J_c(T)$, for the set deposited on Al_2O_3 r-cut substrate (see Figure 4), was derived following the criterion and procedure reported in reference [31]. The J_c values at 0 K (i.e., J_{c0}) were extracted by a least-squares fit using the Ginzburg–Landau equation [31].

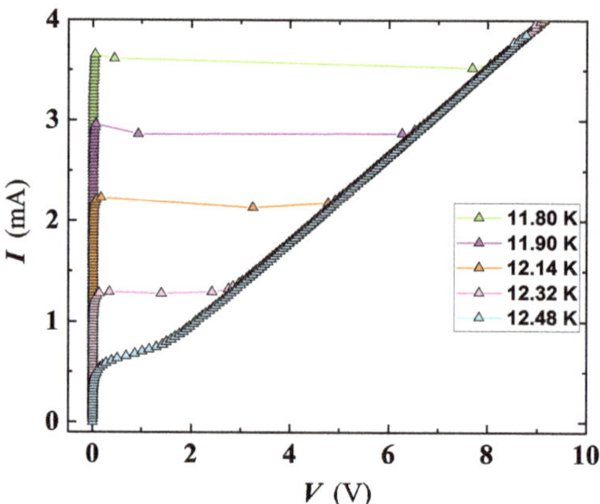

Figure 3. I-V curves of the film SC15, measured at several T close to the SC transition of T_c = 12.73 K. Approaching T_c, the amplitude of the hysteresis between the sweep-up and sweep-down of the current narrows.

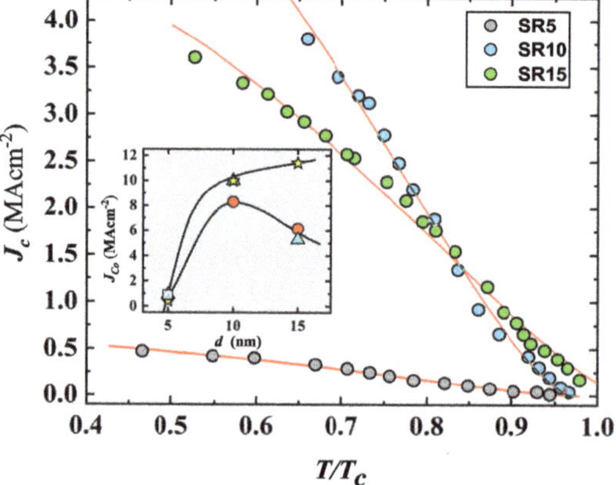

Figure 4. Current density as a function of the temperature, normalized to T_c, for the whole set of NbN films deposited on the Al_2O_3 r-cut substrate. Lines are the result of the least-squares fitting by the Ginzburg–Landau equation (see ref. [31]). Inset: thickness dependence of the critical current density at zero temperature, J_{c0}, for NbN films deposited on Al_2O_3 r-cut (circles), Al_2O_3 c-cut (triangles), MgO (stars), SiO_2 (square).

The thickness dependence of J_{c0} appears related to the substrate types exhibiting a bell shaped behavior on both types of sapphire substrates, and it continues to rise with film thickness on MgO (inset of Figure 4 and Table 1). It is worthwhile noting the drop in J_{c0} at $d = 5$ nm, up to about one order of magnitude, becoming practically independent on the substrate type, and little differences in J_{c0} start to appear from $d = 10$ nm (inset of Figure 4).

The J_{c0} values found in our films are similar to those found in thin films of NbN [42], and the bell shaped behavior was detected also in Nb [31].

3.2. Berezinskii–Kosterlitz–Thouless Transition

Experimental observation of a BKT transition in 2D systems is generally challenging due to the constraints on the film size in relation to the characteristic lengths of the superconductor. Concerning the Pearl length, the condition $\Lambda > w$ must be fulfilled, w being a film dimension. Hence, assuming for NbN films a value of $0.5 \lesssim \lambda \lesssim 0.4$ µm for thicknesses $5 \leq d \leq 15$ nm [43], we get values of $\Lambda \approx 100$ µm, $\Lambda \approx 40$ µm and $\Lambda \approx 20$ µm for $d = 5$ nm, 10 nm and 15 nm. These values must be compared to the maximum physical dimension of the system, which for our films coincides with the width of the Hall bar ($w = 50$ µm): for the Hall bar width of 10 µm, the condition is satisfied at any of the thicknesses here taken into account. Hence, while the 10 nm thick sample can be considered borderline, the 15 nm thick one appears out of range. Regarding the condition $d \ll \xi$, it is worthwhile noting that being $\xi \approx 4$ nm [41], only NbN nanofilms, 5 nm thick, appear to be good candidates to exhibit a well-defined BKT transition.

Experimentally, we have investigated the signature of a BKT superconducting transition in $\rho(T)$ curves and/or in I-V characteristics measured at several fixed Ts [13–16].

We have extracted experimentally the T_c value in a zero magnetic field, T_{c0}, and the sheet resistance in the metallic normal state, $R_{\square N}$, just above T_c, considering a Cooper-pair fluctuation model for a 2D superconducting system developed by Aslamazov and Larkin (AL) [44] and Maki and Thompson (MT) [45,46]. The two parameters were evaluated by a least-squares fitting of the experimental $R_{\square}(T)$ curves (Figure 5), in a T range from T_c to ≈ 15 K, by using the relation [13,14]:

$$R_{\square}(T) = \frac{R_{\square N}}{1 + R_{\square N} \frac{\gamma}{16} \frac{e^2}{\hbar} \left(\frac{T_c}{T - T_c} \right)} \tag{1}$$

where: γ is a numerical factor, $\hbar$ is the reduced Planck constant, e is the electron charge and T_c is here intended as the BCS mean-field transition temperature. Fitting was carried out satisfying the condition $ln(T/T_c) \ll 1$. The result of the $R_{\square}(T)$ fitting of film MO10 is reproduced in Figure 6. Similar findings were found also for MO5a and MO5b.

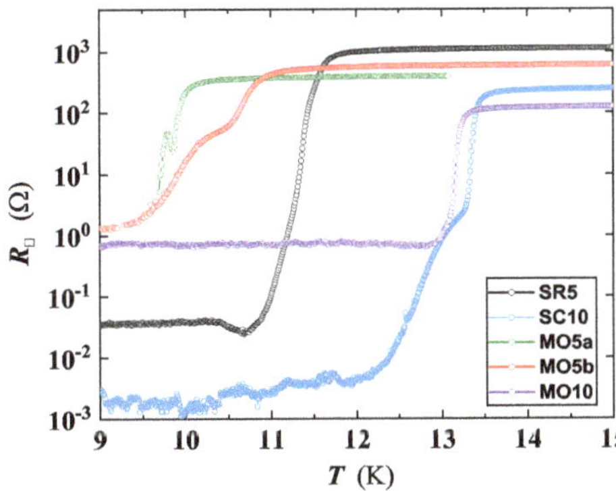

Figure 5. Temperature−dependent sheet resistance of ultrathin NbN films (5 and 10 nm of thickness), deposited on different substrate types, exhibiting quantum effects in 2D. The differences in the curve behavior of MO5a and MO5b (belonging to the same deposition run) are related to specific choices of process parameters during the Hall bar fabrication.

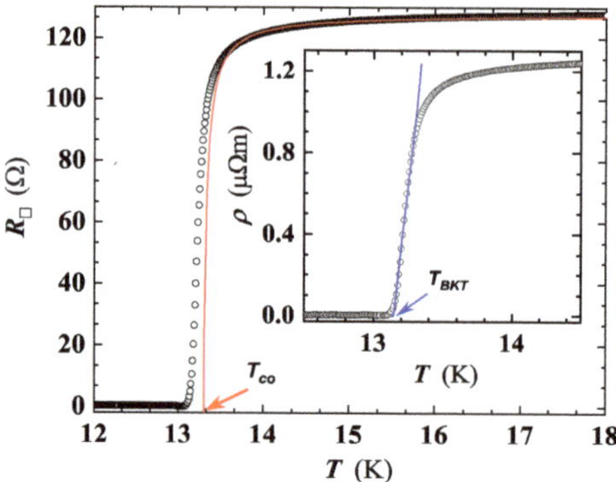

Figure 6. Temperature−dependent sheet resistance around the superconducting transition temperature for MO10 (see Table 2). The red line is the least-squares fit by using Equation (1). The value of $T_{c0} = 13.31$ K (red arrow) is the intercept of the red line with the x-axis. Inset: least-squares fitting of the $\rho(T)$ curve of MO10 by Equation (2). The intercept with the x-axis gives the value of T_{BKT} (13.06 K, blue arrow).

Table 2. Berezinskii–Kosterlitz–Thouless (BKT) parameters derived by the analysis of the resistivity and I-V characteristics curves of some of the thinner NbN films. Column headings, from left: sample acronym *; BKT temperature derived by $\rho(T)$ fitting with Equation (2); SC transition temperature at $B = 0$ (see the text); normal state sheet resistance at 20 K; γ value (see the text and Equation (1)); VAP polarizability.

Sample	T_{BKT} **(K)**	T_{c0} **(K) ****	$R_{\square N}$ **(Ω)**	γ	ϵ
MO5a	9.70 ± 0.03	9.90	502.30 ± 0.01	0.970 ± 0.002	10.28 ± 0.030
MO5b	10.30 ± 0.02	10.60	620 ± 1.06	1.500 ± 0.005	11.49 ± 0.022
MO10	13.060 ± 0.008	13.31	129.8 ± 0.60	1.56 ± 0.09	25.5 ± 0.11

* For meaning of sample acronym see the caption of Table 1. ** The fitting error on T_{co} is of the order of 10^{-4}.

Fitted and experimental $R_{\square N}$ values agree within 2–3%, and T_c and γ exhibit a dependence on d and the substrate type. This is particularly clear for the T_c values for NbN films with 5 nm of thickness (see Figures 5).

In detail, T_c values derived by the fit are close, though systematically lower, than those extracted by the analysis of the SC transition branch of the $\rho(T)$ curves (see the Table 1), and the values of γ, ranging from $\approx$1 to $\approx$2 (Table 2), are in agreement with those reported for NbN films with $d < 10$ nm [13,14], and in general, comparable with those reported in literature for different SC materials.

Based on theoretical studies, below T_{BKT}, all VAPs are bound. While approaching T_{BKT}, thermal fluctuations begin to break VAPs, and under a thermodynamic equilibrium, VAPs and single vortices will coexist [10,13,47]. However, due to the sourced current, single vortices will experience a Lorentz force (neglecting vortex pinning), causing the appearance of a finite voltage drop. The resulting film resistivity, in the temperature range $T_{BKT} < T < T_{c0}$, can be described by the relation:

$$\rho(T) = a\,exp\left(-2\sqrt{b\frac{T_{c0} - T}{T - T_{BKT}}}\right) \tag{2}$$

where a and b are fitting parameters related to the SC material, and the values obtained were less than 1 for all films. The T_{BKT} value derived by the least-squares fitting with

Equation (2) of the $\rho(T)$ curve of film MO10 is shown in the inset of Figure 6. The fitting procedure was extended to other thin films, for which no such sign of BKT transition was detected. It is interesting to estimate the polarizability, ϵ_{BKT}, of a VAP at the BKT-like vortex phase transition in the presence of other VAPs, by using the relation [13]:

$$\frac{T_{BKT}}{T_{c0}} = \frac{1}{1 + 0.173\epsilon_{BKT}R_{\square N}\frac{2e^2}{\pi h}} \tag{3}$$

Equation (3) was applied to NbN films deposited on different substrates, resulting in values of ϵ (Table 2) in close agreement with those found in ref. [13] for 6 nm thick NbN films and also successfully crosschecked with the universal relation $k_B T_{BKT} = A T_{BKT}/4\epsilon_{BKT}$ for topological 2D phase transitions; see Nelson and Kosterlitz [48,49]. All the parameters extracted by fitting (see Table 2) are in excellent agreement with the values reported in the ref. [13,14].

To check the possible BKT-like transition, we carried out further analyses considering that thermal fluctuations occurring in ultra-thin films can excite pairs of vortices, each consisting in a single vortex having supercurrents circulating in opposite directions, then leading to the bound vortex anti-vortex pair state [13]. These VAP pairs lead to the specific signature of a BKT-like transition, consisting in a jump in the superconducting stiffness, J_s, from a finite value, below T_c, to zero above it. In that case, a non linear dependence exists in the I-V characteristics near T_c, since large enough currents may unbind VAPs. Hence, due to this effect, a voltage is generated, depending on the equilibrium density of the free vortices, scaling with the sourced current according to a power law, with an exponent proportional to J_s:

$$V \propto I^\alpha(T) \tag{4}$$

$$\alpha(T) = 1 + \pi\frac{J_s(T)}{T} \tag{5}$$

at the BKT transition, the I-V exponent jumps from $\alpha(T_{BKT}^-) = 3$ to $\alpha(T_{BKT}^+) = 1$, where the unity value signals the metallic ohmic behavior in the normal state of the I-V characteristics. We have extracted the value of α for the MO10 film, by using Equation (4) to carry out a least-squares fitting of the voltage–current curves, measured at several T values (Figure 7). We detected a universal jump from $\simeq 1$ to $\simeq 3$ at T_{BKT} (Figure 8). Our findings evidence a steeper transition (see Figure 8) resulting in agreement with data of Venditti et al. for a thin film of NbN [16]. The behavior here reported for the NbN system was also found by Saito et al. for a completely different system than the MoS_2 [50] (Figure 8), which validates the universal jump in the superfluid density for determining the BKT transition.

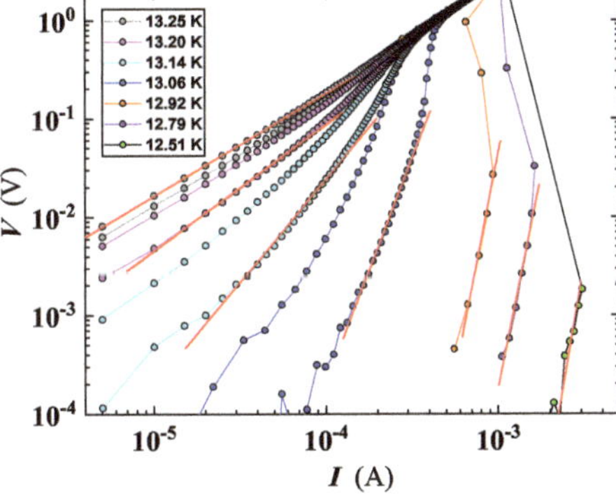

Figure 7. Voltage–current characteristics for the film MO10. Experimental data of the sweep-up curve were fitted by a power law function to extract the value of α (see Equations (4) and (5) and Table 2).

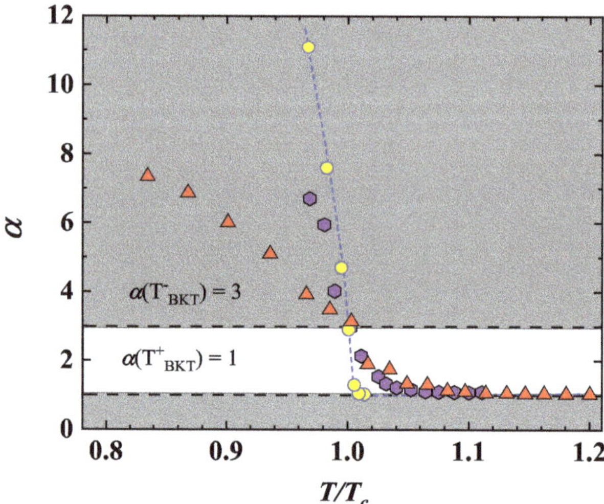

Figure 8. Temperature—dependence of α for MO10 (yellow circle), derived from the power-law fitting of the V-I curves plotted in the Figure 7. A jump in the value of α from $\simeq 1$ to $\simeq 3$ (white region) was detected at T_{BKT}, corresponding to $T/T_c = 0.989$. For comparison, the α values for a 3 nm thick NbN film from the work of Venditti et al. (purple hexagon) [16] and for MoS_2 from the work of Saito et al. (red triangle) [50] were added to the data. The broken line is a guide for the eyes. The temperature was normalized to T_{BKT} for our and Saito et al. data. Venditti et al.'s α values were derived by digitization of Figure (2e) in ref. [16].

3.3. Phase Slips

In addition to the BKT transition, interesting outcomes were found in our NbN films in terms of I-V characteristics by the pulse-mode technique (see the Materials and Methods section). In detail, for two of our thinnest films (SR5 and MO5b), we observed the emergence of resistive states as tailing-like features in I-V curves and a double transition in the $\rho(T)$ curves. These findings were interpreted as possible signatures of phase-slip events, arising due to a discontinuous jump by integer multiple of 2π in the phase of the order parameter of the superconducting state, typically existing in quasi-1D systems as nanowires and nanorings. For a discussion of the resistive states associated with PS events in superconducting quasi-1D nanostructures, see ref. [51].

Nevertheless, the investigated NbN films are a 2D system which is granular in nature. The presence of disorder can lead to a weak localization and inhomogeneous effects, resulting in the appearance of a 1D-like features such as PS events [10,11,24].

Taking into account the above-mentioned scenario, such systems are prone to forming an array of continuous conductive paths, having an effective dimensional size much smaller than the physical dimension of the system. In that case, a region equivalent to the coherence length can create a Josephson-like junction, which shows a PS barrier proportional to the area of quasi-1D SC system. Under these circumstances, Cooper pairs will cross the free energy barrier and the relative phase will jump by 2π, resulting in a measurable voltage drop. This drop, will cause a detectable resistance change at $T < T_c$, giving rise to PS events driven by thermal and quantum activation [24]. To confirm the possible presence of PS in our films, we took into account the theory of Langer, Ambegaokar, McCumber and Halperin (LAMH), accounting for thermally activated phase slips (TAPS) detected as an effective resistance change related to the time evolution of the superconducting phase:

$$R_{TAPS} = \frac{\pi \hbar^2 \Omega_{TAPS}}{2e^2 K_B T} exp\left(\frac{-\Delta F}{K_B T}\right) \tag{6}$$

where ΔF is the energy barrier to Cooper-pair crossing, and the other quantities have the known meanings. Here, Ω_{TAPS} is the attempt frequency defined as:

$$\Omega_{TAPS} = \frac{L}{\xi}\left(\frac{\Delta F}{K_B T}\right)\frac{1}{\tau_{GL}} \tag{7}$$

where τ_{GL} is the relaxation time in the time-independent Ginzburg–Landau equation and ΔF is defined as:

$$\Delta F = 0.83 K_B T_c \frac{L \beta w d R_q}{\rho_n \xi_0} \tag{8}$$

where β is a fitting parameter, w the Hall bar width, d the film thickness, $R_q = \hbar/2e^2 \simeq 6.4 \text{ k}\Omega$ the quantum of resistance and ρ_n the normal state resistivity upon SC transition.

The LAMH theory was originally developed for very long wires, thinner than ξ of the SC material. However, our investigated NbN films had a thickness comparable to ξ. w and L (i.e., the width and the length of the Hall bar, respectively) were orders of magnitude greater then ξ.

In order to check for the presence of TAPS in SR5 and MO5b, least-squares fits were carried out by Equation (6), leaving T_c and β as fitting parameters. For MO5b, good agreement with the LAMH theory was found, leaving T_c and β values comparable to those found in ref. [24] (see Figure 9). Moreover, the T_c value obtained by the LAMH fit coincides with that derived from resistance curve analysis (see Table 1), confirming that fluctuation effects are caused by PS and have a thermal origin.

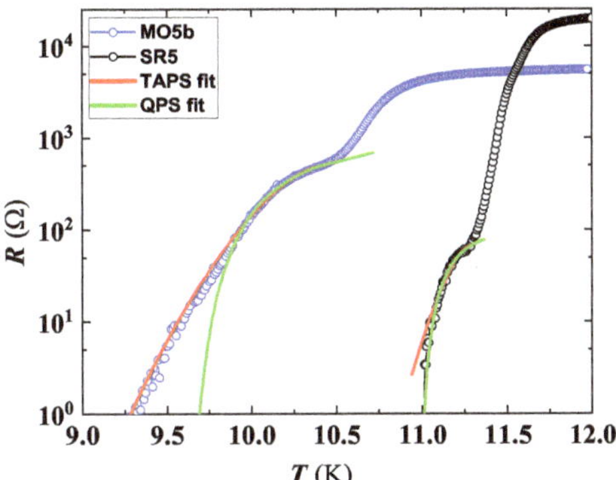

Figure 9. Temperature dependence of resistance for two 5 nm thick NbN films. Both curves showed tailing-like features below about 10.5 and a 11.5 K for MO5b and SR5, respectively. Branches of the $\rho(T)$ curves exhibiting the transition at the lower T range were fitted by using LAMH theory, Equation (6) (red lines), and by Equation (9) (green lines).

On the other hand, for SR5 at lower T, fitting by LAMH theory failed in the first steeper branch of the $\rho(T)$ curve (see Figure 9), suggesting the possibility of a different fluctuation effect present in the same T range. To confirm our hypothesis, we explored the possible contribution by QPS emerging from the quantum tunneling of the order parameter through the same free-energy barrier as in TAPS, which is supposed to dominate at lower T. The dynamics of the order parameter in the quantum fluctuations were first reported by

Giordano [52], suggesting a mechanism similar to TAPS, except that appropriate energy scale $K_B T$ is replaced by $\hbar/\tau_{GL}$, resulting in the equation:

$$R_{QPS} = B \frac{\pi \hbar^2 \Omega_{QPS}}{2e^2 \frac{\hbar}{\tau_{GL}}} exp \left(-a \frac{\Delta F \tau_{GL}}{\hbar} \right) \tag{9}$$

where B and a are numerical factors of ≈ 1 and Ω_{QPS} is defined as:

$$\Omega_{QPS} = \frac{L}{\zeta} \left(\frac{\Delta F}{\frac{\hbar}{\tau_{GL}}} \right) \frac{1}{\tau_{GL}} \tag{10}$$

The result of the fitting by Equation (9) is in agreement with the theoretical predictions for SR5 (see Figure 9), and on the contrary, a progressive deviation from the QPS Equation (9) is evident at lower T for MO5b.

The excellent agreement of our experimental results of $\rho(T)$ with the above-mentioned LAMH and QPS models suggests the existence of a nano-conducting path (NCP) having a lateral size comparable to ζ. To estimate this size, we have used the equation from the model developed by Joshi et al. [53]:

$$d_{NCP} = \sqrt{\left(\frac{12 \rho_{RT} C \zeta_0}{1.76 \sqrt{2} \pi R_q} \right)} \tag{11}$$

where $\zeta_0 \simeq 4$ nm, $C = 8$ (value typically used for quantum systems) [2] and ρ_{RT} is the resistivity value at RT. NCP values calculated by using Equation (11) were 3.2 nm for SR5 and 7.5 nm for MO5b, which are comparable with those found for NbN nanostructures in ref. [53]. In fact, for SR5, the condition $d_{NCP} < \zeta$ (i.e., $\zeta \approx 4$ nm) can be considered as the origin of quantum tunneling, resulting in the formation of QPS detected in our film. On the contrary, for MO5b, $d_{NCP} > \zeta$ TAPS lines will result, as indeed experimentally observed.

An interesting outcome derived by a detailed analysis of I-V characteristics carried out on SR5 and MO5b confirmed the existence of PS events. Figure 10 shows a family of I-V curves measured at different T for SR5, where multiple slanted steps having resistive tailing-like features were detected. The dynamic resistance of these tails rises at increasing V (e.g., along the I-V sweep-up direction, see Figure 11), and the current axis intercept (extending tails slope) occurs for a common current value of $I_{ex} \simeq 0.38$ mA, defined as excess current. These findings can be considered as a proof that detected resistive states are originated by PS [54].

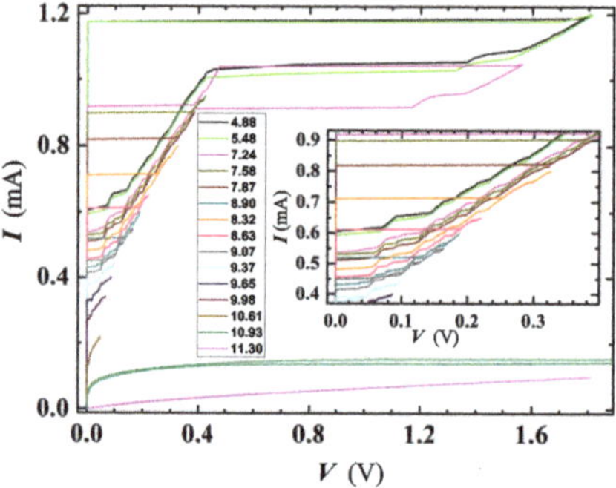

Figure 10. I-V characteristics for the SR5 film, carried out at several T. Both up- and down-current sweeps evidence the presence of slanted steps due to the occurrence of intermediate resistive regimes before the complete transition to the normal state. Inset: magnification of the central part of the plot.

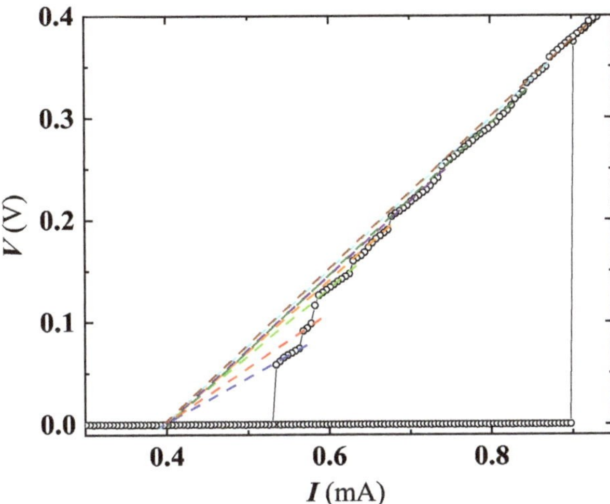

Figure 11. V-I curve at T = 7.48 K is shown for SR5, in which an intercept (represented by dotted lines of different colors for different PSL) on x-axis converge towards a well-defined value of the excess current of $I_{ex} = 0.38$ mA.

Further analysis of the I-V curves evidenced interesting temperature dependence of the number of the resistive states appearing in the sweep-up ($N^{\uparrow}$) and sweep-down ($N^{\downarrow}$) branches of the I-V curves (Figure 12). The $N^{\uparrow}$ and $N^{\downarrow}$ kept an almost constant value up to $T \approx T_c/2$, and then they started to differ. In detail, at $T \gtrsim T_c/2$, $N^{\downarrow}$ jumps to higher values, and then it decreases till about 9 K. On the contrary, $N^{\uparrow}$ continues to gradually increase till 9 K. Onward from 9 K, both curves are perfectly overlapped and continuously rise with T (Figure 12). The behavior exhibited by $N^{\uparrow}$ and $N^{\downarrow}$ allows one to derive additional observations about fluctuation effects involved in our investigated film. Specifically, the unequal distribution of PS in $N^{\uparrow}$ and $N^{\downarrow}$ suggests that different mechanisms are contributing in three distinct T ranges. Below $T_c/2$, the emergence of phase-slip events is driven by quantum tunneling. In the intermediate T range ($T_c/2 < T < 9$ K), the behavior of $N^{\uparrow}$ and $N^{\downarrow}$ can be explained by the competition of QPS and TAPS, the former tending to decrease approaching 9 K. Finally, above 9 K, only TAPS contributes in the observation of PS events (Figure 12). It is interesting to note that in the region where QPS is present, $N^{\uparrow}$ and $N^{\downarrow}$ diverge. This effect can be explained by a presumably different tunneling route followed by the system, during the sweep up and sweep down of the sourced current in the quantum regime. In the intermediate T range, the system self organizes in order to converge towards a condition dominated by TAPS, then leading to a decrease in $N^{\downarrow}$ with an increase $N^{\uparrow}$. Finally, approaching T_c, thermal fluctuations dominate over QPS, and the entire system is driven by the electrons. Since the electrons tend to track the same path while going sweep-up and sweep-down [55], the number of PS becomes equal, followed by a total increase in the number of PSL, due to enhanced fluctuation effects near T_c.

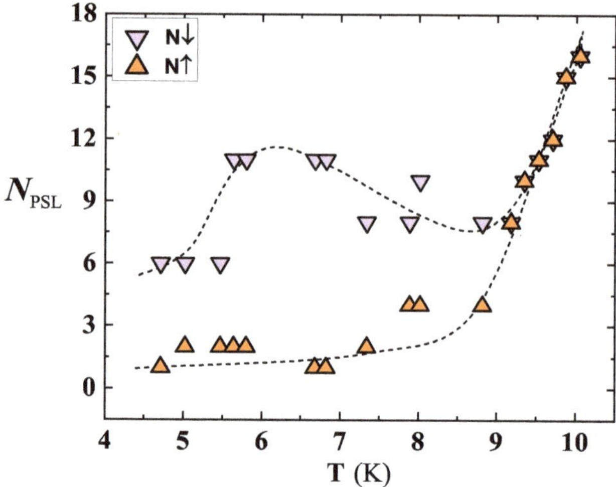

Figure 12. Number of resistive states for SR5, calculated from the I-V curves, during the current sweep-up ($N^\uparrow$) and sweep-down ($N^\downarrow$). Near to $T \simeq 9$ K, $N^\uparrow$ and $N^\downarrow$ suddenly converge, assuming the same value at $T \geq 9$ K. Broken lines are guides for the eye.

In addition to TAPS, BKT-like features were also observed in MO5b (see Figure 13), in the T range where PS disappears and the system still remains in the SC state. Moreover, in the same film, BKT was confirmed by the scaling behavior of Halperin–Nelson equation above T_{BKT} (as explained in the BKT section). Interestingly, with our I-V characteristics, the TAPS completely disappears at 9.9 K, far before reaching T_c (11.02 K, see Table 2), and this PS suppression gives rise to BKT-like transition, i.e., an effect typical of a 2D system, which involves breaking of VAP. For $T > T_{BKT}$, the unbinding of VAP will result in the universal jump in α, as already detected in the I-V characteristics of MO10 (see Figure 7). The parameters derived for this film (see Table 2) are in good agreement with the theory. We believe that the inhomogeneity of the NbN system contributed to the origins of these experimental findings. However, our results suggest that the level of inhomogeneity in the studied system is just sufficient to create a thermal fluctuation which results in the emergence of TAPS but not too strong to destroy the superconductivity of the film. The detection of BKT in the same film seems to confirm this hypothesis, since higher inhomogeneity would have destroyed the BKT effect. The coexistence of BKT and TAPS in the same system, the former being typical in a 2D system and the latter in a quasi-1D system, suggests that the internal structure of our investigated films is on the boundary line of a 2D–1D dimensional crossover.

Similar findings were measured at 5 nm of thickness for NbN films deposited on other types of substrates. However, on SiO_2, resistive tails appeared less pronounced due to the larger width of the Hall bar (see Figure 14).

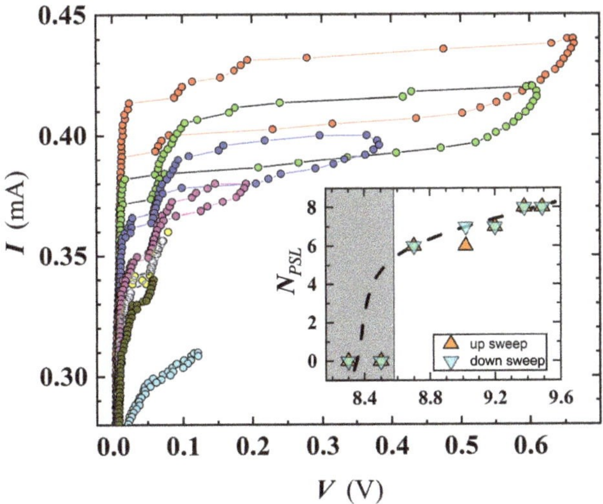

Figure 13. I-V characteristics of MO5b, showing the occurrence of steps in the curves close to T_c, progressively disappearing approaching T_c. The T values for the shown curves are: 8.70 K (red), 9.02 (green), 9.20 K (blue), 9.38 K (magenta), 9.49 K (yellow), 9.60 K (olive), 9.98 K (cyan). Inset: number of TAPS extracted from I-V curves. The gray rectangle defines the T range where no PSL were detected.

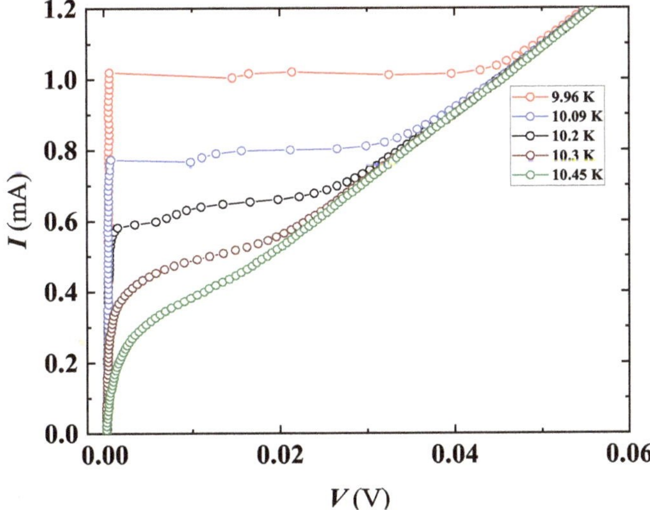

Figure 14. I-V characteristics for SO5, at selected T values close to T_c. Less-marked resistive tails are now visible, whose values are reduced if compared to those detected in the film SR5 (see the Figure 10).

4. Conclusions

In summary, our study of the superconducting properties of NbN nanofilms showed several features associated with complex phase fluctuations of the order parameter. Resistivity and I-V curves showed a well-defined BKT transition to the superconducting state characteristic of 2D systems. In addition to the BKT physics, we also detected and characterized phase-slip events (both quantum and thermal) typical of quasi-1D superconductors. Both effects deman careful fine tuning of the experimental set-up and material system.

To analyze the BKT transition, we used the Cooper pair fluctuation model. We found that the linear in T dependence of the resistivity above T_c is only compatible with 2D fluctuation-conductivity and incompatible with the predictions for 1D and 3D systems, confirming the 2D dimensionality of our NbN films. Our findings of polarizability values of VAP at the BKT transition are in good agreement with the Nelson and Kosterlitz universal relation in two different NbN films. In one case, the polarizability was found to be almost twice the expected value. This evidences the 3D to 2D dimensional crossover at 10 nm, since no BKT transition was detected at 15 nm. Further confirmation is given by the α exponent extracted from I-V curves, exhibiting a steep transition from 1 to 3, close to T_{BKT}, in agreement with theory.

Regarding the PS events, in one sample we detected both quantum and thermal PSs, in different T regimes. QPSs depend on the quantum tunneling route undertaken by the system. These outcomes were explained by the presence of granularity in NbN. A careful analysis of film resistivity suggested the presence of a nano-conductive path, making NbN films equivalent to a quasi-1D system, explaining the presence of PS events. We investigated specific features of PS events as the numbers of PS occurring during the up-sweep and down-sweep sourced current. The distribution of PSs in the quantum regime is uneven, though converging on the same value in the T regime dominated by thermal fluctuations. Moreover, we have evidenced the co-existence of BKT and PS phenomena in the same NbN nanofilm, which has not been reported until now. Considering that BKT and PS events belong to two different dimensionality systems, this means that we have successfully addressed a 2D to quasi-1D dimensional crossover in the same system. Finally, our experimental findings motivate us for in depth study of superconducting NbN nanofilms as a tunable platform to generate and control novel quantum phenomena exploitable for quantum technologies.

Author Contributions: Conceptualization, M.S. (Meenakshi Sharma), A.P., N.P. M.S. (Manju Singh) and R.S.; Film deposition, M.S. (Manju Singh) and R.K.R.; Device fabrication, M.S. (Meenakshi Sharma), M.F. and N.D.L.; investigation, M.S. (Meenakshi Sharma), N.P.; formal analysis, M.S. (Meenakshi Sharma); validation, N.P., S.P.S. and A.P.; writing—original draft preparation, M.S. (Meenakshi Sharma) and N.P.; writing—review and editing, all authors; supervision, N.P.; All authors have read and agreed to the published version of the manuscript.

Funding: This research received no external funding.

Data Availability Statement: Not applicable.

Acknowledgments: We acknowledge the University of Camerino for providing technical and financial support. The School of Science and Technology and the Physics Division are acknowledged for their contribution to the installation of the SEM and the He closed-cycle cryostat equipment. We acknowledge fruitful discussions with Sergio Caprara and Luca Dell'Anna.

Conflicts of Interest: The authors declare no conflict of interest.

Abbreviations

The following abbreviations are used in this manuscript:

SC	Superconducting
BKT	Berezinskii–Kosterlitz–Thouless
PS	Phase slip
TAPS	Thermally activated phase slips
QPS	Quantum phase slips
NCP	Nano-conducting path

References

1. Bezryadin, A.; Lau, C.N.; Tinkham, M. Quantum suppression of superconductivity in ultrathin nanowires. *Nature* **2000**, *404*, 971. [CrossRef]
2. Bezryadin, A. Quantum suppression of superconductivity in nanowires. *J. Phys. Cond. Mat.* **2008**, *20*, 043202. [CrossRef]
3. Zhao, W.; Liu, X.; Chan, M. Quantum Phase Slips in 6 mm Long Niobium Nanowire. *Nano Lett.* **2016**, *16*, 1173–1178. [CrossRef]
4. Lehtinen, J.S.; Sajavaara, T.; Arutyunov, K.Y.; Presnjakov, M.Y.; Vasiliev, A.L. Evidence of quantum phase slip effect in titanium nanowires. *Phys. Rev. B* **2012**, *85*, 094508. [CrossRef]
5. Baumans, X.; Adami, D.C.O.; Zharinov, V.S.; Verellen, N.; Papari, G.; Scheerder, J.E.; Zhang, G.; Moshchalkov, V.V.; Silhanek, A.V.; de Vondel, J.V. Thermal and quantum depletion of superconductivity in narrow junctions created by controlled electromigration. *Nat. Commun.* **2016**, *7*, 10560. [CrossRef]
6. Kim, H.; Gay, F.; Maestro, A.D.; Sacépé, B.; Rogachev, A. Pair-breaking quantum phase transition in superconducting nanowires. *Nat. Phys.* **2018**, *14*, 912–917. [CrossRef]
7. Carbillet, C.; Caprara, S.; Grilli, M.; Brun, C.; Cren, T.; Debontridder, F.; Vignolle, B.; Tabis, W.; Demaille, D.; Largeau, L.; et al. Confinement of superconducting fluctuations due to emergent electronic inhomogeneities. *Phys. Rev. B* **2016**, *93*, 144509. [CrossRef]
8. Mason, N.; Kapitulnik, A. Dissipation Effects on the Superconductor-Insulator Transition in 2D Superconductors. *Phys. Rev. Lett.* **1999**, *82*, 5341–5344. [CrossRef]
9. Breznay, N.; Tendulkar, M.; Zhang, L.; Lee, S.C.; Kapitulnik, A. Superconductor to weak-insulator transitions in disordered tantalum nitride films. *Phys. Rev. B* **2017**, *96*, 134522. [CrossRef]
10. Bell, M.; Sergeev, A.; Mitin, V.; Bird, J.; Verevkin, A.; Gol'tsman, G. One-dimensional resistive states in quasi-two-dimensional superconductors: Experiment and theory. *Phys. Rev. B* **2007**, *76*, 094521. [CrossRef]
11. Gajar, B.; Yadav, S.; Sawle, D.; Maurya, K.K.; Gupta, A.; Aloysius, R.P.; Sahoo, S. Substrate mediated nitridation of niobium into superconducting Nb2N thin films for phase slip study. *Sci. Rep.* **2019**, *9*, 8811. [CrossRef]
12. Kamiński, M.; Schrefler, B. Probabilistic effective characteristics of cables for superconducting coils. *Comp. Meth. Appl. Mechan. Eng.* **2000**, *188*, 1–16. [CrossRef]
13. Bartolf, H.; Engel, A.; Schilling, A.; Il'in, K.; Siegel, M.; Hübers, H.W.; Semenov, A. Current-assisted thermally activated flux liberation in ultrathin nanopatterned NbN superconducting meander structures. *Phys. Rev. B* **2010**, *81*, 024502. [CrossRef]
14. Sidorova, M.; Semenov, A.; Hubers, H.W.; Ilin, K.; Siegel, M.; Charaev, I.; Moshkova, M.; Kaurova, N.; Goltsman, G.N.; Zhang, X.; et al. Electron energy relaxation in disordered superconducting NbN films. *Phys. Rev. B* **2020**, *102*, 054501. [CrossRef]
15. Koushik, R.; Kumar, S.; Amin, K.R.; Mondal, M.; Jesudasan, J.; Bid, A.; Raychaudhuri, P.; Ghosh, A. Correlated Conductance Fluctuations Close to the Berezinskii-Kosterlitz-Thouless Transition in Ultrathin NbN Films. *Phys. Rev. Lett.* **2013**, *111*, 197001. [CrossRef]
16. Venditti, G.; Biscaras, J.; Hurand, S.; Bergeal, N.; Lesueur, J.; Dogra, A.; Budhani, R.C.; Mondal, M.; Jesudasan, J.; Raychaudhuri, P.; et al. Nonlinear $I-V$ characteristics of two-dimensional superconductors: Berezinskii-Kosterlitz-Thouless physics versus inhomogeneity. *Phys. Rev. B* **2019**, *100*, 064506. [CrossRef]
17. Berezinskii, V.L. Destruction of Long-range Order in One-dimensional and Two-dimensional Systems having a Continuous Symmetry Group I. Classical Systems. *Sov. Phys. JETP* **1971**, *32*, 493–500.
18. Berezinskii, V.L. Destruction of Long-range Order in One-dimensional and Two-dimensional Systems Possessing a Continuous Symmetry Group. II. Quantum Systems. *Sov. Phys. JETP* **1972**, *34*, 610–616.
19. Kosterlitz, J.M.; Thouless, D.J. Ordering, metastability and phase transitions in two-dimensional systems. *J. Phys. C Sol. State Phys.* **1973**, *6*, 1181–1203. [CrossRef]
20. Kosterlitz, J.M. The critical properties of the two-dimensional xy model. *J. Phys. C Sol. State Phys.* **1974**, *7*, 1046–1060. [CrossRef]
21. Yong, J.; Lemberger, T.; Benfatto, L.; Ilin, K.; Siegel, M. Robustness of the Berezinskii-Kosterlitz-Thouless transition in ultrathin NbN films near the superconductor-insulator transition. *Phys. Rev. B* **2013**, *87*, 184505. [CrossRef]
22. Giachetti, G.; Defenu, N.; Ruffo, S.; Trombettoni, A. Berezinskii-Kosterlitz-Thouless phase transitions with long-range couplings. *Phys. Rev. Lett.* **2021**, *127*, 156801. [CrossRef]
23. Beasley, M.R.; Mooij, J.E.; Orlando, T.P. Possibility of Vortex-Antivortex Pair Dissociation in Two-Dimensional Superconductors. *Phys. Rev. Lett.* **1979**, *42*, 1165–1168. [CrossRef]
24. Chu, S.L.; Bollinger, A.T.; Bezryadin, A. Phase slips in superconducting films with constrictions. *Phys. Rev. B* **2004**, *70*, 214506. [CrossRef]
25. Benfatto, L.; Perali, A.; Castellani, C.; Grilli, M. Kosterlitz-Thouless vs. Ginzburg-Landau description of 2D superconducting fluctuations. *Eur. Phys. J. B Cond. Mat. Compl. Syst.* **2000**, *13*, 609–612. [CrossRef]
26. Mondal, M.; Kumar, S.; Chand, M.; Kamlapure, A.; Saraswat, G.; Seibold, G.; Benfatto, L.; Raychaudhuri, P. Role of the vortex-core energy on the Berezinskii-Kosterlitz-Thouless transition in thin films of NbN. *Phys. Rev. Lett.* **2011**, *107*, 217003. [CrossRef]
27. Paradiso, N.; Nguyen, A.T.; Kloss, K.E.; Strunk, C. Phase slip lines in superconducting few-layer NbSe2 crystals. *2D Mater.* **2019**, *6*, 025039. [CrossRef]
28. Rezaev, R.; Smirnova, E.; Schmidt, O.; Fomin, V. Topological transitions in superconductor nanomembranes under a strong transport current. *Commun. Phys.* **2020**, *3*, 1–8. [CrossRef]

29. Alfonso, J.; Buitrago, J.; Torres, J.; Marco, J.; Santos, B. Influence of fabrication parameters on crystallization, microstructure, and surface composition of NbN thin films deposited by rf magnetron sputtering. *J. Mat. Sci.* **2010**, *45*, 5528–5533. [CrossRef]

30. Delacour, C.; Pannetier, B.; Villegier, J.C.; Bouchiat, V. Quantum and thermal phase slips in superconducting niobium nitride (NbN) ultrathin crystalline nanowire: Application to single photon detection. *Nano Lett.* **2012**, *12*, 3501–3506. [CrossRef]

31. Pinto, N.; Rezvani, S.J.; Perali, A.; Flammia, L.; Milošević, M.V.; Fretto, M.; Cassiago, C.; De Leo, N. Dimensional crossover and incipient quantum size effects in superconducting niobium nanofilms. *Sci. Rep.* **2018**, *8*, 4710. [CrossRef]

32. Rezvani, S.; Perali, A.; Fretto, M.; De Leo, N.; Flammia, L.; Milošević, M.; Nannarone, S.; Pinto, N. Substrate-Induced Proximity Effect in Superconducting Niobium Nanofilms. *Cond. Matter* **2019**, *4*, 4. [CrossRef]

33. Daire, A. An improved method for differential conductance measurements. *Keithley White Pap.* **2005** . Available online: https://www.tek.com/en/documents/whitepaper/improved-method-differential-conductance-measurements (accessed on 1 November 2022)

34. Keithley Instruments, Inc. *Achieving Accurate and Reliable Resistance Measurements in Low Power and Low Voltage Applications*; Keithley Instruments, Inc.: Cleveland, OH, USA, 2004.

35. Joshi, L.M.; Verma, A.; Gupta, A.; Rout, P.; Husale, S.; Budhani, R. Superconducting properties of NbN film, bridge and meanders. *AIP Adv.* **2018**, *8*, 055305. [CrossRef]

36. Hazra, D.; Tsavdaris, N.; Jebari, S.; Grimm, A.; Blanchet, F.; Mercier, F.; Blanquet, E.; Chapelier, C.; Hofheinz, M. Superconducting properties of very high quality NbN thin films grown by high temperature chemical vapor deposition. *Supercond. Sci. Technol.* **2016**, *29*, 105011. [CrossRef]

37. Soldatenkova, M.; Triznova, A.; Baeva, E.; Zolotov, P.; Lomakin, A.; Kardakova, A.; Goltsman, G. Normal-state transport in superconducting NbN films on r-cut sapphire. *J. Phys. Confer. Ser.* **2021**, *2086*, 012212. [CrossRef]

38. Kang, L.; Jin, B.B.; Liu, X.Y.; Jia, X.Q.; Chen, J.; Ji, Z.M.; Xu, W.W.; Wu, P.H.; Mi, S.B.; Pimenov, A.; et al. Suppression of superconductivity in epitaxial NbN ultrathin films. *J. Appl. Phys.* **2011**, *109*, 033908. [CrossRef]

39. Smirnov, K.; Divochiy, A.; Vakhtomin, Y.; Morozov, P.; Zolotov, P.; Antipov, A.; Seleznev, V. NbN single-photon detectors with saturated dependence of quantum efficiency. *Supercond. Sci. Technol.* **2018**, *31*, 035011. [CrossRef]

40. Senapati, K.; Pandey, N.K.; Nagar, R.; Budhani, R.C. Normal-state transport and vortex dynamics in thin films of two structural polymorphs of superconducting NbN. *Phys. Rev. B* **2006**, *74*, 104514. [CrossRef]

41. Chockalingam, S.P.; Chand, M.; Jesudasan, J.; Tripathi, V.; Raychaudhuri, P. Superconducting properties and Hall effect of epitaxial NbN thin films. *Phys. Rev. B* **2008**, *77*, 214503. [CrossRef]

42. Il'in, K.; Siegel, M.; Semenov, A.; Engel, A.; Hübers, H.W. Critical current of Nb and NbN thin-film structures: The cross-section dependence. *Phys. Stat. Sol.* **2005**, *2*, 1680–1687. [CrossRef]

43. Kamlapure, A.; Mondal, M.; Chand, M.; Mishra, A.; Jesudasan, J.; Bagwe, V.; Benfatto, L.; Tripathi, V.; Raychaudhuri, P. Measurement of magnetic penetration depth and superconducting energy gap in very thin epitaxial NbN films. *App. Phys. Lett.* **2010**, *96*, 072509. [CrossRef]

44. Aslamasov, L.; Larkin, A. The influence of fluctuation pairing of electrons on the conductivity of normal metal. *Phys. Lett. A* **1968**, *26*, 238–239. [CrossRef]

45. Maki, K. Critical Fluctuation of the Order Parameter in a Superconductor. I. *Prog. Theor. Phys.* **1968**, *40*, 193–200. [CrossRef]

46. Thompson, R. Microwave, Flux Flow, and Fluctuation Resistance of Dirty Type-II Superconductors. *Phys. Rev. B* **1970**, *1*, 327–333. [CrossRef]

47. Benfatto, L.; Castellani, C.; Giamarchi, T. Broadening of the Berezinskii-Kosterlitz-Thouless superconducting transition by inhomogeneity and finite-size effects. *Phys. Rev. B* **2009**, *80*, 214506. [CrossRef]

48. Yamashita, T.; Miki, S.; Makise, K.; Qiu, W.; Terai, H.; Fujiwara, M.; Sasaki, M.; Wang, Z. Origin of intrinsic dark count in superconducting nanowire single-photon detectors. *Appl. Phys. Lett.* **2011**, *99*, 161105. [CrossRef]

49. Nelson, D.R.; Kosterlitz, J. Universal jump in the superfluid density of two-dimensional superfluids. *Phys. Rev. Lett.* **1977**, *39*, 1201. [CrossRef]

50. Saito, Y.; Itahashi, Y.M.; Nojima, T.; Iwasa, Y. Dynamical vortex phase diagram of two-dimensional superconductivity in gated Mo S 2. *Phys. Rev. Mat.* **2020**, *4*, 074003.

51. McNaughton, B.; Pinto, N.; Perali, A.; Milošević, M. Causes and Consequences of Ordering and Dynamic Phases of Confined Vortex Rows in Superconducting Nanostripes. *Nanomaterials* **2022**, *12*, 4043. [CrossRef]

52. Giordano, N. Superconductivity and dissipation in small-diameter Pb-In wires. *Phys. Rev. B* **1991**, *43*, 160. [CrossRef]

53. Joshi, L.M.; Rout, P.; Husale, S.; Gupta, A. Dissipation processes in superconducting NbN nanostructures. *AIP Adv.* **2020**, *10*, 115116. [CrossRef]

54. Sivakov, A.G.; Glukhov, A.M.; Omelyanchouk, A.N.; Koval, Y.; Müller, P.; Ustinov, A.V. Josephson Behavior of Phase-Slip Lines in Wide Superconducting Strips. *Phys. Rev. Lett.* **2003**, *91*, 267001. [CrossRef]

55. Kumar, A.; Husale, S.; Pandey, H.; Yadav, M.G.; Yousuf, M.; Papanai, G.S.; Gupta, A.; Aloysius, R. On the switching current and the re-trapping current of tungsten nanowires fabricated by Focussed Ion Beam (FIB) technique. *Eng. Res. Express* **2021**, *3*, 025017. [CrossRef]

MDPI
St. Alban-Anlage 66
4052 Basel
Switzerland
Tel. +41 61 683 77 34
Fax +41 61 302 89 18
www.mdpi.com

Nanomaterials Editorial Office
E-mail: nanomaterials@mdpi.com
www.mdpi.com/journal/nanomaterials

www.ingramcontent.com/pod-product-compliance
Lightning Source LLC
LaVergne TN
LVHW070850170726
843515LV00005B/615